# An Elementary Transition to Abstract Mathematics

## Textbooks in Mathematics

Series editors:
Al Boggess and Ken Rosen

https://www.crcpress.com/Textbooks-in-Mathematics/book-series/CANDHTEXBOOMTH

# An Elementary Transition to Abstract Mathematics

Gove Effinger
Gary L. Mullen

CRC Press
Taylor & Francis Group
Boca Raton London New York

CRC Press is an imprint of the
Taylor & Francis Group, an **informa** business
A CHAPMAN & HALL BOOK

CRC Press
Taylor & Francis Group
6000 Broken Sound Parkway NW, Suite 300
Boca Raton, FL 33487-2742

First issued in paperback 2022

ISBN 13: 978-1-03-247517-2 (pbk)
ISBN 13: 978-0-367-33693-6 (hbk)

DOI: 10.1201/9780429324819

### Library of Congress Cataloging-in-Publication Data

Names: Effinger, Gove W., author. | Mullen, Gary L., author.
Title: An elementary transition to abstract mathematics / Gove Effinger and Gary L. Mullen.
Description: Boca Raton : CRC Press, Taylor & Francis Group, 2020. | Includes bibliographical references and index.
Identifiers: LCCN 2019019500 | ISBN 9780367336936
Subjects: LCSH: Mathematics. | Group theory.
Classification: LCC QA9 .E375 2020 | DDC 512/.2--dc23
LC record available at https://lccn.loc.gov/2019019500

Visit the Taylor & Francis Web site at
http://www.taylorandfrancis.com

and the CRC Press Web site at
http://www.crcpress.com

# Contents

# Preface

Most undergraduate students of mathematics have spent their pre-calculus and elementary calculus years putting a solid majority of their time and effort into *computation*. Here's the Quadratic Formula; use it to solve a polynomial equation. Here's the Power Rule; use it to find a derivative. The primary activity tends to be "turn the handle," or, said another way, "plug and chug." The need to do correct computation never leaves mathematics no matter how far students advance, but as they move upward in the mathematics curriculum the abilities to *conceptualize*, to *generalize* and to *make careful logical arguments* move much more to the foreground. Where does the Quadratic Formula come from anyway, and how can we *prove* it? Why does the Power Rule work, and how can we prove it? What do integers and the real numbers have in common, and how do they differ? How many solutions can a given polynomial equation have, and how can we prove that? And so on.

The aim of this textbook then is to provide a relatively brief and hopefully easily understood transition from the computationally-centered study of pre-calculus and elementary calculus to the more conceptually-centered study of areas like abstract algebra, number theory, real analysis and topology. This transition is never going to be completely smooth, but by providing a platform which is designed to be conversational and supportive, we hope to make it as smooth as possible. We use the term "An *Elementary* Transition" because we do not go into great depth while covering many of the topics that are useful in abstract mathematics, instead attempting to provide enough of a flavor of these topics to spark the students' interest and allow them to move forward into the higher courses.

## To The Student

We have two central ideas which we hope you will fully embrace as you work your way through this book. The first one is:

**"*Why* is it true and how can we *prove* it?" is at least as important a question as "*What* is true and how can we *use* it?"**

In the first two chapters we take a look back at a few ideas from pre-calculus and calculus which should be quite familiar to you, but we look at them in a different way. How can we *prove* that the Quadratic Formula always leads to a solution of the polynomial equation $ax^2 + bx + c = 0$? Why does the Binomial Theorem give us the correct expansion of $(x + a)^n$? Where do the Power Rule and the Product Rule for derivatives come from and how can we prove that they work? What does the Fundamental Theorem of Calculus really say and how can we prove it?

As we progress then to topics with which you may not be as familiar, we continue throughout to emphasize the "why" as well as the "what." This means that you need to learn about various techniques in order to ascertain the "why," so in Chapter 3 we discuss numerous proof techniques which you can try to employ as needed. Knowing which techniques to apply in any given situation is challenging, but we try to help you begin to develop that ability as we go along. We also try to emphasize throughout that if you yourself are trying to recognize a pattern and then prove that it always holds, *looking at data* can often point the direction. We call this the "data to conjecture to proof" method (a "conjecture" is a *guess* at a pattern), and it can be very helpful. So get ready to face the "why is it true?" question many times during this course.

The second central idea is:

**Mathematics is NOT a spectator sport!**

Starting immediately in this course we expect you to have pencil and paper in hand and ready to be used. Mathematics needs to be read and done carefully, including checking each assertion which is made and answering each question which is asked, and this will almost always involve writing things down. For example, if we give an argument which involves a series of equalities, you should focus on *every single equal sign* and be sure that you fully understand why it holds. This kind of *active learning* will quickly deepen your understanding of the material at hand. You of course *must* be active in working through the many examples and exercises we have provided. As with many textbooks, we supply suggested solutions to the odd-numbered exercises (and some of the examples), but you will learn by far the most by not consulting those solutions until you have solved or at least thoroughly struggled with the problem yourself. So, don't read this book with a pizza slice in one hand and with the earbuds cranked to "eleven"; read it and react to it with pencil in hand, working in a quiet place.

So, we hope you enjoy your transition to more advanced, more abstract mathematics. Remember that *math is both important and fun*, so immerse yourself in it and reap the benefits!

# To the Instructor

This textbook is intended to be used in a one-semester course following elementary calculus (and possibly also linear algebra) but prior to any courses which put an emphasis on abstraction or proof. The contents are roughly organized into five parts, as follows:

Chapters 1 though 3 are in a "workshop" form, the first two going back to some familiar topics from pre-calculus and calculus with emphasis on "why is it true?", and then a chapter on proofs and proof techniques. All three are a bit longer than most of other chapters and are highly interactive. They possess fewer exercises at the end but more open-ended examples during the narrative, expecting immediate student input.

Chapters 4 through 12 cover the basic topics of induction, well-ordering, sets, equivalence relations, functions, cardinality, permutations, complex numbers and matrices. Though somewhat less interactive than the three "workshop" chapters, they nonetheless encourage active participation from the student. Each chapter from Chapter 4 on contains at least 10 exercises at the end; there are a total of 414 exercises throughout the book. Following standard practice, we supply suggested solutions to all odd-numbered exercises and to all open-ended examples.

Chapters 13 through 20 introduce various topics from the area of number theory, including some emphasis on congruences and modular arithmetic given that we are definitely living in the age of computers. We include in these chapters discussion of Diffie-Hellman Key Exchange and RSA encryption, which we know from teaching number theory are topics of interest to math/science students. We remark also that as of Chapter 12 (in the previous group), we begin to put an emphasis on *sets with algebraic structure*, which emphasis continues through the remainder of the text.

Chapters 21-32 contain a "light" introduction to the primary objects of abstract algebra, namely groups, rings, fields and vector spaces. These chapters in no way pretend to be a replacement for a full course in abstract algebra, but rather they try to give an overview of some of the main ideas (such as "sub-objects") and try to supply many natural *examples* of each algebraic type.

Chapters 33-35 complete the book with an investigation of single variable polynomials with an emphasis on irreducibility, on the analogy of irreducible polynomials over a field with prime numbers in the integers, and on the various forms irreducible polynomials can take over differing fields. These last chapters try to tie together numerous ideas from the rest of the book, including looping back to the very first example in the book, the Quadratic Formula.

It is our hope that the entire book can be covered in a semester, but of course you may find that to be too much, in which case you can easily omit some of the chapters which don't fit with your particular goals. In any case, we hope you find the book to be a good platform for moving your students toward more serious mathematical abstraction.

We would like to sincerely thank Serge Ballif for his extremely helpful ideas regarding tex files, giving us a very workable framework for the entire book. Thanks are also due James Sellers for his careful reading and constructive thoughts on the first few chapters of our text, and to Robert Ross, senior editor at CRC Press, for encouraging us to pursue this project.

Gove Effinger would like to thank Gary L. Mullen for inviting him to join this exciting venture! He would also like to thank his spouse, colleague and best friend Alice Dean for her support during the writing process and for sharing with him a deep love of mathematics.

Gary L. Mullen would like to thank his wife Bevie Sue Mullen for her patience and understanding during the writing of this textbook.

Gove Effinger
Gary L. Mullen
April 2019

# CHAPTER 1

# A Look Back: Precalculus Math

```
            1
         1     1
      1     2     1
   1     3     3     1
1     4     6     4     1
1   5   10   10   5   1
```

We begin our transition to abstract mathematics by looking backwards in your mathematics education to some ideas from algebra, combinatorics and geometry/trigonometry. The emphasis here will be not on the "what does it say?", but rather on the "why is it true?", as we discussed in the Preface for the Student. By looking at a few familiar ideas from this different point of view, we can begin to develop the kind of abstract thinking required to really understand more advanced mathematics.

In what follows, you are expected to have paper and pencil ready. Mathematics is not a spectator sport; the only way to fully understand an idea is to work with it yourself. In fact, in the whole text, but especially here and in the following two introductory chapters, you should think of the examples really as *exercises* for you to work your way through.

**Example 1.1. (The Quadratic Formula)** Most algebra students have memorized, or at least know how to quickly look up, the Quadratic Formula, which says:

*If $a \neq 0$, the solutions of the equation $ax^2 + bx + c = 0$ are*

$$x = \frac{-b \pm \sqrt{b^2 - 4ac}}{2a}.$$

That's the "what"; let's work out the "why?" now. The derivation of this famous formula comes from the technique for solving quadratic equations known as *completing the square*. This technique is an example of an *algorithm*, which is a finite list of steps to achieve a desired outcome. Here are the steps in this algorithm, given a quadratic equation in standard form (with $a \neq 0$), i.e., set equal to 0 with the 0 on the right-hand side.

1. Divide through by the coefficient of $x^2$ and move the resulting constant term to the right-hand side of the equation.

2. Take half of the coefficient of $x$, square this, add this result to both sides of the equation, and put the right-hand side over a common denominator.

3. The left-hand side is now the square of a linear polynomial. Rewrite the left-hand side in this form.

4. Take the square root of both sides, remembering the plus/minus on the right.

5. Solve for $x$, hence displaying our two solutions.

*Use this algorithm* now to solve the equation $3x^2 + 5x - 1 = 0$. Did you get $x = \frac{-5 \pm \sqrt{37}}{6}$? If so, good.

Now use this algorithm to solve $ax^2 + bx + c = 0$, hence deriving the Quadratic Formula. Having done so, we now know both *what* this formula is and *why* it always works.

Before moving on from the Quadratic Formula, we mention here a very surprising result which was proved by the famous mathematicians Galois and Abel in the early 19th century. It had been known previously that there existed algorithms to solve in radicals every quadratic, cubic (i.e., degree 3) and quartic (i.e., degree 4) polynomial equation with integer coefficients, but no one had been able to find algorithms to solve every such polynomial equation of degree 5 or higher. Galois and Abel proved, using advanced mathematics, that *no such algorithms can exist for degrees* $\geq 5$. So, it's not that no one has as yet discovered them; it's that they are impossible to discover. You might well ask "why?"; to find out, take a course in abstract algebra, more specifically in "field theory" (see Chapter 29).

Next, we move to the area of mathematics called combinatorics, or simply "counting theory".

**Example 1.2. (Binomial or "Choose" Coefficients)** Suppose you have $n$ distinct objects and you wish to select (or "choose") some $k$ of them (note: we are not concerned here with the *order* of the objects, just the number of them). How many choices do you have? The answer is not obvious. We wish to employ here (and elsewhere in this text) an often effective method for finding patterns and then trying to verify that the patterns hold true in general:

1. Gather some data. Compute a few specific examples of what you are studying, enough examples so you might be able to discern a pattern.

2. Conjecture (i.e., guess) at what the general pattern might be.
3. Try to prove your conjecture.

Perhaps we should call this the *Data to Conjecture to Proof Method.*

Let's try this method on our counting question above. Get your paper and pencil ready. First suppose we have 4 distinct objects $a, b, c, d$. There is one way to select none of them. Now work out and fill in below how many ways there are to select 1, then 2, then 3, then all 4. Now do the same with 5 distinct objects $a, b, c, d, e$. (Don't cheat if you already know the answers; work out by hand each of the 6 numbers here). If you are feeling courageous, try 6 objects, getting 7 numbers.

| $n(\downarrow)$ choose $k(\rightarrow)$ | 0 | 1 | 2 | 3 | 4 | 5 | 6 |
|:---:|:---:|:---:|:---:|:---:|:---:|:---:|:---:|
| 4 | 1 | | 6 | | | | |
| 5 | | | | | | | |
| 6 | | | | | | | |

Once filled in, a couple of patterns here are obvious, but others may not be. It is clear, first, that choosing all or none of the objects happens in only one way; second, that choosing 1 or all but 1 (i.e., $n-1$) happens in $n$ ways; and third, the list is palindromic (i.e., in each row the entries read the same left-to-right and right-to-left). What is not obvious from our data is where those "middle" values (e.g., the 6 in the middle for $n = 4$) come from. In this case it is, unfortunately, difficult to form a conjecture about how to compute all the coefficients on the basis of our data, so we go instead to a different counting argument to see how to compute "$n$ choose $k$" (which is denoted by $\binom{n}{k}$).

Let's start by seeing, for example, a different way to get the value of 6 for "4 choose 2". The number of ways to *arrange* 2 of our 4 objects is $(4)(3) = 12$; specifically, the set is $\{ab, ac, ad, ba, bc, bd, ca, cb, cd, da, db, dc\}$. But this set contains each set which is to be *chosen* twice, so we must divide by 2, arriving at our value of 6. Note that this calculation can written as $\frac{4!}{2!2!}$.

Now we can move to the general case. The number of ways to *arrange* $k$ of our $n$ objects is $(n)(n-1)(n-2)\cdots(n-k+1) = \frac{n!}{(n-k)!}$. But now each set to be chosen appears $k!$ times, so we must divide by $k!$. We arrive at:

*The binomial (or choose) coefficient $\binom{n}{k}$ is given by*

$$\binom{n}{k} = \frac{n!}{k!(n-k)!}.$$

Because 0! is defined to be 1, we see, as we already observed, that $\binom{n}{0} = \binom{n}{n} = 1$, that $\binom{n}{1} = \binom{n}{n-1} = n$, and finally that $\binom{n}{k} = \binom{n}{n-k}$, producing the palindromic shape of the list of all coefficients for any given $n$.

For practice, compute the number of ways to select 4 objects from 10, i.e., compute $\binom{10}{4}$. Did you get 210? Want to do this by the first method we used (i.e, "by hand")? Probably not.

What we do not know as yet is why we are calling these "*binomial* coefficients". We now discuss that.

**Example 1.3. (The Binomial Theorem).** The question we wish to address now is: If $a$ and $b$ are variables or numbers, what is the expansion of $(a+b)^n$? Well, you can easily work this out given what we developed in Exercise 1.2. Simply write $(a+b)^n$ in the form $(a+b)(a+b)\cdots(a+b)$; and for each term $a^{n-k}b^k$ in the expansion we seek, ask in how many ways we can select $a$ $(n\text{-}k)$-times and $b$ $k$-times from this string of $n$ binomials (from each binomial $a+b$, we select either $a$ or $b$). It is up to you to write down the details. You have now proved:

**Theorem 1.1. (The Binomial Theorem)** *If $a$ and $b$ are variables or numbers,*

$$(a+b)^n = \sum_{k=0}^{n} \binom{n}{n-k} a^{n-k}b^k.$$

For practice, expand $(a+b)^7$. Now expand $(x-2)^6$.

For our final example in this setting, we consider a question from geometry/trigonometry.

**Example 1.4. (The Interior Angles of a Regular Polygon)** A polygon (from the Greek for "many-sided figure") is called *regular* if all of its (linear) sides and interior angles are the same. The two "smallest" are, of course, an equilateral triangle and a square. A polygon with $n$ sides is called an $n$-gon.

Here is our question: *As a function of $n$, what is the size (in degrees) of an interior angle $A$ of a regular $n$-gon?*

Let us try to have you answer this question using the Data to Conjecture to Proof Method introduced in Exercise 1.2. Here is a (partial) table of information about the first four regular $n$-gons. You fill in the missing data and then see if you can find a pattern:

| $n$ | Interior angle $A$ | Angle $A$ as a fraction of $180°$ |
|---|---|---|
| 3 | | $1/3$ |
| 4 | | $1/2$ |
| 5 | $108°$ | |
| 6 | $120°$ | |

Can you now form a conjecture? *The size of an interior angle $A$ of regular $n$-gon is ... (some function of $n$).*

Having now formulated your conjecture (on your own or in consultation with fellow students and/or your instructor), it's time to move to a proof. Here we give a single hint: Observe that the *exterior* angles must add up to 360°. You take it from there.

In Chapter 2 we will move on to examining some ideas from calculus.

**Exercises**

**1.1.** At the end of Example 1.1 we stated that there cannot exist an algorithm to solve in radicals any given 5th degree polynomial equation with integer coefficients. This does not, of course, say that there aren't *some* 5th degree equations which *can* be solved in radicals. For example, $x^5 - 7 = 0$ is easy enough. See if you can find the solutions (3 real and 2 complex) of $x^5 + 2x^4 - 5x^3 - 8x^2 - 16x + 40 = 0$. *Hint:* Look at this as $x^5 - 8x^2 + 2x^4 - 16x - 5x^3 + 40 = 0$. By the way, any 5th degree polynomial equation with integer coefficients must have at least one *real* solution. Do you know why?

**1.2.** Concerning Example 1.4 on regular $n$-gons, in Chapter 2 we will be examining some ideas from calculus, and so limits will of course be part of our work there. Hence let us ask this question here: What is the limit as $n$ goes to infinity of the angle $A$ in your proof? Why does this make sense?

**1.3.** Try to apply the Data to Conjecture to Proof Method to the following idea about the binomial coefficients. Here are the first 5 rows ($n = 0$ to $n = 4$) of the infinitely tall tree-shaped array known as *Pascal's Triangle*:

$$
\begin{array}{ccccccccc}
 & & & & 1 & & & & \\
 & & & 1 & & 1 & & & \\
 & & 1 & & 2 & & 1 & & \\
 & 1 & & 3 & & 3 & & 1 & \\
1 & & 4 & & 6 & & 4 & & 1
\end{array}
$$

Look at any entry in a given row and the two adjacent entries just above it. State a conjecture as to what, in general, $\binom{n}{k}$ is equal to in terms of entries in the $n-1$ row just above. Now prove your conjecture using the formulas for $\binom{n}{k}$, etc. Note: This is definitely a challenging exercise.

# CHAPTER 2

# A Look Back: Calculus

$$\int_a^b f(x)dx = F(b) - F(a)$$

Moving along from Chapter 1 in reviewing ideas from your mathematics education, we now wish to take a closer look at a few different ideas from calculus. We remark that though the concept of *limit* is central to all of calculus, we will not here be concerned with the precise ($\epsilon$-$\delta$) definition of limit nor with verifying the properties of limits, such as their compatibility with the algebra of functions (e.g., the limit of a sum is the sum of the limits, etc.). We shall also assume that any generic functions we study are continuous and differentiable on their domains (that is, their derivatives exist at every point in their domains). So let's get started.

**Example 2.1. (The Power Rule)** Every student of calculus can tell you immediately that the derivative of the function $x^3$ is $3x^2$; in fact most can say that in general the derivative of the function $x^n$ is $nx^{n-1}$. What they are less likely (unfortunately) to be able to tell you is what this *means* and even less likely to be able tell you *why it is true*. What it means, of course, is that for the basic power function $x^n$, the function's *rate of change* at any point $x$ is given by $nx^{n-1}$. And how about the "why"? Well, if you know the *definition* of the derivative and if you know the Binomial Theorem (see Example 1.3), you can easily prove the result, known of course as the *Power Rule*. We recall the definition of the derivative function $f'(x)$ of the function $f(x)$:

**Definition 2.1.** (**The derivative of** $f(x)$)

$$f'(x) = \lim_{h \to 0} \frac{f(x+h) - f(x)}{h}.$$

The idea is that the derivative of $f$ at a point $x$ in its domain is the limit of the *ratio of the change in the output to the change in the input* as the change in the input goes to 0. You will recall that in order to evaluate this limit for any specific function $f$, you must be able to do algebraic manipulation sufficient to cancel the $h$ from the denominator (since otherwise that denominator is headed toward 0).

Paper and pencil time. Set $f(x) = x^n$ in the definition (so $f(x+h) = (x+h)^n$), expand the binomial, and simplify, arriving (hopefully) at the desired result. If you would like to "warm up" by working through the argument with a specific $n$, like $n = 3$, feel free.

**Theorem 2.1.** (**The Power Rule**) *For every non-negative integer $n$, the derivative function of $x^n$ is $nx^{n-1}$.*

Once you have absorbed the technique of Mathematical Induction (see Chapter 4), you can use it and the Product Rule (see the next example) to do a completely different proof of the Power Rule.

We finish this exercise by remarking that the Power Rule actually holds for any real number $n$, not just non-negative integers. One can use the Quotient Rule to establish the Power Rule for negative integers, then use implicit differentiation to establish it for all rational numbers, and finally use the all-important Chain Rule and the fact that $x^r = e^{r \ln(x)}$ to establish it, once and for all, for every real number $r$. Please see the exercises at the end of this chapter.

**Example 2.2.** (**The Product Rule**) It would be nice if the derivative of the product of two functions $f(x)$ and $g(x)$ were the product of their derivatives, but it just isn't so. For example, try that possibility with the functions $(x)(x) = x^2$. Rather, as most calculus students know, the actual rule is a bit more complicated:

**Theorem 2.2.** (**The Product Rule**)

$$(f(x)g(x))' = f(x)g'(x) + g(x)f'(x).$$

Your task now is to *prove this rule*. Return to the definition of the derivative, apply it to the product $f(x)g(x)$, and see what can be done. Sometimes, unfortunately, proofs involve a "trick" of some sort, and this is one of them. Here we need the so-called "add and subtract trick" wherein you add 0 to an expression by adding some term and its negative. In this case you need to insert a "cross-term"; specifically try inserting $-f(x+h)g(x)$ and its negative. Do this, then do some algebra and evaluate some limits, arriving (again we hope) at the desired result.

As we stated already, once we understand how to use Mathematical Induction (Chapter 4), we can use it and the Product Rule to produce an alternative proof of the Power Rule for non-negative integer exponents $n$.

**Example 2.3.** (**The Fundamental Theorem of Calculus**) Recall that for a given function $f(x)$, an *anti-derivative* $F(x)$ is a *function* whose derivative is $f(x)$. Recall also that the *integral* of $f(x)$ on an interval is (essentially) the sum of its values on that interval, and hence is a *number*. The Fundamental Theorem of Calculus says two things: On the one hand, you can use an anti-derivative of a given function $f(x)$ to compute the integral of $f(x)$ on an interval by evaluating $F(x)$ at the endpoints of the interval; on the other hand, you can produce an anti-derivative of $f(x)$ by turning the integral from a number into a function by freeing up one of the endpoints of the interval. More specifically:

**Theorem 2.3.** (**The Fundamental Theorem of Calculus**)
(1) *If $F(x)$ is an anti-derivative of $f(x)$, then $\int_a^b f(x)dx = F(b) - F(a)$.*
(2) *The function of $x$ given by $\int_a^x f(t)dt$ is an anti-derivative of $f(x)$.*

The first of these is by far the more "popular" since it provides a very quick way to compute the exact value of an integral *provided that you can write down an anti-derivative of your given function*, which may or may not be possible. Its proof depends upon the *Mean Value Theorem* and, though not too complicated, is a bit messy, so we will skip it here.

Your job here is to *prove the other side of the coin*, i.e., (2) above. This proof depends upon the following theorem, which we simply state:

**Theorem 2.4.** (**The Mean Value Theorem for Integrals**) *There exists a point $p$ in the interval $[a, b]$ for which $\int_a^b f(x)dx = f(p)(b - a)$.*

Now, apply the definition of the derivative to the function $\int_a^x f(t)dt$ and go from there, using the above theorem (adjusted to the current setting of course) at some point. Be sure to realize that the integral is a function of $x$, not $t$, so, for example, your proof should include the expression $\int_a^{x+h} f(t)dt$.

We now end this chapter with a look at *numerical integration*; that is, methods of computing the (often approximate) value of an integral when the option of using the Fundamental Theorem is not possible or practical.

**Example 2.4.** (**Simpson's Rule**) The basic technique for approximating the integral of a function $f(x)$ on an interval $[a, b]$ is to use the values of $f$ at some limited number of points in $[a, b]$. In general the more points we use, the better approximation we would hope to get.

We will look at three common such "rules", focusing especially on the third of these, known as *Simpson's Rule*. In all cases we divide the interval $[a, b]$ into $n$ equal subintervals. Here we label the $n + 1$ endpoints of these subintervals as $\{x_0(= a), x_1, x_2, \ldots, x_{n-1}, x_n(= b)\}$, and we label each of the $n$ subinterval midpoints as $m_i(= (x_{i-1} + x_i)/2)$. Also, note that in all cases we must multiply each data-point value by the length of the subinterval, which is, of course $(b - a)/n$ (and is always a common factor in the summations, so can be factored out).

The first rule makes use of $n$ data points:

**The Midpoint Rule.**

$$\int_a^b f(x)dx \approx \frac{b - a}{n} \sum_{i=1}^n f(m_i).$$

The second rule uses $n+1$ data points. Note that here each *internal* endpoint gets used twice since it is counted in two subintervals.

**The Trapezoid Rule.**

$$\int_a^b f(x)dx \approx \frac{b - a}{n} \sum_{i=1}^n \frac{f(x_{i-1}) + f(x_i)}{2}$$

$$= \frac{b - a}{n} \left( \frac{f(x_0) + 2f(x_1) + 2f(x_2) + \cdots + 2f(x_{n-1}) + f(x_n)}{2} \right).$$

The final rule combines these two rules, making use of $2n+1$ data points. *It will be your job to figure out, by looking at some data below, exactly how the two rules were combined.* The result, as you shall see, is a rule which gives really good approximations even when $n$ is small.

**Simpson's Rule.**

$$\int_a^b f(x)dx \approx \frac{b - a}{n} \sum_{i=1}^n \frac{f(x_{i-1}) + 4f(m_i) + f(x_i)}{6}$$

$$= \frac{b - a}{n} \left( \frac{f(x_0) + 4f(m_1) + 2f(x_1) + \cdots + 2f(x_{n-1}) + 4f(m_n) + f(x_n)}{6} \right).$$

Now let's look at some data about how well these approximation techniques do on specific basic functions, for all of which we can get the exact answer using the Fundamental Theorem of Calculus ("FTC"). We fix the interval at $[1, 3]$ and fix $n$ at 4, so the Midpoint Rule uses 4 data points, the Trapezoid Rule uses 5 data points, and so Simpson's Rule uses 9 data points. All results are rounded to at most 4 significant digits. A positive error means the approximation was too high, etc. The data was computed using *Mathematica*.

| Function | FTC | Mid | Error | Trap | Error | Simpson | Error |
|----------|-----|-----|-------|------|-------|---------|-------|
| $x^2$ | 8.667 | 8.625 | $-.042$ | 8.750 | $+.083$ | 8.667 | 0 |
| $x^3$ | 20 | 19.75 | $-.25$ | 20.5 | $+.5$ | 20 | 0 |
| $e^x$ | 17.37 | 17.19 | $-.18$ | 17.72 | $+.35$ | 17.37 | $+.0003$ |
| $\sqrt{x}$ | 2.797 | 2.799 | $+.02$ | 2.793 | $-.04$ | 2.797 | $-.000007$ |
| $\sin(x)$ | 1.530 | 1.546 | $+.016$ | 1.498 | $-.032$ | 1.530 | $+.000003$ |

Looking at such data can be dizzying, but concentrate on the columns showing the Midpoint and Trapezoid *Errors*. Which rule seems to be more accurate? By how much? On the other hand, their errors are always in the opposite direction, so they (sort of) cancel each other out. Using these observations, conclude that Simpson's Rule could be a *weighted average* of these others. What are the appropriate weights? Form a conjecture. Now prove that your conjecture accurately derives the given formula for Simpson from the other two formulas. Have fun!

## Exercises

**2.1.** (a) Use the definition of the derivative to prove that the derivative of $1/g(x)$ is $-g'(x)/g^2(x)$.

(b) Use Part a and the Product Rule to derive the Quotient Rule.

**2.2.** Use the Quotient Rule and the Power Rule for non-negative integer exponents to prove that if $n$ is a positive integer, then the derivative of $x^{-n}$ is $-nx^{-n-1}$. Conclude that the Power Rule holds for *all* integer exponents $n$.

**2.3.** Though not discussed as yet in this chapter, the Chain Rule is arguably the most powerful and useful of the derivative rules. This exercise and the next are two examples of this. Suppose $n$ and $m$ are integers with $m \neq 0$. Let $y = x^{n/m}$. Use implicit differentiation with respect to $x$ and the previous exercise to show that $y' = (n/m)x^{(n/m)-1}$. Conclude that the Power Rule holds for *all rational exponents*.

**2.4.** Let $r$ be an arbitrary real number (rational or irrational), and let $y = x^r$. Taking the natural (i.e., base-e) log of both sides, we get $\ln(y) = \ln(x^r) = r\ln(x)$. Now exponentiating both sides we get $y = e^{r\ln(x)}$. Using this form for $y$ and the Chain Rule, show that the Power Rule holds for *all real numbers*. Note: Though it is true that this quick proof covers every possible case of the form of the exponent in the Power Rule, it is certainly *not* the way one should introduce the Power Rule in the first place. That introduction really must be using the definition of the derivative and the Binomial Theorem to establish the rule for positive integer exponents, as you did at the outset of this chapter.

**2.5.** You probably noticed in the data in Example 2.4 that Simpson's Rule found the exact answer for the integrals of both $x^2$ and $x^3$. Perhaps this was by chance? Set $n = 1$, so that you are using only 3 data points: $f(a)$, $f((a+b)/2)$ and $f(b)$. Prove that Simpson's Rule arrives at the exact value of $\int_a^b f(x)dx$ for both $f(x) = x^2$ and $f(x) = x^3$.

# CHAPTER 3

# About Proofs and Proof Strategies

# QED

Now that we have some experience (from our two chapters on pre-calculus and calculus) with the idea of seeking not just "what's true?" but also *"why is it true?"*, we wish to focus in on the process of doing proofs. Very often students who are new to this process have difficulty knowing how to attack a given problem and in some cases may not really understand what it means to provide a proof. To try to help you through this stage, we first discuss some basic ideas about what constitutes a proof and how to proceed with one you hope to accomplish. After that we'll move to some specific strategies you can try to employ.

**Direct Proofs**. Suppose you are given a mathematical statement of some sort and you are asked to prove that the statement is true. Though the manner in which the statement is expressed can vary, most such statements will contain one or more items which are *assumed to be true* (these are called the *hypotheses* or the *premises*), and usually one item which is to be proved (this is called the *conclusion*). Your job is to find a correct logical argument which leads you from the hypotheses to the conclusion. Hence any proof must start with you clearly identifying the hypotheses and the conclusion. You must ask yourself:

1. What exactly am I trying to prove (i.e., what's the conclusion)?
2. What do I know is true (i.e., what are the hypotheses)?
3. How can I use the hypotheses to get to the conclusion?

Let's look at an example:

**Example 3.1.** Consider the function $f(x) = ax^2 + bx + c$ where $a$, $b$ and $c$ are real numbers and $a \neq 0$. Prove that if $b^2 - 4ac > 0$, then the graph of $f$ crosses the $x$-axis exactly twice.

This statement, by using the words "if" and "then", shows clearly what is the conclusion and what is the hypothesis. Make sure you identify them. Now, how to move from the "if" to the "then"? Well, what does crossing the $x$-axis (a geometric idea) say about $f$ algebraically? Now, how can we use the hypothesis? The expression $b^2 - 4ac$ should "ring a bell"; pretty obviously the Quadratic Formula (as discussed in Chapter 1) is coming into play here. Can you now see how to move directly from the hypothesis to the conclusion using that formula? Write down your proof.

Now suppose the statement was worded differently:

Consider the function $f(x) = ax^2 + bx + c$ with $a \neq 0$. Suppose $b^2 - 4ac > 0$. Show that the graph of $f$ crosses the $x$-axis exactly twice.

This is exactly the same problem, though the words "if" and "then" are not present. Here, of course, "suppose" is a stand-in for "if". Note also that "show" and "prove" mean exactly the same thing.

**If and Only If Proofs**. Now, let's actually change the statement in Example 3.1 :

**Example 3.2.** Consider the function $f(x) = ax^2 + bx + c$ with $a \neq 0$. Prove that the graph of $f$ crosses the $x$-axis exactly twice *if and only if* $b^2 - 4ac > 0$.

Note the words "if and only if." This means that there are *two* proofs you need to provide in which the hypothesis and the conclusion switch rolls. More specifically, this statement is a quick way of saying:

Consider the function $f(x) = ax^2 + bx + c$ with $a \neq 0$. Prove the following:
  (1) If $b^2 - 4ac > 0$, then the graph of $f$ crosses the $x$-axis exactly twice.
  (2) If the graph of $f$ crosses the $x$-axis exactly twice, then $b^2 - 4ac > 0$.

You already did the proof of (1). We leave the proof of (2) as Exercise 3.1. Many statements in mathematics are of this "if and only if" type, and the two proofs show that the two parts are "logically equivalent." The two statements (1) and (2) are called *converses* of each other. Hence to prove an "if and only if" statement, you must prove an "if-then" statement and its converse.

So far we have been discussing what are called *direct* proofs. Much of the time a direct proof is the best option, but there are times another route may be easier. Two such routes are a "contrapositive" proof and a "proof by contradiction". These are similar in nature but a bit different, as we now shall see.

**Contrapositive Proofs.** Given a statement "If A, then B", its *contrapositive* is the statement "If not B, then not A", i.e., we switch the hypothesis with the conclusion and negate both conditions. One can show that these two statements are logically equivalent. Here's a simple example: the contrapositive of "If it's raining, then I'm staying home" is "If I'm not staying home, then it's not raining." Let's look at an example where proving the contrapositive seems easier than a direct proof.

**Example 3.3.** Let $n$ and $m$ be positive integers. We say that $n$ and $m$ are *relatively prime* if their only common divisor is 1. Prove that if there exist integers $x$ and $y$ (one positive and one negative) such that $xn + ym = 1$, then $n$ and $m$ are relatively prime.

Well, it's not obvious how to proceed directly here, so let's try for contrapositive proof. First, of course, we must write down the contrapositive. One tricky thing here is: What is "There do *not* exist $x$ and $y$ such that $xn + ym = 1$ really saying? If you think about it, this is the same as saying "For *every* $x$ and $y$, $xn + ym \neq 1$.

Okay, so our contrapositive is: If $n$ and $m$ are *not* relatively prime, then for every $x$ and $y$, $xn + ym \neq 1$.

So now let's do a *direct* proof of *this* statement. Recall that $n$ and $m$ *not* relatively prime means that there is an integer $d > 1$ which divides them both, i.e., $n = ad$ and $m = bd$ for some positive integers $a$ and $b$. Now you should be able to see that this implies our desired conclusion (write it down!), and we are done.

**Proofs by Contradiction.** As we said above, a similar but somewhat different proof strategy is *proof by contradiction*. Here we *assume that the conclusion is false* and then follow some logical sequence of steps until we arrive at a statement which is itself clearly false. This means that our original assumption that the conclusion is false must itself be false, i.e., the conclusion must be true. (Got it?) Here is a very classic example of a proof by contradiction.

**Example 3.4.** Prove that there are infinitely many prime numbers.

Proof by contradiction (Euclid, about 270 BC): Just suppose there are only a finite number or primes, say $n$ of them, so a complete list of the primes is $\{2, 3, 5, \ldots, p_n\}$. Consider the number $2 \cdot 3 \cdot 5 \cdots p_n + 1$. This number is not divisible by any of the primes in our list (if it were, then 1 would be divisible by that prime also, which it is not), but every number is divisible by a prime (we shall prove this in Example 4.5). This contradiction proves that our assumption of only finitely many primes must have been false, so our conclusion is true. (Isn't that a very nice proof? That Euclid was no dummy.)

We turn now to a proof strategy which is somewhat specialized but very powerful when applied in the right context.

**Proofs by Mathematical Induction.** Suppose you wish to prove that all non-negative integers $n$ which are greater than or equal a smallest integer $n_0$ have a certain property $P$ ($n_0$ is very often 1, but it could be some other non-negative integer instead). A proof by *Mathematical Induction* (or more concisely "proof by induction") can often be used to establish this. The idea is that you do two things:

1. Show that the property $P$ holds for $n_0$ (this is called the "base case").
2. Show that *if* property $P$ holds for an integer $k \geq n_0$, *then* it also holds for $k + 1$ (this is called the "induction step").

Here is an example:

**Example 3.5.** What do you get when you add up the first few powers (including the 0th power) of 2? Try it out. What are $1 + 2$, $1 + 2 + 4$, $1 + 2 + 4 + 8$? Are you seeing a pattern? Let's make a conjecture and try to prove it by induction.

Prove that for all $n \geq 0$, $2^0 + 2^1 + 2^2 + \cdots + 2^n = 2^{n+1} - 1$.

Base case: Because $2^0 = 2^1 - 1$, our conjectured formula works for $n = 0$.

Induction step: *Assume* the formula works for $n = k$, i.e., *assume* that $2^0 + 2^1 + 2^2 + \cdots + 2^k = 2^{k+1} - 1$ (this is called the "induction hypothesis"). But now, moving to $n = k + 1$,

$$
\begin{aligned}
& 2^0 + 2^1 + 2^2 + \cdots + 2^k + 2^{k+1} \\
=\ & (2^0 + 2^1 + 2^2 + \cdots + 2^k) + 2^{k+1} \\
=\ & (2^{k+1} - 1) + 2^{k+1} \quad \text{by the induction hypothesis} \\
=\ & 2(2^{k+1}) - 1 = 2^{k+2} - 1
\end{aligned}
$$

and we have established our conjecture for all $n \geq 0$.

We shall go into a fair amount more detail about Mathematical Induction in Chapter 4, including a discussion of what's called *strong* induction, a technique somewhat different but in the end equivalent to what we introduced here.

We now turn to brief discussions of some ideas which may be proof strategies themselves or may be tools which can be used in the course of a proof. All are discussed in more detail in future chapters. The next three strategies apply primarily to proofs concerning integers.

**The Well-Ordering Principle.** This says that *every non-empty set of non-negative integers has a smallest element.* The usual reaction to this principle from students is "well, of course!", but we must observe that it is *not true* in various other mathematical settings.

**Example 3.6.** Prove that the open interval $(0, 1)$ on the real line has no smallest number (i.e., the interval is *not* well-ordered).

This is a perfect opportunity for a contradiction proof. Just suppose $x > 0$ is the smallest real number in $(0, 1)$. Can you write down a number which is smaller than $x$ but still positive?

We shall discuss this principle in some detail in Chapter 5.

**The Division Algorithm.** This says: *If $b$ is any integer (positive or negative) and if $a$ is a positive integer, then there exist integers $q$ and $r$ such that $b = qa + r$ with $r$ satisfying $0 \le r < a$.* If this just looks like good old long division, that's because it is. The number $b$ is the "dividend", $a$ is the "divisor", $q$ is the "quotient", and $r$ is the "remainder". The important part here is that $r$ will always be non-negative but will be strictly smaller than the divisor $a$.

**Example 3.7.** (a) Given $b = 152$ and $a = 23$, find $q$ and $r$ such that $b = qa + r$ and satisfying the Division Algorithm criteria.

(b) Do the same for $b = -152$ and $a = 23$. (Did you get $r = 9$? Remember that the remainder must be non-negative.)

The Division Algorithm for integers is discussed in more detail in Chapter 5 and Chapter 13, and we'll discuss a version of the Division Algorithm for polynomials in Chapter 33.

**The Euclidean Algorithm.** (Here's Euclid again!) If $n$ and $m$ are two positive integers, their *greatest common divisor $d$* is the largest positive integer which divides both $n$ and $m$. The number $d$ is often denoted $\gcd(n, m)$. *The Euclidean Algorithm* provides a quick procedure for calculating $\gcd(n, m)$. Though we won't go into the algorithm itself here (see Chapter 13), it employs repeated applications of the Division Algorithm, and it is guaranteed to finish because of the Well-Ordering Principle. This algorithm also shows that $d = \gcd(n, m)$ can be written as a linear combination of $n$ and $m$, i.e., we can find integers $x$ and $y$ (one positive and one negative) such that $d = xn + ym$. This can be very useful in numerous settings. Note that we hinted at this idea in Example 3.3.

**Example 3.8.** (a) By any method you wish (probably by factoring), compute the gcd of 105 and 231.

(b) Can you find integers $x$ and $y$ such that $105x + 231y =$ your answer to part a? (Note: it's not easy to see how to do this. The Euclidean Algorithm gives a set procedure for computing one such pair $(x, y)$ in which both $x$ and $y$ are very small.)

As stated previously, we discuss the Euclidean Algorithm in more detail in Chapter 13.

**Case by Case Proofs.** In trying to prove some result, it may be difficult to find a single argument which works. It may however be possible to prove this result for a relatively small number of special cases which, in total, cover the whole desired result.

**Example 3.9.** Prove that the product $n(n+1)(n+2)$ of any three consecutive positive integers is divisible by 6.

By the Division Algorithm, $n = 6q + r$ for some $q$ and $r$ where $0 \leq r < 6$, i.e., $r$ can take on one of the six values 0 through 5. So let's work through those six cases one by one:

Case $r = 0$: Now $n = 6q$, so $n(n+1)(n+2)$ is divisible by 6 since $n$ is.

Case $r = 1$: Here $n = 6q + 1$, so $n + 1 = 6q + 2$ which is divisible by 2. Also $n + 2 = 6q + 3$, which is divisible by 3. Hence $n(n+1)(n+2)$ is divisible by 6.

You get the idea. Work through the remaining four cases yourself.

We shall now discuss two more ideas before bringing this chapter to a close.

**A "Back Door" Counting Technique.** Sometimes you may wish to count the number of elements in a finite set which have a certain property. Keep in mind that if you find it difficult to obtain this count directly, it may be easier to count the number of elements which do *not* have this property. Then of course the count of our desired set will be the count of the whole set minus the count of our "complementary" set. We can think of this as getting our desired result by "coming in the back door."

**Example 3.10.** Suppose $p$ is a prime number and $k$ is a positive integer. How many positive numbers less than or equal to $p^k$ are relatively prime to $p^k$?

See Example 3.3 for the definition of "relatively prime." The count of our whole set is clearly $p^k$. It looks like it would be easier to count the number which are *not* relatively prime to $p^k$. Since $p$ is prime, these are precisely numbers of the form $ap$ where $a$ ranges from 1 to $p^{k-1}$. That set contains $p^{k-1}$ numbers, so, by coming in the back door, we see that the set of numbers between 1 and $p^k$ which *are* relatively prime to $p^k$ has exactly $p^k - p^{k-1}$ numbers, and we are done.

We shall explore these ideas more closely in Chapter 19. Now, one last strategy.

**The Box (or Pigeon Hole) Principle.** This simple but surprisingly helpful principle states that if you have $n$ objects which you wish to distribute into $m$ boxes (or "pigeon holes") with $m < n$ (i.e., there are more objects than boxes), then one of the boxes must end up with at least two objects. For example, if the mail carrier has 11 letters to be distributed into 10 mail boxes, then lots of possible distributions can occur (e.g., all 11 could go into one box, etc.), but one thing we can be sure of is that some box must get at least two letters. Here's a somewhat more interesting example.

**Example 3.11.** Suppose you select 5 (different) numbers from the numbers 1 through 8. Prove that two of the numbers must add up to 9.

Well, we could do a case by case proof, but since there are $\binom{8}{5} = 56$ ways to select the 5 numbers (see Example 1.2), this seems quite impractical. Let's try for a much quicker proof using the Box Principle. Let the boxes be the four *pairs* of numbers from 1 to 8 which add up to 9 (e.g., $1+8$, etc.). Now distribute the 5 selected numbers into those four boxes. Done!

We finish this chapter by remarking that many times (but not in this text) you will see "QED" written at the end of a proof. QED stands for the Latin "quod erat demonstrandum," meaning "what was to be demonstrated." You will have numerous chances to write this down, if you so choose, as you move through this text. We actually prefer to use the symbol at the right to signal that a proof is completed. ∎

### Exercises

**3.1.** Following up on Example 3.2, prove directly that if the graph of $f(x) = ax^2 + bx + c$ $(a \neq 0)$ crosses the $x$-axis exactly twice, then $b^2 - 4ac > 0$. (Hint: Suppose $f$ crosses the $x$-axis at $r_1$ and $r_2$ with $r_1 \neq r_2$, then/newline $(x - r_1)(x - r_2) = 0$. Multiplying out, we see what $a$, $b$ and $c$ are. Now compute $b^2 - 4ac$.)

**3.2.** (a) For a few small positive values of $n$ , say $n = 2$ through 6, write down a little table of $n$ and the value of $2^n - 1$. What seems to happen when $n$ is prime? When $n$ is not prime (i.e., is *composite*, meaning $n = a \cdot b$ with $a > 1$ and $b > 1$)? On the basis of this small amount of data, write down an if and only if conjecture.

(b) It turns out that one direction in your conjecture is true but the other is (unfortunately) false. The statement below is the true direction; see if you can establish it using a contrapositive proof.

Prove that if $2^n - 1$ is prime, then $n$ is prime.

So start by saying "Just suppose $n$ is *not* prime" (i.e., suppose $n = ab$ where both $a$ and $b$ are greater than 1). Now do some algebra, using the fact that for any positive integers $x$ and $b$,

$$x^b - 1 = (x - 1)(x^{b-1} + x^{b-2} + \cdots + x + 1),$$

and (hopefully) conclude that $2^n - 1$ is also not prime.

(c) We claimed that the statement "If $p$ is prime, then $2^p - 1$ is also prime" is false. Can you find a *counter-example*, i.e., a prime $p$ for which the statement is false? (Hint: Such a prime $p$ exists between 10 and 20.)

**3.3.** (a) In relation to Example 3.4, try to use some data to form a conjecture about the sum of the first $n$ powers of 3 (including the 0th power). *Hint:* The resulting sum is similar to the case of powers of 2, but there's now a denominator.

(b) Now prove your conjecture using induction.

**3.4.** (a) Suppose a sold-out theater has 400 seats. Prove that two of the patrons must have the same birthday (month and day, not year).

(b) An interesting fact is that in order for the probability of two matching birthdays occurring to become greater than 50%, we would only need 23 patrons present! One can compute this using a back door approach, i.e., compute the probability that *no* matching birthdays occur. The idea is: there are 365 possible birthdays (leaving out Leap Year). The probability that two people have different birthdays is $365/365 \cdot 364/365 = 99.7\%$. The probability that three people all have different birthdays is $365/365 \cdot 364/365 \cdot 363/365 = 99.2\%$, and so on. If you have access to some math software (or a calculator and some patience), find out how far you have to go before the product goes below 50%. Since we are using the back door, the probability we seek is then above 50%.

**3.5.** (a) Starting with two positive integers $a$ and $b$, suppose we use the Division Algorithm to write $b = aq + r$. Now replace $b$ by $a$ and $a$ by $r$ (i.e., the old divisor becomes the new dividend and the old remainder becomes the new divisor). Apply the division algorithm again obtaining $a = rq_2 + r_2$. Repeat this process over and over. Prove that eventually $r_n$ must be 0 for some $n$.

(b) Try running this procedure on $a = 87$ and $b = 102$. If you do it correctly you should apply the Division Algorithm four times. What is the remainder just preceding of the remainder of 0? What is the $\gcd(87, 102)$? It is not a coincidence that they are the same. You have just used the Euclidean Algorithm to find $\gcd(87, 102)$.

# CHAPTER 4

---

# Mathematical Induction

---

In this chapter we study a very important proof technique, the principle of Mathematical Induction, whose beauty and usefulness stem from the fact that it can be used in many different settings in mathematics to prove a wide array of results.

The basic idea of mathematical induction can be visualized in the following way: suppose you have a set of dominoes standing on end in a row and suppose you want to knock them all down. How can you accomplish this? One way involves two steps:

1. Knock over the first one, and

2. Ensure that *if* any one domino goes over, *then* the one next to it will also go over.

Now suppose that instead of having a set of dominoes, you have the set of all positive integers $\{1, 2, 3, \ldots\}$, and you wish to show that some property holds true for all of them. One method to accomplish this would be:

1. Show that the property holds for 1 (this is called the *base case*), and

2. Show that *if* the property holds for a positive integer $k$, *then* it also holds for $k + 1$ (this is called the *induction step*).

If we can do this, we see that 1 is "knocked over," but then 2 is also knocked over, and 3, and so on. Let's look at a very nice example.

**Example 4.1.** Let us prove that for every positive integer $n$,

$$1 + 2 + \cdots + n = \frac{n(n+1)}{2}.$$

Before tackling the proof itself, it is usually instructive simply to try a few cases (i.e., look at some *data*) to see if the result seems to work. So, substitute $n = 1$, 2 and 3 into this equation and see if it holds. Does it?

So now we proceed. Note that we already have established the base case (in fact, three of them), so we move to the induction step. We now *assume* that our equation above holds for some positive integer $k$; i.e., we assume that

$$1 + 2 + \cdots + k = \frac{k(k+1)}{2}.$$

We need now to show that *given this assumption* our equation holds for $k + 1$. Now we do some algebra:

$$
\begin{aligned}
& 1 + 2 + \cdots + k + (k+1) \\
= \; & (1 + 2 + \cdots + k) + (k+1) \\
= \; & \frac{k(k+1)}{2} + (k+1) \quad \text{by the induction hypothesis} \\
= \; & (k+1)(\frac{k}{2} + 1) \\
= \; & \frac{(k+1)(k+2)}{2}.
\end{aligned}
$$

This is exactly the equation we seek for the case $n = k + 1$, and we have succeeded in proving our result by employing Mathematical Induction. ∎

For example, what is the sum of the first 100 positive integers? Want to figure that out "by hand?" No need - by our equation we have

$$1 + 2 + \cdots + 100 = \frac{100(101)}{2} = 50(101) = 5050.$$

Here now is a formal statement of this method:

**Theorem 4.1. (Mathematical Induction)** *Let $P(n)$ be a mathematical statement about the positive integer $n$. Assume that $P(1)$ holds and whenever $P(k)$ holds so does $P(k + 1)$. Then the statement $P(n)$ holds for all positive integers $n$.*

We note that although the base case in the theorem is the case $n = 1$, we can in fact start induction from *any* non-negative integer $n_0$. Now the base case to check is $P(n_0)$, and the induction step is to assume that $P(k)$ is true for some $k \geq n_0$ and show that the truth of $P(k + 1)$ follows. The assumption that the statement $P(k)$ is true is often called the *induction hypothesis*.

Here is a similar but slightly more complicated example. We note that to begin a proof using Mathematical Induction, you must have an "educated guess" of what property you want to prove (perhaps gotten via the Data to Conjecture to Proof method we have discussed). In this example it would be tricky to come up with the conjectured equation on the basis of a few data items.

**Example 4.2.** We claim that for each positive integer $n \geq 1$,

$$1^2 + 2^2 + \cdots + n^2 = \frac{n(n+1)(2n+1)}{6}.$$

Again, try this equation out for $n = 1$, 2 and 3, just to see if it makes sense at all. If yes, we have already established the base case. Next we assume the statement is true for $n = k$ so that

$$1^2 + 2^2 + \cdots + k^2 = \frac{k(k+1)(2k+1)}{6}.$$

Given this induction hypothesis, for the case $n = k + 1$ we have that

$$
\begin{aligned}
& 1^2 + 2^2 + \cdots + k^2 + (k+1)^2 \\
=\ & \frac{k(k+1)(2k+1)}{6} + (k+1)^2 \quad \text{by the induction hypothesis} \\
=\ & (k+1)[\frac{2k^2 + k + 6(k+1)}{6}] \\
=\ & \frac{(k+1)(2k^2 + 7k + 6)}{6} \\
=\ & \frac{(k+1)(k+2)(2k+3)}{6}
\end{aligned}
$$

which completes the inductive proof. ∎

Here now is an example of a different flavor. In particular, the property we wish to prove is not an equation, as were our first two examples. Also, here, let's try to employ the Data to Conjecture to Proof method.

**Example 4.3.** For a few small positive integers $n$, compute the expression $2^{2n} - 1$ and look for a number which divides all of them. (As will be discussed in greater detail in Chapter 13, we say an integer $d$ *divides* an integer $a$ if there is an integer $m$ for which $a = md$.) If you can identify such a common divisor, form your conjecture: *For every positive integer $n$, the expression $2^{2n} - 1$ is divisible by $\cdots$.*

Hopefully the common divisor you discovered is 3, so now let's try to prove our conjecture by Mathematical Induction. The base case $n = 1$ is established. We *assume* the induction hypothesis, making use of the definition of "divides"; that is, we assume that $2^{2k} - 1 = 3m$ for some integer $m$ (so $2^{2k} = 3m + 1$).

Now we use this to see if our property holds for $n = k + 1$:

$$
\begin{aligned}
2^{2(k+1)} - 1 &= 2^{2k+2} - 1 \\
&= 4 \cdot 2^{2k} - 1 \\
&= 4(3m + 1) - 1 \quad \text{by the induction hypothesis} \\
&= 12m + 3 \\
&= 3(4m + 1).
\end{aligned}
$$

So we see that $2^{2(k+1)} - 1$ is divisible by 3, and the proof is complete. ∎

For our next example, compute the sums $1 + 3$, $1 + 3 + 5$, $1 + 3 + 5 + 7$. See a pattern? Let's prove it. Notice that the $n$th odd number can be denoted by $2n - 1$.

**Example 4.4.** Prove that for all positive integers $n \geq 1$

$$1 + 3 + 5 + \cdots + (2n - 1) = n^2.$$

The base case $n = 1$ clearly works (as do the other cases we checked, of course). We assume the induction hypothesis:

$$1 + 3 + 5 + \cdots + (2k - 1) = k^2.$$

Now, for $n = k + 1$, we have

$$
\begin{aligned}
&(1 + 3 + 5 + \cdots + (2k - 1)) + (2k + 1) \\
&= k^2 + 2k + 1 \quad \text{by the induction hypothesis} \\
&= (k + 1)^2,
\end{aligned}
$$

which completes the (surprisingly easy) proof. ∎

We remark here that most often there are different ways to arrive at a correct proof of a given result. One can think of it as taking different, but all correct, paths to a given destination. Often, for example, one can use something already proved to get a new result. In this "sum of odd numbers" example, we can use Example 4.1 (twice) and the fact that the sum of odd numbers up to some point is the sum of all numbers minus the sum of even numbers up to that point, as follows: (Note that this is an example also of a "back door" approach, as discussed in Chapter 3.)

$$
\begin{aligned}
&1 + 3 + 5 + \cdots + (2n - 1) \\
&= (1 + 2 + 3 + 4 + \cdots + (2n - 1)) - (2 + 4 + \cdots + (2n - 2)) \\
&= \frac{(2n - 1)(2n)}{2} - 2\frac{(n - 1)n}{2} \quad \text{by Example 4.1} \\
&= 2n^2 - n - n^2 + n = n^2.
\end{aligned}
$$

As we said at the start of this chapter, Mathematical Induction can be employed in many different settings. For example, in Chapter 9 we shall use it to prove that if $A$ is a set containing exactly $n$ distinct elements, then the *power set* of $A$, i.e., the set of all subsets of $A$, contains exactly $2^n$ subsets.

Finally in this chapter, we briefly describe a second form of Mathematical Induction, often called *strong induction*. Again $P(n)$ will denote a mathematical statement about the positive integer $n$. We first check that $P(1)$ holds (or that $P(n_0)$ holds for some positive integer $n_0$) to get the induction started. However, in the induction step, instead of assuming that just $P(k)$ holds, we assume that $P(m)$ holds for all $n_0 \leq m \leq k$ and use those (multiple) assumptions to show that $P(k+1)$ holds. Hence this type of induction uses a "stronger" induction hypothesis than the type we have been using to this point. That induction, which we formulated in Theorem 4.1, is often called *weak* induction since the induction hypothesis uses only the assumption that $P(k)$ holds. In fact, it can be shown that the two forms of Mathematical Induction are equivalent, i.e., each implies the other.

Let's illustrate the use of strong induction to get a proof of a fundamental result about integers and prime numbers. Recall that a *prime p* is a positive integer which is divisible by exactly two different values, namely 1 and $p$.

**Example 4.5.** We claim that every positive integer $n \geq 2$ is divisible by a prime.

For the base case, the result certainly holds for $n = 2$ (note here that we will start the induction with $n = 2$ since by the above definition, 1 is not a prime). Now assume the result is true for each $2 \leq m \leq k$, i.e., that every positive integer between and including 2 and $k$ is divisible by a prime. If $k+1$ is a prime, we are done because an integer is always divisible by itself. If $k+1$ is not a prime, then $k+1 = ab$, where $1 < a < k+1$ and $1 < b < k+1$. Now since $a \leq k$ and $b \leq k$, they are, by the strong induction hypothesis, each divisible by a prime. Hence $k+1 = ab$ is also divisible by a prime, and the proof is complete. ∎

Actually, we can use strong induction to show that every positive integer $n \geq 2$ is not only divisible by a prime, but it can be written as a product of primes.

**Example 4.6.** Let $S$ be the set of positive integers greater than 1 which are either primes or can be written as products of primes. Clearly $n = 2$ is in $S$ since 2 is a prime. Now assume that for some positive integer $k$, the set $S$ contains all integers $m$ with $2 \leq m \leq k$. We must show that $k+1 \in S$.

If $k+1$ is a prime, then it is clearly in $S$ because of the definition of the set $S$. On the other hand, if $k+1$ is not a prime, it can be written in the form $k+1 = ab$ where $1 < a < k+1$ and $1 < b < k+1$. From our (strong) induction hypothesis and the fact that $a \leq k$ and $b \leq k$, we know that $a$ and $b$ are in the set $S$. Thus each of the values $a$ and $b$ is either a prime or is a product of primes.

Thus, when we multiply $a$ and $b$ together, the resulting product $ab = k + 1$ will also be a product of primes, and our proof is complete. ∎

As we move on in this text, we will often make use of Mathematical Induction, whether in its weak or strong form.

## Exercises

**4.1.** Define a sequence of positive integers $a_n$ by letting $a_1 = 1$ and then letting $a_{n+1} = 2a_n + 1$ for each $n \geq 2$. List the first five numbers in this sequence. Show, for each positive integer $n$, that $a_n + 1$ is a power of 2.

**4.2.** For each positive integer $n$, consider the number $n^5 - n$. Write down these numbers for $n$ from 1 to 4. Now prove that for all such $n$, these numbers are divisible by 5.

**4.3.** For each non-negative integer $n$, consider the number $3^{2n} - 1$. Write down these numbers for $n$ from 0 to 3. Now prove that for all such $n$, these numbers are divisible by 8.

**4.4.** Is it true, for each positive integer $n \geq 2$, that $2^n - 1$ is a prime number? How about if $n$ is odd? Prove, or disprove by finding counter-examples.

**4.5.** Find a simple "closed" formula (i.e., a formula with no summation in it) for the sum of the first $n$ even positive integers.

**4.6.** In Chapter 2 you proved the Power Rule for positive integer exponents (i.e., if $f(x) = x^n$ with $n$ a positive integer, then $f'(x) = nx^{n-1}$) using the definition of the derivative and the Binomial Theorem. Now do a completely different proof, using Mathematical Induction and the Product Rule.

**4.7.** Consider the summation $1 \cdot 2 + 2 \cdot 3 + 3 \cdot 4 + \cdots + n(n+1)$. Work out this sum for $n = 1$ through 4. Now prove by induction that for any positive integer $n$, this sum is $\frac{n(n+1)(n+2)}{3}$.

**4.8.** Define a sequence of numbers by setting $y_0 = 2$ and $y_1 = 5$ along with the recursion

$$y_{n+2} = 5y_{n+1} - 3y_n$$

for each $n \geq 0$. Show, for each integer $n \geq 0$, that

$$2^n y_n = (5 + \sqrt{13})^n + (5 - \sqrt{13})^n.$$

(Note: This problem requires a fair bit of algebra!)

**4.9.** In this and the next two exercises, $F_n$ will denote the $n$th Fibonacci number. The sequence of Fibonacci numbers is defined by $F_1 = 1, F_2 = 1$ and $F_n = F_{n-1} + F_{n-2}$ for $n \geq 3$. Thus, after the first two Fibonacci numbers, a given Fibonacci number is obtained by adding together the two previous Fibonacci numbers. Write down the first 8 Fibonacci numbers. Now show, for each positive integer $n \geq 1$, that $F_1 + F_2 + \cdots + F_n = F_{n+2} - 1$.

**4.10.** Show, for each positive integer $n \geq 1$, that

$$F_1 + F_3 + \cdots + F_{2n-1} = F_{2n}.$$

**4.11.** Show, for each positive integer $n \geq 1$, that

$$F_2 + F_4 + \cdots + F_{2n} = F_{2n+1} - 1.$$

**4.12.** Let $a_n$ be a sequence defined by $a_1 = 12, a_2 = 30$, and $a_n = 5a_{n-1} - 6a_{n-2}$ for each $n \geq 3$. Prove, for all $n \geq 1$, that $a_n = 2 \cdot 3^n + 3 \cdot 2^n$.

**4.13.** For each positive integer $n \geq 2$, show that

$$\left(1 - \frac{1}{2^2}\right)\left(1 - \frac{1}{3^2}\right)\left(1 - \frac{1}{4^2}\right) \cdots \left(1 - \frac{1}{n^2}\right) = \frac{n+1}{2n}.$$

**4.14.** For each positive integer $n \geq 1$, show that

$$\frac{1}{1 \cdot 3} + \frac{1}{3 \cdot 5} + \cdots + \frac{1}{(2n-1)(2n+1)} = \frac{n}{2n+1}.$$

**4.15.** Show, for each positive integer $n \geq 1$, that

$$1 - 2^2 + 3^2 - 4^2 + \cdots + (-1)^{n-1}n^2 = \frac{(-1)^{n-1}n(n+1)}{2}.$$

**4.16.** Show, for each positive integer $n \geq 1$, that

$$1 \cdot 2 \cdot 3 + 2 \cdot 3 \cdot 4 + 3 \cdot 4 \cdot 5 + \cdots + n(n+1)(n+2) = \frac{n(n+1)(n+2)(n+3)}{4}.$$

**4.17.** Show, for each positive integer $n \geq 1$, that

$$1 \cdot 1! + 2 \cdot 2! + 3 \cdot 3! + \cdots + n \cdot n! = (n+1)! - 1.$$

Recall, for a fixed positive integer $m$, that

$$m! = (m)(m-1)(m-2) \cdots (2)(1)$$

is the product of the first $m$ distinct positive integers.

**4.18.** We know from our previous work (Example 4.1) that the sum of the first $n$ positive integers is a polynomial of degree two in $n$. Similarly (Example 4.2) we know that the sum of the squares of the first $n$ positive integers is a polynomial of degree three in $n$. Let's assume then that the sum of the cubes of the first $n$ positive integers is a polynomial in $n$ of degree four (which is indeed the case), and let's find this polynomial, which must be of the form

$$an^4 + bn^3 + cn^2 + dn + e$$

for some coefficients $a, b, c, d, e$. Here is a possible procedure:

(1) By replacing $n$ in this polynomial by $n+1$ and then gathering the terms, you get one form of the polynomial for the sum of the first $n+1$ cubes.

(2) Get a second form for $n+1$ by simply adding $(n+1)^3$ to the above polynomial and again gathering terms.

(3) By comparing the coefficients of each term in the two forms, you get four (useful) equations in the five values $a$ through $e$. Solve those equations, obtaining the value of each one except $e$.

(4) Use the fact that our polynomial must be correct for the case $n=1$ to nail down the value of $e$, and you are done.

(Note: This exercise requires some very careful calculations. To make sure you are on the right track, we reveal that $a = 1/4$.)

# CHAPTER 5

# The Well-Ordering Principle

# $\{\ldots, 5, 4, 3, 2, 1, \text{Go!}\}$

In this chapter we focus on the *Well-Ordering Principle*, which we first introduced in Chapter 3. It turns out that this simple-sounding principle is actually *equivalent* to the Principle of Mathematical Induction which we studied in the previous chapter. This means that each principle holds if and only if the other holds.

As you work through this and subsequent chapters, you will notice that the style becomes somewhat more formal as we continue our transition to higher mathematics. This is to be expected, but it does not mean that it's time to put the pencil and paper aside. No matter what the level of mathematics you are engaging, you should always be ready to look for examples, write down some data, see if you can find or at least finish off a proof, and so on. Always try to be an active learner.

Before moving on, we wish to state here that throughout the book we shall use the term "integers" to refer to the set of *all* integers (positive, negative and 0), "non-negative integers" for the set $\{0, 1, 2, 3, \ldots\}$, and "positive integers" for the set $\{1, 2, 3, \ldots\}$. If we refer to these sets by symbols, those will be, respectively, $\mathbb{Z}$, $\mathbb{Z}^+ \cup \{0\}$, and $\mathbb{Z}^+$.

So now we state the principle:

**Theorem 5.1. (The Well-Ordering Principle)** *Let $Y$ be a non-empty set of non-negative integers. Then $Y$ has a smallest element.*

As we remarked in Chapter 3, this principle seems obvious, but it does not hold for many sets. Clearly it does not hold for (all) the integers, and in Example 3.5 you were asked to prove that the open interval $(0, 1)$ on the real line has no smallest number.

How could anything that sounds so simple be useful? We illustrate by using it to prove the Division Algorithm, which we introduced but did not prove in Chapter 3. In Example 3.6 you were asked to compute two specific examples of this algorithm; you might want to look back at that now. Again, keep in mind that it is important to look at some *data* to make abstract concepts more concrete. Below then is the statement of this result followed by a proof which assumes the Well-Ordering Principle.

One word first on notation: we shall frequently use "set-builder notation", whose form is {elements|condition(s)}. For example, the set $D$ below is read "the set of all integers of the form $b - ak$ such that $b - ak$ is positive."

**Theorem 5.2. (Division Algorithm)** *Let $a$ and $b$ be integers with $a > 0$. Then there are integers $q$ and $r$ with $0 \leq r < a$ so that $b = aq + r$.*

**Proof.** If $a$ divides $b$, i.e., if $b = ak$ for some integer $k$, then we clearly we have $r = 0$ and we're done. Hence we assume that $a$ does not divide $b$. Let

$$D = \{b - ak \mid b - ak > 0 \text{ with } k \text{ an integer}\}.$$

(Note that if $b$ is negative, then so is $k$.) The set $D$ certainly consists of positive integers, and it is non-empty because if we select any $k$ less than the rational number $b/a$, we have $b > ak$.

Hence, by the Well-Ordering Principle, $D$ contains a smallest element, say $r$. Since $r \in D$, $r$ must be of the form $r = b - aq$ for some integer $q$. To finish the proof we must now show that $r < a$. Suppose not; that is, just suppose $r \geq a$. First, $r$ cannot equal $a$ since then we would have $b = a(q + 1)$, i.e., $b$ would be divisible by $a$. Also, though, if $r > a$, then $r - a = b - aq - a = b - a(q+1) > 0$; that is, $r - a$ is in $D$. This is a contradiction, since $r$ is the smallest element in $D$. We must conclude that $r < a$, and we are done (Whew!) ∎

As we observed in Chapter 3, the Division Algorithm is the theoretical basis of ordinary long division; that is, given an integer dividend $b$ and a positive integer divisor $a$, we are always guaranteed to find an integer quotient $q$ and a non-negative remainder $r$ which is strictly less than the divisor $a$.

Here comes another important result which follows from the Well-Ordering Principle. You should recall that we actually proved this result using induction in the previous chapter (Example 4.5). We shall see shortly why it's not surprising that we could use either induction or well-ordering to achieve a proof.

**Theorem 5.3.** *Every positive integer larger than 1 is divisible by a prime number.*

**Proof.** Let $b \geq 2$ be a positive integer. Let $S$ be the set of positive integers $\geq 2$ which divide $b$. The set $S$ is non-empty since $b$ itself is in $S$. By the Well-Ordering Principle, the set $S$ has a least positive integer, say $d$. If $d$ is not a prime, then $d$ has a smaller divisor $\geq 2$, which of course also divides $b$. This contradicts the fact that $d$ is the smallest element in the set $S$. Thus $d$ must be a prime, and the proof is complete. (That was pretty easy.) ∎

As a final example of the important role the Well-Ordering Principle plays, we show that assuming it, the Principle of Mathematical Induction follows.

**Theorem 5.4.** *The Well-Ordering Principle implies the Principle of Mathematical Induction (Theorem 4.1).*

**Proof.** Let $P(n)$ be a mathematical statement about a positive integer $n$. We assume that
(i) $P(1)$ is true and
(ii) for any $n \geq 1$, if $P(n)$ is true, then $P(n+1)$ is true.
We wish to show that $P(n)$ is true for all $n \geq 1$.

Let $S$ be the set of positive integers $n \geq 1$ for which $P(n)$ is *false*. The set $S$ must be either empty or non-empty. If the set $S$ is empty, we are done.

Just suppose then that the set $S$ is non-empty, and so, by the Well-Ordering Principle, $S$ must contain a smallest number $s$. Since $P(1)$ holds, we know that 1 is not in the set $S$. Hence $s \geq 2$, and so $s - 1$ is positive.

The definition of $s$ as the smallest number in $S$ implies that $P(s-1)$ must hold. But now by our assumption (ii) above, we must have that $P(s)$ also holds, which contradicts the fact that $s$ is in the set $S$. The element $s$ cannot be both in the set $S$ and not in the set $S$ at the same time, and hence we must conclude that the set $S$ is empty. Thus $P(n)$ holds for every positive integer $n$, as desired. ∎

It turns out that the two principles of Well-Ordering and Mathematical Induction are indeed equivalent, i.e., each implies the other. In Exercise 5.6 below you will be asked to prove the converse of Theorem 5.4.

## Exercises

**5.1.** Consider the two numbers 30 and 72. An example of a *common multiple* of them is $(30)(72) = 2160$, but it is probably not their *least* common multiple. By any method, probably factoring, find the least common multiple of 30 and 72.

**5.2.** Show that any two positive integers $a$ and $b$ have a **least common multiple**; i.e., an integer $m$ which is a common multiple of both $a$ and $b$, and which is less than or equal to any other common multiple of $a$ and $b$. (Hint: Let $M$ be the set of all common multiples of $a$ and $b$. Establish that $M$ is non-empty, and then apply the Well-Ordering Principle.)

**5.3.** In Example 3.5 you were to show that the open interval $(0,1)$ is not well-ordered. Intervals on the real line can be open, closed, or half-open half-closed. Describe exactly which such intervals are well-ordered and which are not.

**5.4.** An infinite subset of the real numbers is called *countable* if one can display a list of all its elements. It turns out that the positive rational numbers are countable, as we shall prove Chapter 9. One might hope that since this set is countable, it might be well-ordered as well. Show that this is not the case.

**5.5.** The *Gaussian integers* are complex numbers of the form $a + bi$ where $a$ and $b$ are integers. The *norm* $N$ of a Gaussian integer is its distance from the origin in the complex plane, so (by the Pythagorean Theorem, everyone's favorite), we have $N(a + bi) = \sqrt{a^2 + b^2}$.
(a) Prove that the set of norms of any non-empty set $S$ of Gaussian integers is well-ordered. (Hint: Look at the set of *squared* norms in $S$.)
(b) What is the smallest norm of the set all non-zero Gaussian integers?
(c) What is the smallest norm of the set of all Gaussian integers lying strictly outside of the circle centered at the origin of radius 5? Write down three Gaussian integers which have this norm. How many are there in total?

**5.6.** Prove that the Principle of Mathematical Induction implies the Well-Ordering Principle. (Hint: Let $S$ be a set of positive integers which contain no least element. We must show that $S$ is empty. Define a set $T$ to be the set of all positive integers $n$ such that $n$ is not greater than or equal to any element of $S$. Show by induction that $T$ is the set of all positive integers, and hence the set $S$ is empty, to complete the proof.—

**5.7.** In Example 4.1 we used induction to prove that for all positive integers $n$,

$$1 + 2 + 3 + \cdots + n = \frac{n(n+1)}{2}.$$

In light of Exercise 5.6 and Theorem 5.4, it should not be a surprise that one can use the Well-Ordering Principle to prove this result. Do so.

**5.8.** The rational number $b/a$ (where $b$ and $a$ are positive integers) is said to be *in lowest terms* provided that $b$ and $a$ are relatively prime (i.e., $b$ and $a$ have no common divisor besides 1). Use the Well-Ordering Principle to prove that every rational number $b/a$ can be put in lowest terms. (Hint: Let $S$ be the set of all numerators $b$ of rational numbers $b/a$ which *cannot* be put into lowest terms and assume that $S$ is non-empty. By well-ordering, let $b_0$ be the smallest number in $S$. We have then that for some $a_0$, $b_0/a_0$ cannot be put into lowest terms, so there exists an integer $d > 1$ such that $d$ divides both $b_0$ and $a_0$. But now

$$\frac{b_0/d}{a_0/d} = b_0/a_0.$$

You finish this contradiction argument.)

**5.9.** Prove that $\sqrt{2}$ is irrational; that is, prove that $\sqrt{2}$ cannot be written in the form $b/a$ where $b$ and $a$ are positive integers. (Hint: Just suppose $\sqrt{2}$ *can* be so written, i.e., just suppose $\sqrt{2} = b/a$. By Exercise 5.8, we may assume $b/a$ is in lowest terms. Multiply both sides by $a$ and square, so we're now looking at an integer equation. Inspecting the divisors of each side, argue your way to a contradiction.)

**5.10.** Prove that if $p$ is a prime number, then $\sqrt{p}$ is irrational.

**5.11.** Suppose that the only postage stamps available are 5 cent and 3 cent. Can we form any postage amount of 8 cents or greater using these two types? Prove that we can. (Hint: Try using well-ordering. Let $S$ be the set of non-negative integers $n$ which have the property that $n + 8$ *cannot* be formed using these stamps. Let $n_0$ be the smallest number in $S$. Show that $n_0 \geq 3$, so $n_0 - 3$ is non-negative and cannot be in $S$ by definition of $n_0$. Carry on.)

**5.12.** Consider the following procedure: Start with a positive integer $n_0$. If $n_0$ is even, let $n_1 = n_0/2$. If $n_0$ is odd, let $n_1 = 3n_0 + 1$. Now repeat this procedure on $n_1$, obtaining $n_2$, and so on. Let $S(n_0)$ be the set of positive integers you obtain by repeating this procedure starting with $n_0$. By the Well-Ordering Principle, $S(n_0)$ has a least element.
(a) Find the least element in $S(13)$. How many steps did it take?
(b) Find the least element in $S(7)$. How many steps did it take?
Note: This procedure is called either the "$3n + 1$ Problem" or the "Collatz Conjecture." It has a rich history (try Googling it), but the bottom line is that no one knows whether or not, regardless of what $n_0$ you pick, you will always arrive eventually at that least element you found above. The famous 20th century mathematician Paul Erdös said "Mathematics may not be ready for such problems."

# CHAPTER 6

---

## Sets

---

$$\mathbb{Q} = \{b/a \,|\, a, b \in \mathbb{Z}, a \neq 0\}$$

In this chapter we consider one of the most basic concepts in all of mathematics, namely that of a set. You are likely to be familiar with many properties of sets, but we will, at least briefly, review many of these basic but very important properties.

A *set* is a collection of objects. The objects are called **elements** or **members** of the set. We indicate that an element $a$ is a member of the set $A$ by writing $a \in A$. In Chapter 5, just prior to Theorem 5.2, we introduced "set-builder notation," which we use below and often in this text.

There are various ways to form new sets from old ones. For example, the *union* of two sets, denoted by $A \cup B$, is the set

$$\{x \,|\, x \in A \text{ or } x \in B\}$$

We remark here that in mathematics, unless otherwise stated, the word "or" means the "inclusive or;" that is, it means "one or the other or *both*." Similarly, the *intersection* of the sets $A$ and $B$, denoted by $A \cap B$, is the set

$$\{x \,|\, x \in A \text{ and } x \in B\}.$$

The *empty set* $\emptyset$ is the set which does not contain any elements.

**Example 6.1.** If $A = \{1, 2, 3, 4\}$ and $B = \{2, 4, 6\}$ then $A \cup B = \{1, 2, 3, 4, 6\}$ and $A \cap B = \{2, 4\}$.

35

**Example 6.2.** Pencil and paper time. Let $A = \{6n|n \in \mathbb{Z}\}$ and $B = \{8n|n \in \mathbb{Z}\}$. Write down a few elements of each of these (infinite) sets. Now use set-builder notation to describe $A \cap B$.

Two sets are **equal** if they contain the same elements. A set $A$ is a **subset** of a set $B$ if each element in the set $A$ is also an element in the set $B$. We denote this by writing $A \subseteq B$. A subset $A$ is a **proper** subset of $B$ if $A$ is a subset of $B$ and there is at least one element $b \in B$ with $b \notin A$. We will denote that the set $A$ is a proper subset of the set $B$ by writing $A \subset B$.

A good technique for showing that two sets $A$ and $B$ are equal is to show that each is a subset of the other, i.e., we can show that $A \subseteq B$ and $B \subseteq A$. We illustrate and try out this technique in the following two examples.

**Example 6.3.** If $A$ is the set of even integers and

$$B = \{b|b = 2k + 2 \text{ for some integer } k\},$$

then to show $A = B$, we proceed as follows.

Let $a \in A$. Then $a = 2m$ for some integer $m$. Hence, we also have that $a = 2(m-1)+2$ so that $a \in B$. Thus $A \subseteq B$. Similarly, if $b \in B$ then $b = 2k+2$ for some integer $k$. Thus, $b = 2(k + 1)$, so that $b$ is even which means $b \in A$. Hence $B \subseteq A$, and thus $A = B$. ∎

**Example 6.4.** Pencil and paper again. Consider the set $C$ to be the set of all odd integers and let $D = \{d|d = 2k - 3 \text{ for some integer } k\}$. Show that $C$ and $D$ are equal by showing that each is a subset of the other.

Two sets $A$ and $B$ are **disjoint** if they have no elements in common, i.e., if $A \cap B = \emptyset$.

**Example 6.5.** Let $P$ be the set of all prime numbers and let $B = \{5n|n \in \mathbb{Z}^+\}$. Are $P$ and $B$ disjoint? If not, why not? Are the sets $P$ and the set $C = \{6n|n \in \mathbb{Z}^+\}$ disjoint?

The **power set** of a set $A$, denoted by $\mathcal{P}(A)$, is the set consisting of all subsets of $A$. Hence $\mathcal{P}(A)$ is a *set of sets*, and we include the empty set $\emptyset$ and the set $A$ itself in this collection.

**Example 6.6.** Let $A = \{1, 2, 3\}$. Then

$$\mathcal{P}(A) = \{\emptyset, \{1\}, \{2\}, \{3\}, \{1, 2\}, \{1, 3\}, \{2, 3\}, \{1, 2, 3\}\}$$

Note that the set $A$ has 3 elements and $\mathcal{P}(A)$ has 8 elements. Why do we get 8 subsets from 3 elements?

**Example 6.7.** Have a little courage and write down the power set $\mathcal{P}(B)$ of the set $B = \{1, 2, 3, 4\}$. Does $\mathcal{P}(B)$ have 16 elements? If so, you should now be able to formulate a conjecture about the number of elements in the power set of a finite set with $n$ elements. Do so.

There are several methods one could use to try to prove your conjecture. Here we choose to use proof by induction.

**Theorem 6.1.** *If a set $A$ has $n$ elements, then the power set $\mathcal{P}(A)$ of $A$ contains $2^n$ elements.*

**Proof.** We prove this result by induction on the number $n$ of elements in the set $A$.

For the base case, if $n = 1$, assume that the set $A = \{a_1\}$. The subsets of $A$ are the empty set $\emptyset$ and the set $\{a_1\}$, so the result is true for $n = 1$. We note also that if $n = 0$, i.e., if $A$ is the empty set, then $\mathcal{P}(A)$ has $2^0 = 1$ element.

For the induction step, we now assume the result is true for any set that contains $n = k$ distinct elements. Suppose $A = \{a_1, a_2, \ldots, a_{k+1}\}$ is a set with $k + 1$ elements. Let $A^*$ be the subset of $A$ in which the element $a_{k+1}$ has been removed, so that $A^*$ has $k$ elements. By the induction hypothesis, $A^*$ has exactly $2^k$ distinct subsets, which we label as $T_1, T_2, \ldots, T_{2^k}$. In addition, the $2^k$ subsets

$$T_1 \cup \{a_{k+1}\}, T_2 \cup \{a_{k+1}\}, \ldots, T_{2^k} \cup \{a_{k+1}\}$$

of $A$ are all distinct from each other as well as distinct from the collection $\{T_i\}$, so we have identified $2^k + 2^k = 2(2^k) = 2^{k+1}$ subsets of $A$. Finally, suppose $S$ is an arbitrary subset of $A$, then $a_{k+1}$ either is or is not in $S$. If not, then $S$ is a subset of $A^*$, so $S = T_i$ for some $i$. If so (i.e., if $a_{k+1}$ is in $S$), then $S$ is the disjoint union $S^* \cup \{a_{k+1}\}$ and now $S^*$ is a subset of $A^*$ and hence $S^* = T_j$ for some $j$. We conclude that the $2^{k+1}$ subsets identified are all the subsets of $A$, and we are done. ∎

Another method which we can employ to prove Theorem 6.1 involves functions. In Chapter 8, we will see that the number of distinct functions from a finite set containing $n$ elements to a finite set containing $m$ elements is $m^n$. Now suppose $A = \{a_1, a_2, \ldots, a_n\}$ is a set with $n$ elements and let $S$ be a subset of $A$. We define a function $f_S$ from $A$ to $\{0, 1\}$ by $f_S(a_i) = 1$ if $a_i$ is in $S$ and $f_S(a_i) = 0$ if $a_i$ is not in $S$. Then each distinct subset corresponds to a distinct function, of which there are $2^n$.

We now finish this chapter with one more way to combine functions. Given two sets $A$ and $B$, the **Cartesian product** of the sets $A$ and $B$ is the set of ordered pairs defined by

$$A \times B = \{(a, b) | a \in A, b \in B\}.$$

**Example 6.8.** A Cartesian product with which you are very familiar from single variable calculus is the case of $A = B = \mathbb{R}$ ($\mathbb{R}$ being the infinite set of real numbers), where now $A \times B = \mathbb{R}^2$ is the usual $xy$-plane.

**Example 6.9.** Looking at finite sets instead, if $A = \{1, 2\}$ and $B = \{a, b, c\}$ then

$$A \times B = \{(1, a), (1, b), (1, c), (2, a), (2, b), (2, c)\}.$$

We note that in general for finite sets, if the set $A$ has $m$ elements and the set $B$ has $n$ elements where $m$ and $n$ are positive integers, then the Cartesian product set $A \times B$ has $mn$ elements. This multiplicative property explains why we use the term Cartesian "product."

The fundamental notion of sets runs through all of mathematics, as we have seen and shall continue to see as we move forward.

### Exercises

**6.1.** If $A = \{a, b\}$ and $B = \{1, 2, c\}$, write down the sets $A \cup B, A \cap B$, and $A \times B$.

**6.2.** Write down all of the $8 = 2^3$ subsets of the set $A = \{2, 4, 6\}$.

**6.3.** Let $A$ be the set in the previous exercise. List the elements of the Cartesian product $A \times A$. How many elements are in this Cartesian product? How many elements are there in the power set $\mathcal{P}(A \times A)$?

**6.4.** If $S$ is any set and $\emptyset$ is the empty set, what is $S \times \emptyset$? (Note: This is a further example of why the Cartesian product is called a product.)

**6.5.** Let $A$ and $B$ be arbitrary sets. Prove that if $A \cap B = B$, then $B \subseteq A$.

**6.6.** Suppose $S = \{x \in \mathbb{R} | x^2 = p \text{ for some prime number } p\}$. What is $S \cap \mathbb{Q}$? (Hint: See Exercise 5.10.)

**6.7.** If a set $A$ contains 10 distinct elements, how many elements are in $A \cup A, A \cap A, A \times A$, the power set $\mathcal{P}(A)$, and the power set $\mathcal{P}(A \times A)$? (Note: This last number has 31 decimal digits!)

**6.8.** Consider the following sets:

$$A = \{1, 2, 3, 4, 5\} \quad B = \{0, 1, 2, 4\} \quad C = \{1, 3, 5\}$$
$$D = \emptyset \qquad E = \{2\} \qquad F = \{0, 2, 4\}$$

List all true statements of the form $X \subseteq Y$ and $X \subset Y$ for these six sets. (Note: If $X \subset Y$ is true, then $X \subseteq Y$ is automatically true.)

**6.9.** Suppose $A = \{1, 2, 3\}$, $B = \{1, 4\}$ and $C = \{a, b\}$. Show that $(A \cup B) \times C = (A \times C) \cup (B \times C)$.

**6.10.** Prove that for *all* sets $A$, $B$ and $C$, $(A \cup B) \times C = (A \times C) \cup (B \times C)$. (Hint: Recall that a good technique for proving that two sets are equal is to show that each is a subset of the other. So start by saying: "Suppose $(x, y) \in (A \cup B) \times C$" and show that $(x, y) \in (A \times C) \cup (B \times C)$, proving that $(A \cup B) \times C \subseteq (A \times C) \cup (B \times C)$. Finally, go the other direction.)

**6.11.** The following statement looks similar to the one in the previous two exercises: $(A \times B) \cup C = (A \cup C) \times (B \cup C)$. Prove or disprove this statement. (Note: To disprove a general statement like this, you need only display a single *counterexample*, i.e., an example for which the statement is false. Try, for example, the three sets in Exercise 6.9.)

**6.12.** Suppose $U$ is a set and $A$ is a subset of $U$. The *complement of $A$ in $U$*, denoted $U\backslash A$, is $\{x \in U | x \notin A\}$. If the "universal" set $U$ is fixed, the complement of $A$ can also be denoted by $A^C$.

A very standard way to visualize the relationship among sets is a *Venn diagram*. In the two-subset Venn diagram below, identify the areas representing the following sets:

$A \cup B$, $A \cap B$, $U\backslash A$, $U\backslash(A \cup B)$, and $(U\backslash A) \cap (U\backslash B)$.

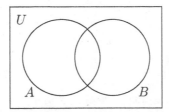

**6.13.** In the previous exercise, the last two sets, $U\backslash(A\cup B)$ and $(U\backslash A)\cap(U\backslash B)$, appear to be the same. This is true in general and is one of the two *De Morgan Laws*. Prove that if $A$ and $B$ are subsets of $U$, then

$$U\backslash(A \cup B) = (U\backslash A) \cap (U\backslash B),$$

or, if $U$ is fixed, the simpler looking statement

$$(A \cup B)^C = A^C \cap B^C.$$

**6.14.** State and prove the "other" De Morgan Law (i.e., reverse the roles of $\cap$ and $\cup$). You might want to visualize these two (equal) sets in a Venn diagram. (Note: The De Morgan Laws also apply in logic. If $P$ and $Q$ are statements, then the laws say "not $(P \ or \ Q)$ = not $P$ *and* not $Q$," and the reverse.)

---

# CHAPTER 7

# Equivalence Relations

---

$$[4] = \{..., -6, -1, 4, 9, ...\}$$

We now consider another important concept in mathematics, namely, that of relations between elements within a set and in particular a special type of relation called an equivalence relation.

Assume that $A$ is a non-empty set. We say an element $a \in A$ *is related to* an element $b \in A$ if the elements $a$ and $b$ have some common property. A simple example of a relation $R$ is equality; that is, $a$ is related to $b$ if $a = b$. In general, we will denote the fact that an element $a$ is related to an element $b$ by writing $aRb$. Alternately, we may choose to write a relation $R$ as a set of ordered pairs, where $(a, b) \in R$ means the same thing as $aRb$ (see, for example, Exercises 7.5 and 7.14).

There are three important properties which a relation $R$ may or may not possess:
**Reflexivity**: $R$ is *reflexive* if for every $a \in A$, $aRa$.
**Symmetry**: $R$ is *symmetric* if for every $a, b \in A$, $aRb$ if and only if $bRa$.
**Transitivity**: $R$ is *transitive* if for every $a, b, c \in A$, if $aRb$ and $bRc$, then $aRc$.

**Example 7.1.** Consider four basic relations on the set $\mathbb{Z}$ of integers. Let $R_1$ be "is equal to" ($=$), $R_2$ be "is less than or equal to" ($\leq$), $R_3$ be "is less than" ($<$) and $R_4$ be "is not equal to" ($\neq$). For each relation, determine which of the three properties above they possess. You should find that whereas $R_3$ and $R_4$

possess only one and $R_2$ possesses two, $R_1$ possesses all three. Hence equality is an example of a special class of relations we now define.

A relation is called an *equivalence relation* if it possesses all three of the above properties, i.e., it is reflexive, symmetric and transitive.

**Example 7.2.** Let $R$ be the following relation on the set $\mathbb{Z}$ of integers: we say $aRb$ provided that $b - a$ is divisible by 5. To see what's going on, write down, say, six integers $a$ for which $aR4$. Now verify that $R$ is in fact an *equivalence* relation on the elements of $\mathbb{Z}$ by checking that it satisfies the three properties.

If $R$ is any equivalence relation on the elements of a set $S$ and if $a \in S$, the set of all elements which are related to $a$ will be called the *equivalence class of* $a$. We will denote the equivalence class of $a$ by $[a]$. Hence $[a] = \{b \mid bRa\}$ (or, since symmetry holds, $[a] = \{b \mid aRb\}$).

In Example 7.2 above, you wrote down six elements which are in the equivalence class [4] (upon division by 5) within the set $\mathbb{Z}$. The equivalence relation we studied there can, of course be generalized to any positive integer $n$.

**Example 7.3.** Let $n$ be a fixed positive integer. For two integers $a$ and $b$, define $aRb$ if $b - a$ is divisible by $n$, or, said another way, if $b - a = nk$ for some integer $k$. You should check that this relation, like the one in Example 7.2, is indeed an equivalence relation. We will denote this relation by writing

$$a \equiv b \pmod{n}$$

where $a, b \in \mathbb{Z}$. The equivalence class $[a]$ of an element $a \in \mathbb{Z}$ is given by

$$[a] = \{b \in \mathbb{Z} \mid b \equiv a \pmod{n}\}$$
$$= \{b \in \mathbb{Z} \mid b = a + kn \text{ for some integer k}\}.$$

The equivalence classes of this relation will be called *congruence classes modulo* $n$ and will be discussed in more detail in Chapter 15.

For another example beyond Example 7.2, if $n = 6$, the equivalence class $[2] = \{\dots, -10, -4, 2, 8, 14, 20, \dots\}$.

Here is an example with a geometric flavor.

**Example 7.4.** Verify that the following relation $R$ on the ordered pairs of the Cartesian product $\mathbb{R} \times \mathbb{R}$ (i.e., the $xy$-plane) is an equivalence relation: $(a, b)R(c, d)$ if $\sqrt{a^2 + b^2} = \sqrt{c^2 + d^2}$. Describe, geometrically, the equivalence class $[(0.6, 0.8)]$. More generally, describe, geometrically, the equivalence class $[(a, b)]$ of any given ordered pair $(a, b) \in \mathbb{R} \times \mathbb{R}$.

Finally, we consider an example from calculus.

**Example 7.5.** Suppose $S$ is the set of all real-valued functions $f(x)$ which are continuous and differentiable (i.e., their derivatives exist) on all of $\mathbb{R}$. Let $R$ be the following relation: $(f(x))R(g(x))$ if $f'(x) = g'(x)$. Verify that $R$ is an equivalence relation. Describe the equivalence class $[x^2]$. More generally, describe the equivalence class $[f(x)]$ of an arbitrary function $f(x) \in S$.

We shall encounter other equivalence relations, as well as the important one of congruences introduced in Examples 7.2 and 7.3 above, as we move forward.

**Exercises**

**7.1.** If $a$ and $b$ are integers, define $a$ to be related to $b$ (i.e., define $aRb$) if $b - a$ is even. Is this relation reflexive, symmetric, and transitive? Does the relation form an equivalence relation? If so, describe the equivalence classes. How many classes are there?

**7.2.** If $a$ and $b$ are integers, define $aRb$ if $b - a$ is a multiple of 3. Now answer the same questions as in the previous exercise.

**7.3.** If $a$ and $b$ are integers, define $a$ to be related to $b$ if $a + b$ is an odd integer. Is this relation reflexive? Is it symmetric? Is it transitive? Does the relation form an equivalence relation? (Note: Always remember that looking at specific examples can clarify a concept.)

**7.4.** If $a$ and $b$ are integers, define $aRb$ if $b - a$ is a multiple of a fixed positive integer $m$. Does this relation form an equivalence relation? If so, describe the equivalence classes? How many classes are there?

**7.5.** Let $A = \{a, b, c, d\}$. Let a relation $R$ be defined by

$$R = \{(a, a), (a, b), (b, b), (b, a), (c, d), (c, c), (d, d), (b, c), (a, c), (b, d)\}.$$

Is the relation $R$ reflexive? Is $R$ symmetric? And finally, is this relation $R$ transitive? Give a reason for each answer.

**7.6.** If $a$ and $b$ are real numbers, define $a$ to be related to $b$ if $b = 3a$. Is this relation reflexive? Is it symmetric? Is it transitive? Does it form an equivalence relation?

**7.7.** In the set $\mathbb{Q}$ of all rational numbers $\frac{a}{b}$ (i.e., with $a$ and $b$ integers and $b \neq 0$) define the fractions $\frac{a}{b}$ and $\frac{c}{d}$ to be related if $\frac{a}{b} \leq \frac{c}{d}$. Does this relation form an equivalence relation? Why or why not?

**7.8.** In the set $\mathbb{Q}$ of all rational numbers, define the fractions $\frac{a}{b}$ and $\frac{c}{d}$ ($b \neq 0$, and $d \neq 0$) to be related if $ad = bc$. Does this relation form an equivalence relation? Why or why not? If so, write down, say, four fractions which are in the equivalence class $[\frac{2}{5}]$. More generally, describe fractions which are in the equivalence class $[\frac{a}{b}]$. (Note: This equivalence relation is the reason, for example, that the fractions $\{1/2, 2/4, \ldots\}$ all have the same value.)

**7.9.** The set $\mathbb{C}$ of complex numbers is $\{a + bi \,|\, a, b \in \mathbb{R}, i = \sqrt{-1}\}$. On $\mathbb{C}$, define $(a + bi)R(c + di)$ if $a = c$. Does this relation form an equivalence relation? Why or why not? If yes, can you describe the equivalence class of, say, $3 + 4i$ in the complex plane, where the "real" coordinates are horizonal and the "imaginary" coordinates are vertical? (Note: For more on the set $\mathbb{C}$, see Chapter 11.)

**7.10.** Consider the set $\mathbb{Z}^+ \cup \{0\}$ of non-negative integers. Let $R$ be the relation "is greater than". How many elements in this set are related to, say, 3? How about 8? How about $n$ for any $n$? How about 0? This exercise shows us that a given element of a set can be related to any number of elements in that set, from infinitely many at one extreme (not for this relation, but for, say, "is less than" instead) to *none* at the other extreme (as for 0 in this relation).

**7.11.** The following "theorem" is false, and hence the given "proof" must contain an error:
"Theorem." If a relation $R$ on a set $X$ is both symmetric and transitive, then it must also be reflexive.
"Proof." Suppose $x \in X$. If $xRy$ holds, then by symmetry $yRx$ also holds, and then by transitivity $xRx$. "QED"
(a) Find a counterexample to the "theorem," hence denying its validity. (Hint: Use the idea in the previous exercise about the possibility of elements having no other elements related to them. Let $X = \{a, b\}$ and create an $R$ with only one relation in it. Then verify that $R$ satisfies the hypotheses of the "theorem" but not the conclusion.)
(b) Identify the exact place where the above "proof" goes astray.

**7.12.** For the relation in Example 7.4, write down all the elements in the set $(\mathbb{Z} \times \mathbb{Z}) \cap [0.6, 0.8]$. (Hint: There are less than 10.)

**7.13.** A little more calculus: Let $S$ be the set of continuous non-negative real-valued functions defined on the closed interval $[0, 1]$. We say $f$ and $g$ in $S$ are related if $\int_0^1 f(x)dx = \int_0^1 g(x)dx$. Verify that this is an equivalence relation. Describe, in terms of area under a curve, the equivalence class $[x^2]$. More generally, describe the equivalence class $[f(x)]$ of an arbitrary function $f(x) \in S$.

**7.14.** This exercise requires some careful thought. Let $A = \{a, b, c\}$. Denote reflexivity by R, symmetry by S, and transitivity by T. It will be easiest to use the ordered pair notation for your relation $R$ (i.e., $R = \{(a, a), \ldots\}$). In each part, write down a (non-empty) relation on $A$ which possesses: (a) all three properties; (b) R and S only; (c) R and T only; (d) S and T only; (e) R only; (f) S only; (g) T only; (h) none of the properties. (Note: These can be accomplished using as few as one ordered pair and at most seven ordered pairs.)

# CHAPTER 8

# Functions

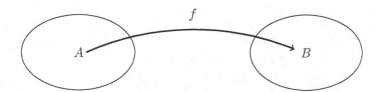

In the previous chapter we discussed relations between the elements of a given set. We now expand this idea to cover relations between the elements of (often but not necessarily) two different sets. The key idea here is the ultra-important concept of a function.

A *function* $f$ from a set $A$ (the *domain*) to a set $B$ (the *co-domain*) is a rule which assigns to each element $a \in A$ a unique element $f(a) \in B$. We will denote a function $f$ from the set $A$ to the set $B$ by writing

$$f : A \to B.$$

We note that for a function to be fully defined, all three of the domain, the co-domain and the rule must be specified.

There are several ways to describe what a function is besides its definition as a rule. One way is to think of a function $f$ from $A$ to $B$ as being a *machine* which takes in any element $a \in A$ and puts out exactly one element $f(a) \in B$. Another way is to think of a function $f$ as a set of ordered pairs $(a, f(a)) \in A \times B$. If we now consider the special case where $A = B$, i.e., where $f$ is a function from $A$ to itself, then the function $f$ is simply an example of the relations we studied in

Chapter 7, but this relation $f$ has the property that *every* $a \in A$ is the left-hand entry in *exactly one* ordered pair from $A \times A$.

You are of course familiar with functions from at least your courses on pre-calculus and calculus. We note that in those courses the functions usually had the set $\mathbb{R}$ of real numbers as both the domain and co-domain. (In multivariable calculus, the domain and/or co-domain might be $\mathbb{R} \times \mathbb{R}$, etc.) Here, however, we wish to look at functions from a broader perspective, in particular allowing the domain and co-domain to be much more general.

We now define some terms and concepts about functions. The ***image*** or ***range*** of a function $f : A \to B$ is the subset $\{f(a)|a \in A\}$ of $B$. We note that the image may be all of the co-domain $B$ or may be a proper subset of $B$. In the former case (i.e., when the image is equal to the co-domain), the function $f$ is said to be ***onto*** or ***surjective***. Hence a given function $f : A \to B$ is onto if and only if for *every* $b \in B$, there exists an $a \in A$ such that $b = f(a)$. Finally, a function $f : A \to B$ is said to be ***one-to-one*** or ***injective*** if for every $b$ in the *image* of $f$, there exists *exactly one* $a \in A$ such that $f(a) = b$.

It is time for some examples.

**Example 8.1.** Consider the two functions $f : \mathbb{R} \to \mathbb{R}$ and $g : \mathbb{Z} \to \mathbb{Z}$ which have the same rule: $f(x) = g(x) = 2x$. We claim that both of them are one-to-one, but only one of them is onto.

In general, to check if a function $h$ is one-to-one, take two general elements $h(a)$ and $h(b)$ from the image, assume that they are equal, and show that this implies that $a$ and $b$ must be equal. For our example $f$ above, if $f(a) = f(b)$, then $2a = 2b$, and so $a = b$, and the exact same argument shows that $g$ is also one-to-one.

To check if a given function $h$ is onto, we must show that every element of the co-domain is the image of some element in the domain. For $f$ above, if we take any $b \in \mathbb{R}$, then $0.5b$ is a real number and $f(0.5b) = b$, establishing that $f$ is onto. However, $g$ is not onto, and to show this we need only find a single counter-example, i.e., a single element of $\mathbb{Z}$ which is not the image of another integer under this rule. Obviously 1, or 3, or any odd integer provides such an example.

**Example 8.2.** Consider the function $f : \mathbb{R} \to \mathbb{R}$ given by the rule $f(x) = x^2$ and the function $g : \mathbb{R} \to \mathbb{R}$ given by the rule $g(x) = x^3$. We claim that one of these functions is both one-to-one and onto, whereas the other is neither one-to-one nor onto. Using the techniques outlined in the previous example, establish our claim. In the case of the function which is not one-to-one, it is sufficient to display a single counter-example, i.e., find two different real numbers $a$ and $b$ whose images are the same. In the case of the function which is not onto, describe the image of that function within $\mathbb{R}$.

In reference to the above example, we remind you from pre-calculus that there is a nice characterization of both one-to-one and onto in terms of the graph of a function from $\mathbb{R}$ to $\mathbb{R}$ called the *horizontal line test*. It is: A function is one-to-one if any horizontal line touches the graph in *at most* one place; a function is onto if any horizontal line touches the graph in *at least* one place.

**Example 8.3.** We claim that $f : \mathbb{R} \to \mathbb{R}^+$ (i.e., the positive real numbers) given by the rule $f(x) = x^2$ is "not quite" a function. Why not? (Recall: *Every* element of the domain must have an image in the co-domain.) "Fix" the given co-domain by adding a single element. Is $f$ now one-to-one? Is it onto?

**Example 8.4.** Let $A = \{a, b, c, d, e\}$ and consider the following relation $R$ on $A$: $\{(a, b), (b, b), (d, e), (e, c), (a, c)\}$. We claim that $R$ fails to be a function from $A$ to $A$ for two different reasons. What are they? Turn $R$ into a function by changing one entry in one ordered pair.

**Example 8.5.** Suppose $A = \{1, 2, 3\}$ and $B = \{a, b\}$. An example of one function from $A$ to $B$ is $\{(1, a), (2, a), (3, a)\}$. Note that by the definition of a function it must contain exactly three ordered pairs. Since each of the three right-hand slots has two options, there must be $2^3 = 8$ possible functions from $A$ to $B$. Write down the other seven.

A particularly important class of functions are those which are both one-to-one and onto. Such a function $f$ is called a **bijection**. In the above examples of functions from $\mathbb{R}$ to $\mathbb{R}$, we discovered that $f(x) = 2x$ and $g(x) = x^3$ are bijections. In fact it is not hard to show that for functions from $\mathbb{R}$ to $\mathbb{R}$, all linear functions $f(x) = mx + b$ are bijections provided $m \neq 0$, and all power functions of the form $g(x) = x^k$, where $k$ is an odd positive integer, are bijections (see Exercises 8.6 and 8.7).

We remark here that we can sometimes *create* bijections by restricting the domain and/or co-domain of a given function. For example, we saw above that $f : \mathbb{R} \to \mathbb{R}$ given by $f(x) = x^2$ is neither one-to-one nor onto, but the new function $h : \mathbb{R}^+ \cup \{0\} \to \mathbb{R}^+ \cup \{0\}$ given by $f(x) = x^2$ is indeed a bijection.

Concerning functions from a finite set to another finite set, note that none of the eight functions you wrote down in Example 8.5 is a bijection. Why is this? Can you see what must be true of two finite sets in order for the possibility of a bijection to occur? In Chapter 10 we shall discuss *permutations*, which are bijections from a set to itself.

A key reason that bijections are an important class of functions is that they possess *inverses*. Suppose $f : A \to B$ is a bijection (so every element of $B$ is of the form $f(a)$ for some unique $a \in A$) then the **inverse** of $f$, denoted $f^{-1}$, is the function from $B$ to $A$ defined by the rule $f^{-1}(f(a)) = a$. The function $f^{-1}$ is also a bijection which simply "reverses" $f$. Thinking of $f$ as a machine, $f^{-1}$ runs the machine backwards.

**Example 8.6.** As already remarked, the functions from $\mathbb{R}$ to $\mathbb{R}$ given by $f(x) = 2x$ and $g(x) = x^3$ are bijections and hence possess inverses. When a function like these is defined by a formula, a standard technique to compute the formula for the inverse is to replace $f(x)$ (or $g(x)$, etc.) by $y$, solve for $x$, and finally switch $x$ and $y$. For our two given functions above the obvious formulas are $f^{-1}(x) = x/2$ and $g^{-1}(x) = \sqrt[3]{x}$. Try a slightly trickier example: What is the formula for the inverse of $h(x) = 2x^3 - 4$?

For our last concept of this chapter we discuss a natural way, called *composition*, to combine two different functions into a third single function. Suppose $f : A \to B$ and $g : B \to C$ are functions which have the property that the co-domain of the former function is equal to the domain of the latter function. Then their **composition**, denoted $g \circ f$, is defined as

$$(g \circ f)(x) = g(f(x))$$

for each $x$ in the set $A$. Here we operate from right to left, applying the function $f$ (the "inner" function) first, then applying the function $g$ (the "outer" function) to the result. Note that the above function $g \circ f$ has domain $A$ and co-domain $C$.

**Example 8.7.** Suppose that $f(x) = x^2$ and $g(x) = 2x + 1$ are functions whose domain and co-domain are the real numbers. Then we can compute.

$$(g \circ f)(x) = g(f(x)) = 2(x^2) + 1 = 2x^2 + 1.$$

Because these functions have fully matching domains and co-domains, the composition $f \circ g$ can also be computed. Please complete the following calculation:

$$(f \circ g)(x) = f(g(x)) = f(\cdots) =$$

If you did this right, you see that the two compositions are different functions. In general, functions do not "commute" under the operation of composition; that is, if $f \circ g$ and $g \circ f$ are both defined, it is most often the case that $(f \circ g)(x) \neq (g \circ f)(x)$.

**Example 8.8.** Suppose $A = \{a, b, c\}$, $B = \{1, 2, 3\}$ and $C = \{x, y, z\}$. Suppose further that the function $f : A \to B$ is given by $\{(a, 2), (b, 3), (c, 2)\}$ and the function $g : B \to C$ is given by $\{(1, y), (2, x), (3, z)\}$.
(a) Is the composition $f \circ g$ defined? Why or why not? If so, write down its ordered pairs.
(b) Answer the same questions for $g \circ f$.

We note that in defining the inverse function $f^{-1}$ of a bijection $f$ above (just before Example 8.6), we actually used their composition $f^{-1} \circ f$. Please look back at that now. The composition $f \circ f^{-1}$ can also be defined and in both cases they result in an *identity* function, i.e., a function which takes every element of its domain back to itself.

Functions, like sets, play a central role in all of mathematics. Understanding their properties is essential as we move further toward more advanced topics.

**Exercises**

**8.1.** Let $A = \{a, b\}$ and $B = \{1, 2, c\}$.
(a) Using ordered pair notation, give an example of a function $f : A \to B$ which is one-to-one. How many such one-to-one functions are there from $A$ to $B$?
(b) Are there functions from $A$ to $B$ which are onto? If so, write one down. If not, why not?

**8.2.** Using the data from the previous exercise as a guide,
(a) suppose the set $A$ has 4 elements and the set $B$ has 7 elements. How many one-to-one functions are there from $A$ to $B$?
(b) suppose the set $A$ has $n$ elements and the set $B$ has $m$ elements where $n < m$. How many one-to-one functions are there from $A$ to $B$?

**8.3.** List *all* of the functions
(a) from the set $\{1, 2\}$ to the set $\{a, b\}$;
(b) from the set $\{1, 2, 3\}$ to the set $\{a, b\}$;
(c) from the set $\{a, b\}$ to the set $\{1, 2, 3\}$.

**8.4.** Using the data from the previous exercise as a guide,
(a) how many functions (in total) are there from a set of 4 elements to a set of 7 elements? Why?
(b) how many functions (in total) are there from a set of $n$ elements to a set of $m$ elements? Why?

**8.5.** Suppose that the set $A$ has 7 elements and the set $B$ has 4 elements.
(a) How many functions (in total) are there from $A$ to $B$?
(b) How many of these functions are one-to-one?
(c) How many of these functions are onto? (Note: This part is challenging. You might try a case-by-case approach. There are three possibilities for how the 7 images are distributed to the 4 co-domain elements: 4-1-1-1, 3-2-1-1, and 2-2-2-1. Try to count each of these and then add them up. The answer is between 5,000 and 10,000.)

**8.6.** Suppose $f$ is the linear function from $\mathbb{R}$ to $\mathbb{R}$ given by $f(x) = mx + b$ where $m, b \in \mathbb{R}$ and $m \neq 0$. Prove that $f$ is a bijection. (Note: Recall (from Example 8.1) that to prove one-to-one, assume that $f(x_1) = f(x_2)$ and show that implies that $x_1 = x_2$. To prove onto, take an arbitrary element $y \in \mathbb{R}$ and work out (by algebra) an element of $\mathbb{R}$ whose image under $f$ is $y$.)

**8.7.** Suppose $g$ is the function from $\mathbb{R}$ to $\mathbb{R}$ given by $g(x) = x^k$ where $k$ is an odd positive integer. Prove that $g$ is a bijection.

**8.8.** Let $f$ and $g$ be the functions in the previous two exercises. Write down the formulas for their inverses $f^{-1}$ and $g^{-1}$ (which they possess since they are bijections).

**8.9.** Suppose $f$ is a function from a set $A$ to itself, and suppose that $f$ is also an equivalence relation on $A$. Describe $f$ explicitly.

**8.10. A little algebra and calculus.** Consider the function $f$ from $\mathbb{R}$ to $\mathbb{R}$ whose rule is $f(x) = x^3 - x^2 - 12x + 5$. Show that $f$ is onto by showing that $\lim_{x \to -\infty} = -\infty$ whereas $\lim_{x \to \infty} = \infty$, and use the fact that $f$ is continuous. However, show that $f$ is *not* one-to-one by considering the horizontal line $y = 5$.

**8.11.** Prove that the composition of two one-to-one functions is one-to-one.

**8.12.** Prove that the composition of two onto functions is onto.

**8.13.** Give an example of a one-to-one function from the set of positive integers $\mathbb{Z}^+$ to itself which is not onto.

**8.14.** Give an example of an onto function from $\mathbb{Z}^+$ to itself which is not one-to-one. (Note: This is a bit trickier than the previous exercise.)

**8.15.** Let $a$ and $b$ be fixed positive integers with the integer $a$ dividing the integer $b$. Find a bijection between the set $A = \{as \mid s \in \mathbb{Z}\}$ and $B = \{bt \mid t \in \mathbb{Z}\}$.

**8.16.** Let $f : [0, \infty) \to [0, \infty)$ be the function defined by $f(x) = \sqrt{x}$. Is $f$ one-to-one? Is $f$ onto? In each case, prove your answer.

**8.17.** Suppose the set $A$ is $\{1, 2, 3\}$ and the set $B$ is $\{0, 2, 4\}$
(a) How many functions are there from the Cartesian product set $A \times B$ to itself?
(b) How many functions are there from the Cartesian product set $A \times B$ to the set $A$?

**8.18.** (a) If a function $f : A \to A$ is one-to-one, is the function $f^3 = f \circ f \circ f$ also $1 - 1$? Here, as usual, $\circ$ denotes the composition of functions. (Note that $f \circ f \circ f = f \circ (f \circ f) = (f \circ f) \circ f$.)
(b) Generalize this result.

# CHAPTER 9

## Cardinality of Sets

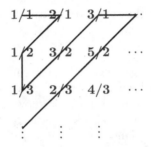

In Chapter 8 we observed that two finite sets can have a bijection between them if and only if they have the exact same number of elements. With infinite sets such as $\mathbb{Z}$, $\mathbb{R}$, etc., we obviously cannot talk about "number of elements," but it would be nice if we at least had a precise way of measuring the "size" of these sets and be able to compare those sizes. Well, it turns out that the crucial concept of a *bijection* is just the ticket here.

Two sets $A$ and $B$ (finite or infinite) are said to have the **same cardinality** if there is a bijection between them. If no such bijection exists, the sets are said to have **different cardinality**. We note then that for *finite* sets, cardinality mean exactly what we think it should mean (i.e., the two sets have the same number of elements).

The concept of cardinality when applied to *infinite* sets is much more interesting and at times may seem counter-intuitive. An excellent illustration of this is the following: If a set $B$ is a proper subset of a finite set $A$, then it is obvious that $A$ and $B$ have different cardinalities. However, *among infinite sets, it is perfectly possible for a set and a proper subset of itself to have the same cardinality!* Time for an example or two.

**Example 9.1.** Let $\mathbb{Z}$ be the set of integers, let $E$ be the set of *even* integers, and define the function $g : \mathbb{Z} \to E$ by $g(x) = 2x$. We already analyzed almost the same function (only the co-domains were different) in Example 8.1, observing that it is one-to-one and, though not onto $\mathbb{Z}$, it *is* onto $E$. Hence $g$ is a bijection, and so $\mathbb{Z}$ and its proper subset $E$ have the same cardinality.

**Example 9.2.** Find a bijection from the set $\mathbb{Z}^+$ of positive integers to the set $\mathbb{Z}$ of all integers. Note that we can describe such a bijection by producing a *list* (i.e., {first entry, second entry, $\cdots$}) which clearly contains all the elements of the co-domain. (Hint: How about starting the list with $\{0, 1, -1, \cdots\}$?)

Because of this "make a list" idea, the set $\mathbb{Z}^+$ plays a special role in the concept of cardinality. In fact, we make the following definition: An infinite set $A$ is called ***countable*** if it has the same cardinality as $\mathbb{Z}^+$, i.e., if there exists a bijection from $\mathbb{Z}^+$ to $A$ (or, said somewhat less formally, if we can make a list of all the elements of $A$).

So at this point we can see that the integers $\mathbb{Z}$ and all of its infinite subsets are countable. Is it possible that any sets which at least appear to be "bigger" than $\mathbb{Z}$ are also countable? Again, this would seem counter-intuitive, but the answer is *yes!*

**Example 9.3.** We now show the very surprising fact that *the positive rational numbers* $\mathbb{Q}^+$ *are countable.* (Hence it would follow that *all* the rationals $\mathbb{Q}$ are also countable - why?) "All we need to do" is show how to make a list of all the positive rationals, but it is not at all obvious how to do that. Well, they say that a picture is worth a thousand words, so below is a picture of a list we can produce. In the picture, each row consists of all fractions in lowest terms (i.e., the greatest common divisor of the numerator and denominator is 1) with a fixed denominator, and the denominators increase by one from each row to the next. The line starting at $1/1$ shows a way to eventually "cover" every fraction, hence making the desired list.

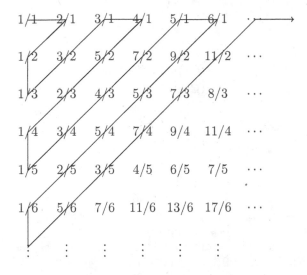

**Example 9.4.** An almost identical argument can be used to show that $\mathbb{Z}^+$ and its Cartesian product with itself $\mathbb{Z}^+ \times \mathbb{Z}^+$ have the same cardinality. Hence if the infinite set $A$ is countable, then the Cartesian product $A \times A$ is also countable. We note that this is definitely different for *finite* sets; in fact, we saw in Chapter 6 that if the set $A$ has $n$ elements, then the set $A \times A$ has $n^2$ elements, which is larger than $n$ as long as $n > 1$.

Okay, so at this point you may believe that all infinite sets are countable, and hence that it's not a very interesting idea for infinite sets. This, however, is definitely not the case. In 1891 the mathematician Georg Cantor published a lovely proof, known as the **Cantor diagonalization argument**, that *there are infinite sets which are not countable*, and the real numbers $\mathbb{R}$ are one of those. Such sets are called **uncountable**.

**Theorem 9.1.** *The real numbers $\mathbb{R}$ are uncountable.*

**Proof by Cantor Diagonalization.** We actually show that the open interval $(0,1) \subset \mathbb{R}$ is uncountable, which then proves that $\mathbb{R}$ itself is uncountable. The proof is by contradiction. Just suppose that we are able to produce a list of all real numbers strictly between 0 and 1, written in their decimal representation. Recall that all real numbers can be so written with infinitely many decimal places, even if those places are eventually all 0 (for example, $2/5 = 0.4000\ldots$). We show a listing of the first 8 numbers in one such (supposed) list, in a randomly chosen order, and notice that we have enlarged the "diagonal" entries:

$$0.5\overset{\cdot}{4}789321\ldots$$
$$0.3\overset{\cdot}{0}978234\ldots$$
$$0.04\overset{\cdot}{4}47800\ldots$$
$$0.950\overset{\cdot}{6}4789\ldots$$
$$0.1637\overset{\cdot}{0}321\ldots$$
$$0.84702\overset{\cdot}{5}89\ldots$$
$$0.061854\overset{\cdot}{4}2\ldots$$
$$0.7921076\overset{\cdot}{7}\ldots$$
$$\vdots$$

Now consider the real number $x \in (0,1)$ whose first decimal digit is *not* 5, second digit is not 0, third digit is not 4, and so on all the way down the diagonal. The number $x$ definitely does not appear in our list, but this contradicts our assumption that *every* real number between 0 and 1 *is* in the list. We conclude that no such list can be built, and so $(0,1)$ is uncountable. ∎

We finish this chapter by looking at one instance in which a property of cardinality matches up for finite and infinite sets. We shall say that a set $A$ **has cardinality less than** the cardinality of a set $B$ if there does not exist a function from $A$ onto $B$. It turns out that for *every* set $A$, $A$ has cardinality less than its power set $\mathcal{P}(A)$. We proved in Theorem 6.1 that if the finite set $A$ has $n$ elements, then $\mathcal{P}(A)$ has $2^n$ elements, and $n < 2^n$ for all $n$. Now, Cantor proved that $\mathbb{Z}$ has cardinality less than the cardinality of $(0,1)$ (and hence less than that of $\mathbb{R}$). In Exercise 9.10 below, we explore the idea that $(0,1)$ can be viewed as the power set $\mathcal{P}(\mathbb{Z}^+)$, and so Cantor's theorem becomes an example of an infinite set (namely $\mathbb{Z}^+$) having cardinality less than its power set.

Finally, we mention an interesting idea about cardinalities of infinite sets. No one has been able to identify a set $I$ (for "in between") such that $\mathbb{Z}^+$ has cardinality less than $I$ but $I$ has cardinality less than $(0,1)$ (and hence less than $\mathbb{R}$). On the other hand, no one has been able to prove that such a set $I$ cannot exist. The unsolved problem in mathematics of proving that $I$ cannot exist is called the **Continuum Hypothesis**. This is a nice reminder that there are many important ideas in mathematics which remain mysterious to this day.

**Exercises**

**9.1.** Suppose the set $A$ is all positive divisors of 24 (including 1 and 24) and the set $B$ is all positive divisors of 54. Prove that $A$ and $B$ have the same cardinality. What is that cardinality?

**9.2.** Prove that the set $E$ of all even integers has the same cardinality as the set $O$ of all odd integers.

**9.3.** Prove that the set $E$ of all even integers has the same cardinality as its proper subset $S$, the set of all multiples of 14.

**9.4.** Let $m \geq 2$ be a fixed positive integer. Show that the set of integers divisible by $m$ has the same cardinality as the set of integers divisible by $mk$ for any positive integer $k$.

**9.5.** Suppose that $A$ and $B$ are disjoint countable sets. Prove that $A \cup B$ is also countable.

**9.6.** In the previous exercise, the hypothesis that the sets are disjoint is not necessary. Prove that if $A$ and $B$ are countable sets, then $A \cup B$ is also countable. (Hint: Work with the 3 disjoint sets $A \cap B$, $A \backslash B$ and $B \backslash A$.)

**9.7.** Prove that the open interval $(0, 1)$ on the real line has the same cardinality as the open ray $\{x \in \mathbb{R} | x > 1\}$.

**9.8.** Prove that the set of all positive real numbers whose decimal representations *terminate* (i.e., eventually all decimal places are 0) is countable. (Hint: For example, suppose that like the number 231.56, the total number of non-zero places is 5. Observe that there are only a finite number of such numbers, in this case $9 \cdot 10 \cdot 10 \cdot 10 \cdot 9 = 81,000$. This should give you an idea about how to create a list of all these numbers.)

**9.9.** Perhaps an easier way to answer the question in the previous exercise is the following: show that all the real numbers described above are actually rational, so we are looking at an infinite subset of $\mathbb{Q}^+$.

**9.10.** The real numbers, and in particular the open interval $(0, 1)$, can be expressed in *binary* (as opposed to decimal) form. Hence, for example, the decimal number $0.3333 \cdots$ is the same as the binary number $0.010101 \cdots$. (To see this, let $x = 0.010101 \cdots$, then $2^2 x = 4x = 1.010101 \cdots$, and subtracting these two expressions, we get that $3x = 1$, i.e., $x = 1/3$.) Hence the Cantor diagonalization argument can be run on an array of numbers which consist entirely of 0's and 1's. But now each real number (i.e., each row in the array) can represent a subset of $\mathbb{Z}^+$, where 0 means do not put this integer in this subset and 1 means *do* put this integer in this subset. For example, the real number $0.010101 \cdots$ represents the set of all even positive integers. Now you complete the argument that $\mathcal{P}(\mathbb{Z}^+)$ is uncountable.

**9.11.** Prove that the open unit interval $(0, 1)$ on the real line and the open unit square $(0, 1) \times (0, 1)$ in the real plane have the same cardinality. (Note: This is another counter-intuitive idea.)

# CHAPTER 10

# Permutations

$$1 \quad 2 \quad 3 \quad 4 \quad 5$$
$$\downarrow \quad \downarrow \quad \downarrow \quad \downarrow \quad \downarrow$$
$$2 \quad 4 \quad 1 \quad 3 \quad 5$$

In the previous two chapters we discovered that among types of functions, bijections (i.e., functions which are both one-to-one and onto) play a particularly important role. In this chapter we will concentrate on bijections *from a set to itself*. Such functions are called permutations.

Assume that $X$ is a non-empty set. A bijection $f : X \to X$ is called a **permutation of** $X$.

**Example 10.1.** The functions $f(n) = -n$ and $g(n) = n+1$ are both permutations of the set $\mathbb{Z}$ of integers.

While permutations can be defined on any non-empty set, in this chapter we will focus on permutations of *finite* sets. Hence we let $X$ denote a finite set, which we may then assume to be the set

$$X = \{1, 2, \ldots, n\}.$$

In this text we first encountered permutations in the very first chapter on pre-calculus when we studied binomial coefficients (see Example 1.2). In that context we noted that a permutation is simply a way of *arranging* the elements of a set $X$. If $X$ is a finite set with $n$ elements, then we have $n$ choices for which element goes in the "first" slot, $n-1$ choices for which goes in the second slot, and so on. This tells us then the total number of permutations on a set of $n$ elements is

$$n! = n(n-1)\cdots(2)(1),$$

this number being called "n factorial." In Example 1.2 we showed that the binomial coefficients are "built out of" permutations; specifically

$$\binom{n}{m} = \frac{n!}{m!(n-m)!}.$$

Looking ahead a bit, if we let $S_n$ denote the set of all permutations of a set $X$ having $n$ elements, then, as just observed, $S_n$ contains exactly $n!$ different permutations. In Chapter 21 the set $S_n$ will be called the **symmetric group on $n$ elements** because this set of all the permutations forms what is called a "group" using the operation of composition of functions. We will discuss this idea in some detail in that chapter.

Unlike more general functions which are commonly denoted by $f$, $g$, etc., permutations are often denoted by small Greek letters such as sigma ($\sigma$) and rho ($\rho$). So let us assume that $\sigma$ and $\rho$ are two permutations in $S_n$. As in Chapter 8, $\sigma \circ \rho$ is their composition, computed for each element $x$, as expected, as $\sigma(\rho(x))$ (i.e., from right to left, first apply $\rho$, then apply $\sigma$). By Exercises 8.11 and 8.12, this composition is indeed itself a permutation in $S_n$.

**Example 10.2.** Suppose $X = \{1, 2, 3, 4\}$ and suppose
$$\sigma(1) = 3, \quad \sigma(2) = 1, \quad \sigma(3) = 2, \quad \sigma(4) = 4,$$
$$\rho(1) = 4, \quad \rho(2) = 3, \quad \rho(3) = 2, \quad \rho(4) = 1.$$
Then $\sigma \circ \rho(1) = \sigma(\rho(1)) = \sigma(4) = 4$,
$\sigma \circ \rho(2) = \sigma(\rho(2)) = \sigma(3) = 2$,
$\sigma \circ \rho(3) = \sigma(\rho(3)) = \sigma(2) = 1$, and
$\sigma \circ \rho(4) = \sigma(\rho(4)) = \sigma(1) = 3$.

For practice, compute $\rho \circ \sigma$, and observe, as we have before, that compositions in the opposite order are, in general, not equal.

It is evident from this example that it would be nice to have some less cumbersome notations to describe the permutations of the finite set $X = \{1, 2, \cdots, n\}$. We introduce two such notations, the first of which is called **two row notation**. Using this notation for a permutation $\sigma$, in the first row we simply list the values $1, 2, \ldots, n$ in order, whereas in the second row, for each $i = 1, 2, \ldots, n$, underneath the value $i$ we place the image $\sigma(i)$. We thus have

$$\sigma = \begin{pmatrix} 1 & 2 & \cdots & n \\ \sigma(1) & \sigma(2) & \cdots & \sigma(n) \end{pmatrix}.$$

**Example 10.3.** For the permutation $\sigma$ in Example 10.2, the two row notation is

$$\sigma = \begin{pmatrix} 1 & 2 & 3 & 4 \\ 3 & 1 & 2 & 4 \end{pmatrix}.$$

For practice, redo the calculation from Example 10.2 of $\rho \circ \sigma$, this time using the two row notation for both permutations and for the answer. As usual with compositions, you will work from right to left, i.e., for each element apply $\sigma$ first, then $\rho$.

Because every permutation is a bijection, we know from Chapter 8 that it must possess an inverse function which is itself a permutation. In two row notation, it is easy to calculate this inverse in two steps, namely:

1. Switch the two rows, and
2. Rearrange the columns so that the top row is in ascending order.

**Example 10.4.** Looking again at our permutation $\sigma$ from Example 10.3, we see that

$$\sigma^{-1} = \begin{pmatrix} 3 & 1 & 2 & 4 \\ 1 & 2 & 3 & 4 \end{pmatrix} = \begin{pmatrix} 1 & 2 & 3 & 4 \\ 2 & 3 & 1 & 4 \end{pmatrix}.$$

We observe, as must be the case, that

$$\sigma \circ \sigma^{-1} = \begin{pmatrix} 1 & 2 & 3 & 4 \\ 3 & 1 & 2 & 4 \end{pmatrix} \begin{pmatrix} 1 & 2 & 3 & 4 \\ 2 & 3 & 1 & 4 \end{pmatrix} = \begin{pmatrix} 1 & 2 & 3 & 4 \\ 1 & 2 & 3 & 4 \end{pmatrix},$$

i.e., the identity permutation. Verify that $\sigma^{-1} \circ \sigma$ is also the identity permutation.

We now turn to a second notation for permutations, called **cycle notation**, which has the advantage of directly showing important properties of the given permutation, but the disadvantage that the notation of a given permutation need not be unique.

Suppose we are given a permutation $\sigma$ on the set $X = \{1, 2, \ldots, n\}$. We shall denote the composition $\sigma \circ \sigma$ by $\sigma^2$, $\sigma \circ \sigma \circ \sigma$ by $\sigma^3$, and so on. Now for a specific element $x \in X$, we can compute the "cycle" of $x$ by calculating the elements

$$x, \sigma(x), \sigma^2(x), \sigma^3(x), \ldots,$$

continuing until we find the first power, say $r$, of the permutation $\sigma$ so that $\sigma^r(x) = x$. In Exercise 10.9 you will be asked to prove that since our set $X$ is finite and since $\sigma$ is one-to-one, such a value of $r$ must exist. We then call the ordered tuple

$$(x \quad \sigma(x) \quad \sigma^2(x) \quad \cdots \quad \sigma^{r-1}(x))$$

the **cycle** of the element $x$ under the permutation $\sigma$. Since this cycle contains exactly $r$ distinct elements, we call the cycle of length $r$ an **r-cycle**. Cycles of length one are called **fixed points** while cycles of length two are called **transpositions**.

**Example 10.5.** A cycle notation for the permutation $\sigma$ in Example 10.3 is $(132)(4)$, which is read "1 goes to 3, 3 goes to 2, 2 goes back to 1, and 4 is fixed." Note then that $\sigma$ is composed of a 3-cycle and a fixed point. Note also that we say "*A* cycle notation" since $\sigma$ could also be written as $(321)(4)$, or $(4)(213)$, etc.

Write down a cycle notation for the permutation $\rho$ in Example 10.2. Note that $\rho$ consists of two transpositions.

A procedure then for writing down a cycle notation for an arbitrary permutation $\tau$ ("tau") of the set $X = \{1, 2, 3, \ldots, n\}$ is:

1. Start with 1 and build its cycle,
2. identify the smallest element which is *not* in this first cycle and build its cycle, and
3. continue until all elements appear in some cycle.

**Example 10.6.** Suppose in two row notation, we have

$$\tau = \begin{pmatrix} 1 & 2 & 3 & 4 & 5 & 6 & 7 & 8 \\ 5 & 7 & 3 & 6 & 8 & 4 & 2 & 1 \end{pmatrix}$$

Use the above procedure to write down a cycle notation for $\tau$, showing that it consists of a 3-cycle, two transpositions, and a fixed point.

Because any permutation of a set $X$ is a bijection of $X$ to itself, each element of $X$ must appear once and only once in the cycle notation produced by the above procedure. It follows then that any two cycles in the notation are **disjoint**, i.e., they contain no elements in common. Thus we see that every permutation of $X$ has a representation as a "product" of disjoint cycles, and that representation is unique except for the order in which the cycles appear (and the starting element in each cycle, as noted above).

**Example 10.7.** The permutation given in two row notation by

$$\begin{pmatrix} 1 & 2 & 3 & 4 & 5 & 6 & 7 & 8 & 9 \\ 4 & 1 & 5 & 7 & 9 & 8 & 2 & 6 & 3 \end{pmatrix}$$

can be represented in disjoint cycles as $(1472)(359)(68)$, $(359)(68)(1472)$, $(593)$ $(7214)(86)$, and so on. In any case, however, we observe that these disjoint cycle notations give a very clear "picture" of this permutation, i.e., it consists of a 4-cycle, a 3-cycle, and a transposition.

How do we compose two permutations which have been represented in their disjoint cycle notations? Although disjoint cycles can appear in any order, this is definitely *not* the case when cycles fail to be disjoint, which will automatically be the case when we compose permutations. An example should make this clear.

**Example 10.8.** Let $\sigma = (132)(4)$ and $\rho = (14)(23)$ be the two permutations on $X = \{1, 2, 3, 4\}$ studied in several examples above. Let us compute the composition $\sigma \circ \rho$ using cycle notation. Recall that we compose from right to left, i.e., $\rho$ first, then $\sigma$. We are computing then a "product" of *non-disjoint* cycles:

$$\sigma \circ \rho = (132)(4) \circ (14)(23) = (132)(4)(14)(23).$$

It is not obvious how to proceed here. Try this procedure:

1. Find the rightmost occurrence of 1. In that cycle, 1 goes to 4. Now (immediately) move left to find the next occurrence of 4. In that cycle, 4 is fixed. Hence we have that 1 goes to 4 in this composition.

2. Now where does 4 go? In the rightmost occurrence of 4, 4 goes to 1. Now move left and find 1. Here 1 goes to 3. Hence 4 goes to 3.

3. Now where does 3 go? In the rightmost occurrence of 3, 3 goes to 2. Now move left and find 2. Here 2 goes to 1. Hence 3 goes to 1, which closes off the new cycle as (143)

4. It's good to check the "missing" element 2. Rightmost for 2, 2 goes to 3. Moving left, 3 goes to 2. Hence 2 is indeed fixed by this composition, and we are done, i.e., $\sigma \circ \rho = (143)(2)$. Note that this matches what we got all the way back in Example 10.2.

Compute $\rho \circ \sigma$ using cycle notation and confirm your answer with what you got in Example 10.2. Note once again that $\rho \circ \sigma$ and $\sigma \circ \rho$ are not the same.

Working through a product of non-disjoint cycles to produce a disjoint cycle representation (as must always be done with compositions, for example) takes some practice. You will get some more when you do this chapter's exercises.

We remark that frequently fixed points (i.e., 1-cycles) are omitted from the cycle notation. For example, $\sigma = (132)(4)$ may be written simply as $(132)$.

Finally, cycle notation lends itself nicely to writing down the *inverse* of a given permutation. Specifically, the inverse of any $r$-cycle $(k_1 k_2 \cdots k_{r-1} k_r)$ is simply the cycle $(k_r k_{r-1} \cdots k_2 k_1)$, so if the permutation $\tau$ is written in disjoint cycle notation, then the cyclic notation for $\tau^{-1}$ results from reversing each of $\tau$'s cycles.

**Example 10.9.** The inverse of the permutation $(1472)(359)(68)$ of Example 10.7 is $(2741)(953)(86)$. This of course can also be written as $(1274)(395)(68)$, and so on. Again, the order in which the individual cycles appear does not matter since the cycles are disjoint. You are asked to consider the case of inverses of non-disjoint cycles (for example, the inverse of a composition) in Exercise 10.6.

We shall encounter permutations as we move on in this text, especially in Chapter 21 where we encounter the "group" $S_n$ of all permutations of a set containing $n$ elements.

**Exercises**

**10.1.** Suppose $\theta$ (theta) and $\tau$ (tau) are permutations of the set $X = \{1, 2, 3, 4, 5, 6\}$ (i.e., they are in $S_6$) which are defined by
$$\theta(1) = 4, \quad \theta(2) = 3, \quad \theta(3) = 2, \quad \theta(4) = 6, \quad \theta(5) = 5, \quad \theta(6) = 1,$$
$$\tau(1) = 6, \quad \tau(2) = 5, \quad \tau(3) = 1, \quad \tau(4) = 3, \quad \tau(5) = 2, \quad \tau(6) = 4.$$
(a) Write down $\theta$ and $\tau$ in their two row notations.
(b) Write down $\theta$ and $\tau$ in disjoint cycle notations.
(Note that disjoint cycle notation gives the clearest pictures of these permutations.)

**10.2.** Write down, in disjoint cycle notation, the $3! = 6$ permutations in $S_3$.

**10.3.** For the two permutations in Exercise 10.1, write down both $\theta^{-1}$ and $\tau^{-1}$ in both two row and disjoint cycle notations. (Note that a permutation and its inverse have the same "cycle structure.")

**10.4.** Prove that a permutation $\sigma$ and its inverse are identical if and only if $\sigma$ consists entirely of transpositions and fixed elements.

**10.5.** For the two permutations in Exercise 10.1, compute the composition $\theta \circ \tau$, first using two row notation and then using cycle notation. Be sure to check that your answers match up.

**10.6.** Do this exercise with cycle notation:
(a) Compute $(\theta \circ \tau)^{-1}$ given your solution to the previous exercise.
(b) Compute both $\theta^{-1} \circ \tau^{-1}$ and $\tau^{-1} \circ \theta^{-1}$ using your solution to Exercise 10.3. One of these should match your answer to Part (a). (Note: This exercise is an example of a general fact about the relationship between inverse functions and compositions. Specifically, it is the case that if $f$ and $g$ are bijections for which $f \circ g$ is defined, then $(f \circ g)^{-1} = g^{-1} \circ f^{-1}$. This fact is often called "socks and shoes", since to undo putting on your socks and then your shoes, you need to first take off the shoes, then the socks. See Chapter 22 for more on this idea.)

**10.7.** (a) For the permutation $\sigma = (423)$ in $S_4$, compute $\sigma^3 (= \sigma \circ \sigma \circ \sigma)$.
(b) For the permutation $\rho = (1423)$ in $S_4$, compute $\rho^4 (= \rho \circ \rho \circ \rho \circ \rho)$.
(c) Make a conjecture on the basis of your data: If the permutation $\alpha$ is an $r$-cycle, then $\alpha^r$ is $\cdots$. Can you prove this conjecture?

**10.8.** For the permutation $\sigma = (423)$ in $S_4$, compute the permutation $\sigma^{-2}$. (This can be calculated in two different ways: as $(\sigma^{-1})^2$ or as $(\sigma^2)^{-1}$.) Given the result of Exercise 10.7(a), why does it make sense that $\sigma^{-2} = \sigma$?

**10.9.** As mentioned prior to Example 10.5, prove that if $X$ is a finite set and if $\sigma$ is a permutation of $X$, then for every $x \in X$, there exists a positive integer $r$ such that $\sigma^r(x) = x$ (i.e., we must eventually cycle back to $x$).

**10.10.** In Chapter 23 we shall define the *order* of an element $\beta$ of a "group" to be the smallest integer $r$ such that $\beta^r =$ the "identity element." As an example of this, in Exercise 10.7(c) you conjectured that the order of an $r$-cycle is simply $r$ (and this is in fact true). Here we ask what happens in the case of a permutation which consists of at least two disjoint cycles of length at least 2.

(a) Suppose $\rho \in S_6$ is given by $(1452)(36)$. Compute the order of $\rho$. (Note: the easiest way is to compute $\rho^2$, then from that $\rho^3$, and so on until you arrive at the identity permutation.)

(b) Suppose $\tau \in S_6$ is given by $(145)(26)$. Compute the order of $\tau$.

(c) Make a conjecture on the basis of your data: If a permutation $\sigma$ consists of an $r$-cycle and an $s$-cycle, then the order of $\sigma$ is $\cdots$.

# CHAPTER 11

# Complex Numbers

$$i = \sqrt{-1} \notin \mathbb{R}$$

At the very beginning of this text, in Example 1.1, you derived the Quadratic Formula using the technique of completing the square. Again, that formula says that, if $f(x) = ax^2 + bx + c$ with $a \neq 0$, the solutions (or "roots") of $f(x) = 0$ are given by

$$x = \frac{-b \pm \sqrt{b^2 - 4a}}{2a}.$$

Let us assume here that the coefficients $a$, $b$ and $c$ are real numbers. Let us also call the quantity $b^2 - 4ac$ the *discriminant* of the expression. A key idea here is that if this discriminant is negative, the equation has no real roots (since for all real numbers $r$, $r^2 \geq 0$). We see then that the equation $f(x) = 0$ has exactly two real solutions if the discriminant is positive, exactly one real solution if the discriminant is 0, but *no* real solutions if the discriminant is negative.

Having no solutions to a simple equation in real numbers was long since a signal to mathematicians that we needed to move beyond real numbers and widen the domain in which possible solutions to real number equations can exist. This is accomplished by simply allowing the number $\sqrt{-1}$ to be defined as an "imaginary" number and then building a wider domain using that single number. For ease of notation, we denote $\sqrt{-1}$ by the symbol $i$. This wider domain is the set of complex numbers.

The ***complex numbers***, denoted by $\mathbb{C}$, are the set defined by

$$\mathbb{C} = \{a + bi \mid a, b \in \mathbb{R}\}.$$

The value $a$ is called the **real part** of the complex number $a + bi$ while the value $b$ is called the **imaginary part** of $a + bi$.

**Example 11.1.** Let us solve the equation $f(x) = x^2 - 4x + 7 = 0$. By the Quadratic Formula we get

$$x = \frac{4 \pm \sqrt{16 - 28}}{2} = \frac{4 \pm \sqrt{-1}(2\sqrt{3})}{2} = 2 \pm i\sqrt{3}.$$

Hence we have no real solutions but two complex solutions, $2 + i\sqrt{3}$ and $2 - i\sqrt{3}$.

The two solutions in this example have a special relationship; they are called complex conjugates. For the complex number $a + bi$, its **complex conjugate** is the number $a - bi$. The conjugate of $a + bi$ is often denoted by

$$\overline{a + bi} = a - bi.$$

Complex conjugates have the special property that their product is real. Since $i^2 = -1$, we see that $(a + bi)(a - bi) = a^2 - b^2 i^2 = a^2 + b^2$. This will come into play when we divide complex numbers (see below), and in Chapter 35 when we examine "irreducible" polynomials with real coefficients.

Two complex numbers $a + bi$ and $c + di$ are **equal** if $a = c$ and $b = d$. The sum or difference of two complex numbers $a + bi$ and $c + di$ is found by simply combining the real parts and the imaginary parts, i.e., $(a + bi) + (c + di) = (a + c) + (b + d)i$ and $(a + bi) - (c + di) = (a - c) + (b - d)i$. Multiplication is done as in the previous paragraph, i.e.,

$$(a + bi)(c + di) = (ac - bd) + (ad + bc)i.$$

Finally, what about division? We wish to write the answer in the form $s + ti$ with $s$ and $t$ real numbers, so we use the old trick from algebra of multiplying the top and bottom by the conjugate of the denominator. Hence division goes like this:

$$\frac{a + bi}{c + di} = \frac{(a + bi)(c - di)}{(c + di)(c - di)} = \frac{ac + bd}{c^2 + d^2} + i\frac{bc - ad}{c^2 + d^2}.$$

**Example 11.2.** For practice, add, subtract, multiply and divide the two complex numbers $3 + 4i$ and $5 - 2i$.

Looking ahead, we note that the set $\mathbb{C}$ of all complex numbers forms what is called a "field" in mathematics. Roughly speaking, a field is a set of numbers in which you can add, subtract, multiply and divide (except by 0) and obtain an answer which is still in the set. For example, the rational numbers $\mathbb{Q}$ and the real numbers $\mathbb{R}$ are both fields, but the integers $\mathbb{Z}$ are not (why not?). The sets $\mathbb{Q}$ and $\mathbb{R}$ form "subfields" of $\mathbb{C}$; that is, they are both subsets of $\mathbb{C}$ which are themselves fields. In Chapter 29 this will all be examined in some detail.

We complete this short chapter by addressing the following question: Does the expansion of the real numbers using the square root of $-1$ help us find solutions of polynomial equations beyond quadratic (i.e., degree 2) equations? The answer is yes. In fact, the so-called Fundamental Theorem of Algebra, which we shall introduce formally in Chapter 35, guarantees us that every polynomial equation "over $\mathbb{C}$" (i.e., with complex coefficients, which of course can be real since $\mathbb{R}$ is the set of elements of $\mathbb{C}$ whose imaginary parts are 0) has at least one complex solution $\alpha$. After dividing out the factor $(x-\alpha)$ and continuing, we see that every polynomial of degree $n$ over $\mathbb{C}$ actually has $n$ (not necessarily distinct) complex solutions. This same statement with "real solutions" substituted for "complex solutions" is, as we saw at the outset of this chapter, definitely not true.

**Example 11.3.** Let us find all the solutions of the equation $x^3 + 8 = 0$. Obviously $x = -2$ is one of the solutions. Using factorization (or in a harder example long division), we see that $x^3 + 8 = (x+2)(x^2 - 2x + 4)$. Now, by the Quadratic Formula, we get that the other two solutions are $1 + i\sqrt{3}$ and $1 - i\sqrt{3}$. For practice and to verify, check that $(1 + i\sqrt{3})^3 + 8 = 0$.

**Example 11.4.** Trigonometry has only come up once in this text (in Example 1.4), but here's another appearance. Let us solve the polynomial equation $x^4 + 1 = 0$. Well, solving for $x^2$ using the Quadratic Formula, we get that $x^2 = \pm i$. Applying the formula again to these two equations, we get $x = \pm\sqrt{i}$ or $x = \pm\sqrt{-i}$. So, in the desired form of $a + bi$ with $a, b \in \mathbb{R}$, what are $\sqrt{i}$ and $\sqrt{-i}$ ? We now make two observations using trigonometry:
(1) $i = \cos(\pi/2) + i\sin(\pi/2)$ (check this).
(2) Recall that for any real number $\theta$, $\cos(2\theta) = \cos^2(\theta) - \sin^2(\theta)$ and $\sin(2\theta) = 2\sin(\theta)\cos(\theta)$. Setting $\theta = \pi/4$, we see that

$$(\cos(\pi/4) + i\sin(\pi/4))^2 = \cos^2(\pi/4) + 2i\cos(\pi/4)\sin(\pi/4) - \sin^2(\pi/4)$$

$$= \cos(\pi/2) + i\sin(\pi/2) = i.$$

Since the same calculation works for $(-(\cos(\pi/4) + i\sin(\pi/4)))^2$, we conclude that

$$\sqrt{i} = \pm(\cos(\pi/4) + i\sin(\pi/4)) = \pm(\sqrt{2}/2 + i\sqrt{2}/2).$$

By a very similar argument, we get

$$\sqrt{-i} = \pm(\cos(\pi/4) - i\sin(\pi/4)) = \pm(\sqrt{2}/2 - i\sqrt{2}/2),$$

and we have identified the four solutions of $x^4 + 1 = 0$.

The complex numbers play a central role in all of mathematics, not least of all because they contain all the possible solutions to every polynomial equation over $\mathbb{C}$, and hence over $\mathbb{R}$. We shall discuss the set $\mathbb{C}$ further in Chapter 35, where we look more closely at solutions to polynomial equations.

**Exercises**

**11.1.** Write each of the following in the form $a + bi$ with $a$ and $b$ in $\mathbb{R}$:
(a) $(2 + 3i) + (6 - 2i)$,
(b) $(2 - 3i) + (6 + 2i)$,
(c) The complex conjugate $\overline{(2 + 3i) + (6 - 2i)}$.

**11.2.** Prove that "complex conjugacy is compatible with complex addition," i.e., prove that
$$\overline{(a + bi) + (c + di)} = \overline{a + bi} + \overline{c + di}.$$

**11.3.** Compute:
(a) $(2 + 3i)(6 - 2i)$,
(b) $(2 - 3i)(6 + 2i)$,
(c) The complex conjugate $\overline{(2 + 3i)(6 - 2i)}$.

**11.4.** Prove that complex conjugacy is compatible with complex multiplication, i.e., prove that
$$\overline{(a + bi)(c + di)} = \overline{a + bi} \cdot \overline{c + di}.$$

**11.5.** Compute:
(a) $(2 + 3i)^3$,
(b) $(2 - 3i)^3$,
(c) The complex conjugate $\overline{(2 + 3i)^3}$.

**11.6.** Prove that complex conjugacy is compatible with raising to positive integer powers, i.e., prove that if $n \in \mathbb{Z}^+$,
$$\overline{(a + bi)^n} = (\overline{a + bi})^n.$$

(Hint: Use induction on $n$ and Exercise 11.4.)

**11.7.** Suppose $c + di \neq 0$. The ***multiplicative inverse*** of $c + di$ is $1/(c + di)$. Compute (in the form $a + bi$):
(a) the multiplicative inverse of $6 - 2i$,
(b) the multiplicative inverse of $6 + 2i$.

**11.8.** Prove that complex conjugacy is compatible with multiplicative inverses, i.e., prove that if $c + di \neq 0$,
$$\overline{1/(c + di)} = 1/(\overline{c + di}).$$

**11.9.** Compute (in the form $a + bi$):
(a) $(2 + 3i)/(6 - 2i)$,
(b) $(2 - 3i)/(6 + 2i)$.

**11.10.** Prove that complex conjugacy is compatible with complex division. (Hint: This an excellent opportunity to use what you've already proved. See Exercises 11.4 and 11.8.)

**11.11.** Simplify $\frac{(2+4i)^2(3-2i)}{5-i}$ into the form $a + bi$.

**11.12.** Calculate the powers $i^n$ for $n = 1, 2, \ldots, 8$.

**11.13.** Find the two complex solutions (or "roots") of the polynomial equation $x^2 + 4x + 5 = 0$.

**11.14.** Find the three complex roots of the polynomial equation $x^3 - 5x^2 + 7x = 0$.

**11.15.** Find the four complex solutions of the polynomial equation $x^4 - 16 = 0$.

**11.16.** Find the three complex roots of the polynomial equation $x^3 - 2x^2 + 5x - 10 = 0$.

**11.17.** In Example 11.4 we found that the two complex square roots of $i$ are $\pm(\cos(\pi/4) + i\sin(\pi/4)) = \pm(\sqrt{2}/2 + i\sqrt{2}/2)$. Let's discover the three complex *cube* roots of $i$.
(a) Show that $(-i)^3 = i$. So there is one of our cube roots.
(b) Since $i$ can be written in terms of the sine and cosine of $\pi/2$, and its square root can be written in terms of $\pi/4$, perhaps we can get a cube root by looking at $\pi/6$. Again using the Binomial Theorem, prove that $(\cos(\pi/6) + i\sin(\pi/6))^3 = i$. So we have found a second cube root of $i$.
(c) The third cube root of $i$ is of the form $(\cos(k\pi/6) + i\sin(k\pi/6))$ where $k$ satisfies $2 \leq k \leq 5$. Guess which $k$ works and then prove that your guess is correct.

**11.18.** (a) On the basis of what we learned in Example 11.4 and Exercise 11.17, make a conjecture about the form, in terms of sines and cosines, of one of the $n$-th roots of $i$.
(b) It is a fact, given your hopefully correct conjecture in Part (a), that the other $n - 1$ complex $n$th roots of $i$ are of the form $\cos(k\pi/(2n)) + i\sin(k\pi/(2n))$ where $k$ is of the form $4m + 1$ with $1 \leq m \leq n - 1$. Using your conjecture and this, write down the five 5th roots of $i$.
(c) The second root you wrote down in Part (b) should "look familiar." What is it and why does this make sense? (See Exercise 11.12.)

# CHAPTER 12

## Matrices and Sets with Algebraic Structure

$$
\begin{pmatrix}
1 & 0 & -1 \\
2.3 & 1/5 & 4 \\
\pi & 1 & 2
\end{pmatrix}
$$

We begin this chapter with some general remarks about sets. If $S$ is a set, then a **binary operation** on $S$ is a way to "combine" two elements of $S$, obtaining another element of $S$. More formally, it is a function from the Cartesian product $S \times S$ to $S$. If the set $S$ has one or more binary operations defined on it, we say that $S$ is a **set with algebraic structure**. Here is a chart of some of the sets we have studied (or will soon study) listed with their binary operations (see the Notations page if needed):

| Set(s) | Binary operation(s) |
|---|---|
| $\mathbb{Z}^+$ | addition, multiplication |
| $\mathbb{Z}$ | addition, subtraction, multiplication |
| $\mathbb{Q}^+, \mathbb{R}^+$ | addition, multiplication, division* |
| $\mathbb{Q}, \mathbb{R}, \mathbb{C}$ | addition, subtraction, multiplication, division* |
| permutations ($S_n$, etc.) | composition |
| $n \times n$ matrices | addition, subtraction, multiplication |

*No division by 0.

71

The following three properties are key in the study of sets with algebraic structure:

(1) A binary operation $*$ on a set $S$ is said to be **commutative** if for every $a, b \in S$, $a * b = b * a$.

(2) A binary operation $*$ on a set $S$ is said to be **associative** if for every $a, b, c \in S$, $(a * b) * c = a * (b * c)$.

(3) If a set $S$ has an addition "+" and a multiplication "·", those two binary operations are said to be **distributive** if for every $a, b, c \in S$, $a \cdot (b+c) = a \cdot b + a \cdot c$ and $(a + b) \cdot c = a \cdot c + b \cdot c$.

**Example 12.1.** (a) It is clear that on any of our sets of numbers ($\mathbb{Z}$, $\mathbb{R}$, etc.) addition and multiplication are both commutative and associative. However, show via examples that subtraction and division on $\mathbb{Q}$ (say) are neither commutative nor associative.

(b) We have observed previously that composition of permutations (in fact, of functions) is *not* commutative. Do you think composition of permutations is associative?

(c) You have been using the distributive property on real numbers ever since Algebra I when you "factored out" or "multiplied through"

(e.g., $3(x - 7) = 3x - 21$). The distributive property does not apply to permutations since they possess only one binary operation. We will see below that it *does* apply to $n \times n$ matrices.

We now introduce a new collection of sets with algebraic structure. A **matrix over a set** $S$ is a rectangular array of elements of $S$. An $n \times m$ matrix has $n$ rows and $m$ columns. You were likely introduced to matrices in a course on linear algebra, wherein a matrix is a way of writing down a "linear transformation" from some $m$-dimensional vector space to some $n$-dimensional vector space. In this chapter we will concentrate on *square* matrices (hence representing linear transformations from some vector space to itself), and we will concern ourselves just with the algebraic structure of such matrices. Thus here, if $n \in \mathbb{Z}^+$, we are looking at the set of all $n \times n$ matrices with entries in $S$. For a closer look at vector spaces, see Chapter 30.

We now assume that $A$ is an $n \times n$ matrix whose entries are real numbers. We may write $A$ in the form $A = (a_{ij})$ where $a_{ij}$ denotes the element of $A$ that lies in row $i$ and column $j$ for $1 \le i, j \le n$. Assume that $B = (b_{ij})$ is another $n \times n$ matrix. The matrices $A$ and $B$ are **equal** if $a_{ij} = b_{ij}$ for each $1 \le i, j \le n$; i.e., if the corresponding entries of the two matrices are the same.

Now we get to the algebraic structure of such matrices. We can add two matrices in the rather obvious way of adding the corresponding entries; that is:

$$A + B = (a_{ij}) + (b_{ij}) = (a_{ij} + b_{ij}), 1 \le i, j \le n.$$

Not surprisingly, we subtract matrices by replacing the plus signs in the above line with minus signs.

**Example 12.2.** If

$$A = \begin{pmatrix} 1 & 0 & 1 \\ 2 & 1 & 3 \\ 1 & 1 & 2 \end{pmatrix}$$

and

$$B = \begin{pmatrix} 2 & 1 & 1 \\ 1 & 2 & 1 \\ 1 & 1 & 1 \end{pmatrix}.$$

then

$$A + B = \begin{pmatrix} 3 & 1 & 2 \\ 3 & 3 & 4 \\ 2 & 2 & 3 \end{pmatrix},$$

and similarly

$$A - B = \begin{pmatrix} -1 & -1 & 0 \\ 1 & -1 & 2 \\ 0 & 0 & 1 \end{pmatrix}.$$

Another simple (but not binary) operation on matrices is *scalar multiplication*. If $A = (a_{ij})$ is an $n \times n$ matrix, we can multiply the matrix $A$ by a constant value $c$ to obtain a new matrix

$$cA = c(a_{ij}) = (ca_{ij}), 1 \leq i, j \leq n.$$

Now, how can we multiply two $n \times n$ matrices? You might think that we simply multiply the corresponding entries as we did for addition and subtraction, but such a process turns out to be not very useful. What *is* very useful is to, again, view an $n \times n$ matrix as a linear transformation on a set of $n$-dimensional column vectors. If $A$ and $B$ are $n \times n$ matrices and $x$ is an $n$-dimensional column vector, then we need to define the product of $A$ and $B$ so that $(AB)x = A(Bx)$. The details here are tricky, but the upshot is that multiplication of matrices needs to employ an "across/down" strategy; that is, we go across the rows of $A$ and down the columns of $B$, multiplying and adding as we go. For the details here see, for example, [1]. Some specific examples should be very helpful here.

**Example 12.3.** (a) Consider the $2 \times 2$ matrices

$$A = \begin{pmatrix} 1 & -4 \\ -2 & 3 \end{pmatrix} \text{ and } B = \begin{pmatrix} 3 & 1 \\ 1 & -4 \end{pmatrix}.$$

Then

$$AB = \begin{pmatrix} (1 \times 3) + (-4 \times 1) & (1 \times 1) + (-4 \times -4) \\ (-2 \times 3) + (3 \times 1) & (-2 \times 1) + (3 \times -4) \end{pmatrix} = \begin{pmatrix} -1 & 17 \\ -3 & -14 \end{pmatrix}.$$

(b) Using this procedure, calculate $BA$. Given the answer, what have you just showed about whether matrix multiplication (which is really composition of functions) is commutative?

(c) Let's run this procedure on a couple of $3 \times 3$ matrices $C$ and $D$

$$CD = \begin{pmatrix} 1 & -1 & 0 \\ 2 & -2 & 1 \\ 1 & 2 & 2 \end{pmatrix} \begin{pmatrix} 3 & -1 & 2 \\ 0 & 2 & 0 \\ 1 & 3 & -4 \end{pmatrix} =$$

$$\begin{pmatrix} 1 \times 3 + -1 \times 0 + 0 \times 1 & 1 \times -1 + -1 \times 2 + 0 \times 3 & 1 \times 2 + -1 \times 0 + 0 \times -4 \\ 2 \times 3 + -2 \times 0 + 1 \times 1 & 2 \times -1 + -2 \times 2 + 1 \times 3 & 2 \times 2 + -2 \times 0 + 1 \times -4 \\ 1 \times 3 + 2 \times 0 + 2 \times 1 & 1 \times -1 + 2 \times 2 + 2 \times 3 & 1 \times 2 + 2 \times 0 + 2 \times -4 \end{pmatrix}$$

$$= \begin{pmatrix} 3 & -3 & 2 \\ 7 & -3 & 0 \\ 5 & 9 & -6 \end{pmatrix}.$$

(d) Calculate the matrix product $DC$ and compare your answer to that of $CD$.

Although we are concentrating here on square matrices, we should mention that non-square matrices can be multiplied under the right conditions. Recall from Chapter 8 that functions can be composed provided that the co-domain of the first (or right-most) function matches the domain of the second (or left-most) function. With matrices this translates to the number of rows of the right-most matrix must match the number of columns of the left-most matrix. Hence if $A$ is a matrix of size $n \times m$ and $B$ is a matrix of size $m \times k$, then the product matrix $AB$ can be calculated and the result will be a matrix of size $n \times k$. For example, if $A$ is a $2 \times 4$ matrix and $B$ is a $4 \times 3$ matrix, then we can compute the resulting product matrix $AB$, whose size will be $2 \times 3$. However, we cannot compute $BA$.

**Example 12.4.** (a) Suppose that

$$A = \begin{pmatrix} 1 & 2 & -3 & 5 \\ 2 & -1 & 4 & -2 \end{pmatrix} \text{ and } B = \begin{pmatrix} 1 & 3 & 4 \\ -2 & 1 & 1 \\ -1 & 3 & 2 \\ 5 & -2 & 1 \end{pmatrix}$$

then verify that the product matrix

$$AB = \begin{pmatrix} 25 & -14 & 5 \\ -10 & 21 & 13 \end{pmatrix}.$$

(b) Write down the matrix $B$ on the left and the matrix $A$ on the right. Try to multiply them by this "across/down" procedure, thus seeing why the left-hand column count must match the right-hand row count.

We note here that matrix addition and matrix multiplication do have the following properties which binary operations may possess. If $A, B, C$ are $n \times n$ matrices with real entries, then

(1) $A + B = B + A$ and $A + (B + C) = (A + B) + C$, i.e., addition is both commutative and associative (because these properties hold in $\mathbb{R}$).

(2) $A(BC) = (AB)C$, i.e., multiplication is associative (but *not* commutative). See Exercises 12.2 and 12.3.

(3) $A(B + C) = AB + AC$ and $(A + B)C = AC + BC$, i.e., multiplication distributes over addition on both the left and the right. See Exercises 12.4 and 12.5.

We now end this chapter by discussing two further concepts which apply to any set $S$ which has algebraic structure, and then see how those concepts apply to matrices in particular. Suppose that a set $S$, having a binary operation $*$, has an element $e$ which satisfies that for every $a \in S$, $e * a = a * e = a$; then $e$ is called an **identity element** for $*$ in $S$. In words, an identity element does not change any elements when combined with them.

**Example 12.5.** (a) Among the number sets in the chart at the outset of this chapter (i.e, $\mathbb{Z}^+$, $\mathbb{Z}$, $\mathbb{Q}^+$, $\mathbb{Q}$, $\mathbb{R}^+$, $\mathbb{R}$ and $\mathbb{C}$), state whether the set has an identity element for addition (i.e., an **additive identity**), and if so what it is.

(b) Among the number sets in the chart at the outset of this chapter, state whether the set has an identity element for multiplication (i.e., a **multiplicative identity**), and if so what it is.

(c) Does the set $S_n$ of permutations of a set of $n$ elements possess an identity element for composition? If so, what is it?

Now suppose that the set $S$ has a binary operation $*$ and an identity element $e$. If $a$ is an arbitrary element of $S$, there may or may not be an element $b \in S$ which has the property that $a * b = b * a = e$ (i.e., combining $b$ with $a$ brings us back to the identity). If such an element $b$ exists, it is called an **inverse of** $a$ with respect to $*$, and can be denoted $a^{-1}$. If $a^{-1}$ exists, we say that $a$ is **invertible** (with respect to $*$).

**Example 12.6.** (a) Among the number sets in the chart at the outset of this chapter (i.e, $\mathbb{Z}^+$, $\mathbb{Z}$, $\mathbb{Q}^+$, $\mathbb{Q}$, $\mathbb{R}^+$, $\mathbb{R}$ and $\mathbb{C}$), state which elements (if any) of the set are invertible with respect to addition, and if so what the "additive inverses" look like.

(b) Among the number sets in the chart at the outset of this chapter, state which elements (if any) of the set are invertible with respect to multiplication, and if so what the "multiplicative inverses" look like.

(c) Recall that we established in Chapter 10, and more generally in Chapter 8, that every permutation, being a bijection, is invertible with respect to composition of functions.

In the above definitions, we have said "an" identity element and "an" inverse element. We shall ask you in Exercises 12.6 and 12.8 to show that in fact the identity element is always unique and, provided that the binary operation is associative, that inverses are also unique.

Turning back to matrices then, we wish to describe identity elements and inverse elements for the two binary operations addition and multiplication on the set of $n \times n$ matrices. Three of these are pretty obvious, but the fourth is not. An additive identity is the **zero matrix**, i.e., the matrix all of whose entries are 0, and an additive inverse of any matrix is its "negative", i.e., each of its entries is negated. Concerning matrix multiplication, it is easy to verify that an identity element is the (appropriately named) **identity matrix**, often denoted $I_n$, which is the matrix with 1 as each of its main diagonal entries and zeros otherwise (see Exercise 12.10).

The matter of multiplicative inverses of square matrices is more complicated. If you took a course in Linear Algebra, you will recall that a finite set of $n$-dimensional vectors (i.e., $n$-tuples) with entries from $\mathbb{R}$ is called **linearly independent** if none of them can be written as a linear combination of the others (another way to say this is that no non-zero linear combination of all them can equal 0). Now, in the matrix $A$, if the $n$ rows (each of which is an $n$-dimensional vector) are linearly independent, then $A$ is said to be **non-singular** (otherwise $A$ is called **singular**). Finally then, $A$ possesses a multiplicative inverse, i.e., is invertible with respect to matrix multiplication, *if and only if it is non-singular*.

Okay, given that an $n \times n$ matrix $A$ is invertible, how do we compute its inverse? This is not as easy as in all of the other cases (number sets and permutations) we have considered. For the case $n = 2$, one can simply memorize the inverse:

$$\text{If } A = \begin{pmatrix} a & b \\ c & d \end{pmatrix}, \text{ then } A^{-1} = \begin{pmatrix} \frac{d}{ad-bc} & \frac{-b}{ad-bc} \\ \frac{-c}{ad-bc} & \frac{a}{ad-bc} \end{pmatrix}.$$

The quantity in the denominator of the inverse's entries is called the **determinant** of $A$, which will be non-zero if and only if $A$ is invertible. When $n > 2$, there is a process for computing inverses involving "elementary row operations" which is called Gauss-Jordan Elimination. We shall not discuss this procedure here; for a full discussion of this idea and numerous other aspects of matrix theory, see, for example, [1].

We shall return to an extended look at different types of sets with algebraic structure starting with Chapter 21. For example, the set of $n \times n$ matrices over $\mathbb{R}$, which we have examined here, will belong to a class of sets called **rings**. A ring has binary operations with which you can add, subtract and multiply the elements but not necessarily divide them (or, put another way, additive inverses exist, but multiplicative inverses need not exist). The study of the classes of sets with algebraic structure is generally called *Abstract Algebra*.

## Exercises

**12.1.** If $A = \begin{pmatrix} 1 & 4 \\ 2 & -3 \end{pmatrix}$ and $B = \begin{pmatrix} -3 & 2 \\ 1 & 4 \end{pmatrix}$, determine $A + B, A - B, AB,$ and $BA$.

**12.2.** If

$$A = \begin{pmatrix} 1 & 2 \\ 2 & 1 \end{pmatrix}, B = \begin{pmatrix} 2 & 1 \\ -1 & 4 \end{pmatrix}, \text{ and } C = \begin{pmatrix} 2 & 3 \\ -1 & -2 \end{pmatrix},$$

verify the associativity of matrix multiplication for these three matrices, i.e., show that $A(BC) = (AB)C$.

**12.3.** For the case of $2 \times 2$ matrices over $\mathbb{R}$, do a partial proof that matrix multiplication is associative by showing that the upper left-hand entries of the two matrices being computed below are equal.

$$\begin{pmatrix} a & b \\ c & d \end{pmatrix} \left( \begin{pmatrix} e & f \\ g & h \end{pmatrix} \begin{pmatrix} j & k \\ m & n \end{pmatrix} \right) \text{ and } \left( \begin{pmatrix} a & b \\ c & d \end{pmatrix} \begin{pmatrix} e & f \\ g & h \end{pmatrix} \right) \begin{pmatrix} j & k \\ m & n \end{pmatrix}.$$

(Note: These entries are each a sum of four terms.)

**12.4.** For the three matrices $A$, $B$ and $C$ in Exercise 12.2, verify that matrix addition and multiplication are distributive for these three matrices. That is, show that $A(B + C) = AB + AC$.

**12.5.** For the case of $2 \times 2$ matrices over $\mathbb{R}$, do a partial proof that matrix addition and multiplication are distributive by showing that the upper left-hand entries of the two matrices being computed below are equal.

$$\begin{pmatrix} a & b \\ c & d \end{pmatrix} \left( \begin{pmatrix} e & f \\ g & h \end{pmatrix} + \begin{pmatrix} j & k \\ m & n \end{pmatrix} \right) \text{ and } \begin{pmatrix} a & b \\ c & d \end{pmatrix} \begin{pmatrix} e & f \\ g & h \end{pmatrix} + \begin{pmatrix} a & b \\ c & d \end{pmatrix} \begin{pmatrix} j & k \\ m & n \end{pmatrix}.$$

(Note: Again, these entries are each a sum of four terms.)

**12.6.** Prove that if a set $S$ has a binary operation $*$ and an identity element $e$ for $*$, then $e$ is unique. (Note: A standard technique to establish uniqueness is to assume that two elements (say $e_1$ and $e_2$) both have the given property (in this case, being an identity), and show that they are in fact equal. So what is $e_1 * e_2$?)

**12.7.** Let

$$A = \begin{pmatrix} 1 & 2 \\ 1 & -1 \end{pmatrix} \text{ and } B = \begin{pmatrix} 2 & 3 & 1 \\ -1 & 4 & 1 \end{pmatrix}$$

Of the three combinations $A + B$, $AB$ and $BA$, only one is possible to compute. Which one is it, and what is the answer?

**12.8.** Prove that if a set $S$ has an *associative* binary operation $*$ with identity element $e$, and if $a \in S$ has an inverse $b$, then $b$ is unique. (Note: As in Exercise 12.6, suppose both $b$ and $c$ have the given property (in this case, being an inverse for $a$). Show that they are in fact equal. Start with $b = b * e = \cdots$.)

**12.9.** Let

$$A = \begin{pmatrix} 2 & 1 & -1 \\ -3 & 4 & 1 \\ 2 & -1 & 3 \end{pmatrix} \text{ and } B = \begin{pmatrix} 1 & -1 & 3 \\ 2 & -1 & 2 \\ 3 & -1 & -2 \end{pmatrix}.$$

Calculate $5A$, $AB$, $BA$ and $A^2$.

**12.10.** For the matrices $I_3 = \begin{pmatrix} 1 & 0 & 0 \\ 0 & 1 & 0 \\ 0 & 0 & 1 \end{pmatrix}$ and $A = \begin{pmatrix} a & b & c \\ d & e & f \\ g & h & j \end{pmatrix}$, verify that $I_3$ is the identity element for the set of $3 \times 3$ matrices under multiplication, i.e., verify that $I_3 A = A$ and $A I_3 = A$.

**12.11.** Consider the two matrices $A = \begin{pmatrix} 2 & 1 \\ -3 & 4 \end{pmatrix}$ and $B = \begin{pmatrix} 4 & 6 \\ 2 & 3 \end{pmatrix}$. By computing their determinants (see the discussion following Example 12.6), determine which of these matrices is invertible, and if so, write down the inverse(s).

**12.12.** If you computed an inverse matrix in the previous exercise, verify now that it is indeed the inverse by multiplying it by the original invertible matrix.

**12.13.** Though we have not covered the computation of determinants and multiplicative inverses for $n \times n$ matrices with $n > 2$, you may have learned how to do one or both of these computations in a previous course. If so, compute the inverse of the matrix $A = \begin{pmatrix} 1 & 2 & 1 \\ -1 & 1 & 1 \\ 1 & -1 & 1 \end{pmatrix}$.

(Recall: Write the matrix $A$ on the left and separate it from the matrix $I_3$ using a vertical line. Now perform the elementary row operations on the whole display until the three left-hand columns are $I_3$. The three right-hand columns will now be the inverse of $A$.)

**12.14.** As loosely described at end of this chapter, a **ring** is a set with algebraic structure consisting of an associative (and commutative) addition and an associative multiplication (which together are distributive), and for which all elements possess additive inverses but not necessarily all non-zero elements possess multiplicative inverses. A **field** is a ring whose multiplication is commutative and which has the property that *all* non-zero elements possess multiplicative inverses. Hence every field is automatically a ring, but not conversely.

For each of the sets in the opening chart of this chapter, classify that set as a field, a ring but not a field (why?), or not even a ring (why?).

---

# CHAPTER 13

# Divisibility in $\mathbb{Z}$ and Number Theory

---

$$\gcd(858, 1092) = 78$$

In this and the next few chapters, we turn our attention to the set $\mathbb{Z}$ of integers. As we observed in Chapter 12, this set has algebraic structure consisting of addition and multiplication, the former possessing an additive identity (0) and additive inverses (negatives), the latter possessing a multiplicative identity (1) but multiplicative inverses for only two of its elements (1 and $-1$). Thus $\mathbb{Z}$ falls in the category of *rings* (see Exercise 12.14).

The study of the set $\mathbb{Z}$ is generally known as *number theory*. Number theory is one the oldest and most fascinating areas of mathematics because it attempts to understand the most basic set among the sets of numbers (e.g., $\mathbb{Q}, \mathbb{R}, \mathbb{C}$). One of its beauties is that most often the *questions* we ask are easy to understand, but the *answers* to those questions can often be very difficult to find. Here are two famous examples, discussed very briefly:

**Example 13.1. Fermat's Last Theorem.** Every student of geometry knows about the famous 3-4-5 triangle, which is a *right* triangle because those numbers satisfy the Pythagorean Theorem: $3^2 + 4^2 = 5^2$. It turns out that there are infinitely many such "Pythagorean triples" (e.g., 5-12-13, 8-15-17, etc.); that is, there are infinitely many solutions in positive integers to the equation $a^2 + b^2 = c^2$. However, in 1637, the mathematician Pierre de Fermat claimed to have found (but never wrote down) a proof of the following quite startling statement:

If $n > 2$, then there are *no* solutions in positive integers to the equation $a^n + b^n = c^n$.

Making a long story very short, 358 years later, in 1995, Andrew Wiles finally confirmed Fermat's statement. A tremendous amount of work by many mathematicians went on during those 358 years in trying to solve this problem. Wiles's proof, by the way, uses extremely high-level mathematics involving the complex numbers $\mathbb{C}$ and much more.

**Example 13.2. The Twin Primes Conjecture.** In Chapter 3 we wrote down Euclid's beautiful proof that there are infinitely many prime numbers in $\mathbb{Z}$ (Example 3.4). Looking a little more closely, two prime numbers are called *twins* if they differ by 2. For example, 3 and 5 are twins, as are 5 and 7. In Exercise 13.13 you will be asked to identity all the twin prime pairs starting with 3-5 and going up to 100. Does the list keep on going after that? How far?

**Conjecture**: There are infinitely many twin prime pairs.

At the present, there is no proof of this conjecture. No one has been able to prove that there are, or that there are *not*, infinitely many such pairs. Hence this conjecture is what's known as an "open problem" in mathematics. This is another of very many topics in number theory where it is easy to ask the question but hard to find an answer.

Returning to the algebraic structure of the integers $\mathbb{Z}$, we know that nearly all its elements do not possess multiplicative inverses, but we can still investigate *divisibility* in $\mathbb{Z}$, a topic which has been studied in depth for more than 2000 years. In fact, Euclid ($\sim$325BC – 265BC) dealt with this topic in Book VII of his *Elements*. So here we go.

Suppose $b$ is an integer and $a$ is a non-zero integer. We say that $a$ *divides* $b$ if there is an *integer* $q$ so that $b = aq$. If there are such integers, we denote the fact that $a$ divides $b$ by using the notation $a|b$. Note that the notation $a|b$ is a *sentence* with the verb being "divides." Contrast this with the notation $a/b$, which is an element of the rational numbers $\mathbb{Q}$, not a sentence.

**Example 13.3.** Clearly $2|8$ since $8 = 2(4)$; $36|108$ since $108 = 36(3)$; $3|(-36)$ since $-36 = 3(-12)$; and for any integer $m$, $3|(15m + 3)$ since $15m + 3 = 3(5m + 1)$. On the other hand, 3 does not divide 13 as there is no integer $q$ with $13 = 3q$.

In the following lemma, we provide a few basic properties involving the divisibility of integers. (Note: A "lemma" is a "helping theorem", i.e., an often easily proved result which is then used to establish bigger results.)

**Lemma 13.1.** *Let $a, b, c, d$ be integers with $a > 0$ and $d > 0$.*
(i) *If $a|b$ and $a|c$, then $a|(b + c)$;*
(ii) *If $a|b$ and $a|c$, then $a|(b - c)$;*

(iii) *If $a|b$ and $a|c$, then $a|(mb + nc)$ for any integers $m$ and $n$;*

(iv) *If $d|a$ and $a|b$, then $d|b$.*

Proof. To prove Part (i), we may assume that $b = aq$ and $c = as$ where $q$ and $s$ are integers. Then $b + c = aq + as = a(q + s)$ so that $a$ divides $b + c$ (since $q + s$ is an integer). The proof of Part (ii) is similar and hence omitted. Proofs of the remaining parts are left to Exercise 13.2. ∎

Now, what if a positive integer $a$ does *not* divide an integer $b$? Here is where the seemingly simple but very important ***Division Algorithm*** comes into play when doing computations in $\mathbb{Z}$. We introduced this result in Chapter 3 and then proved it in Theorem 5.2, but for completeness we state it again here:

**Theorem 13.2. (Division Algorithm)** *Let $a$ and $b$ be integers with $a > 0$. Then there are integers $q$ and $r$ with $0 \le r < a$ so that $b = aq + r$.*

As pointed out in Chapter 3, this is just the long division process you learned early in your math education. The integer $b$ is often called the *dividend*, $a$ the *divisor*, $q$ the *quotient*, and $r$ the *remainder*. The key is that the remainder $r$ must be non-negative and must be less than the divisor $a$. It is clear that $a|b$ if and only if $r = 0$. We asked you to show examples of this computation twice in Example 3.7.

We now apply the Division Algorithm to another important concept in the algebraic structure of $\mathbb{Z}$. Having introduced the following ideas in Chapter 3, we now look at them more closely. Given two positive integers $a$ and $b$, we define the ***greatest common divisor*** of $a$ and $b$ to be the largest positive integer that divides them both. This is often denoted by writing $\gcd(a, b)$. (Note: $\gcd(a, b)$ is sometimes denoted by $(a, b)$, but we shall avoid this notation since it can easily be confused with the open interval $(a, b)$ in $\mathbb{R}$ or the point $(a, b)$ in $\mathbb{R} \times \mathbb{R}$.) We say that two positive integers are ***relatively prime*** if their greatest common divisor is 1, i.e., if the only factor they have in common is 1. For example, 5 and 16 are relatively prime as are 24 and 37, but 6 and 28 are not relatively prime since they have a common factor of 2. You should convince yourself that a prime $p$ is relatively prime to any integer $n$ as long as $p$ does not divide the integer $n$.

So, given two positive integers $a$ and $b$, how do we compute their greatest common divisor? If $a$ and $b$ are relatively small, one way is to factor them both into their prime number factorizations and gather what those factorizations have in common. However, if at least one of the integers $a$ and $b$ is large, this method can be difficult and relatively inefficient. Euclid, in Book VII of his *Elements*, now comes to the rescue again with a beautiful and very efficient procedure for computing greatest common divisors. Before stating his algorithm (which we first introduced in Chapter 3), let us look at a couple of examples.

**Example 13.4.** (a) What is gcd(420, 378)? Well, as suggested above, one way is to factor them both: $420 = 2^2 \cdot 3 \cdot 5 \cdot 7$ and $378 = 2 \cdot 3^3 \cdot 7$. These factorizations have in common $2, 3$ and $7$, so gcd(420, 378) $= 2 \cdot 3 \cdot 7 = 42$.
(b) What is gcd(858, 1092)? We could factor again, but it does not look easy (try it if you like). We are seeing that the larger the numbers become, the more difficult the factorization method becomes. We could use a better algorithm, and here it is.

**Theorem 13.3. (Euclidean Algorithm)** *Let $a$ and $b$ be positive integers. If $a$ divides $b$ then the gcd$(a, b) = a$. Otherwise, there exists a strictly decreasing sequence of positive integers $r_1, \ldots, r_n$ so that*

$$
\begin{aligned}
b &= aq_1 + r_1 \\
a &= r_1q_2 + r_2 \\
r_1 &= r_2q_3 + r_3 \\
&\ \vdots \\
r_{n-2} &= r_{n-1}q_n + r_n \\
r_{n-1} &= r_nq_{n+1} + 0.
\end{aligned}
$$

*Then gcd$(a, b) = r_n$.*

At first glance this procedure looks complicated, but all it really says is that in order to calculate the greatest common divisor of two positive integers, we repeatedly apply the Division Algorithm, replacing at each step the previous dividend with the previous divisor and the previous divisor with the previous remainder. In doing this, each divisor is strictly smaller than the previous divisor, and hence each remainder is strictly smaller than the previous remainder. It follows that the sequence $\{r_1, \ldots, r_n\}$ is a strictly decreasing sequence of nonnegative numbers, and so by the Well-Ordering Principle we must eventually get to a remainder of 0. Finally, you will be asked to prove in Exercise 13.7 that (a) $r_n$ is indeed a divisor of $a$ and $b$, and (b) it is in fact the *greatest* such divisor.

It is time for some examples.

**Example 13.5.** (a) Following up on Example 13.4, we now use the Euclidean Algorithm to find the greatest common divisor of $a = 378$ and $b = 420$:

$$
\begin{aligned}
420 &= 378(1) + 42 \\
378 &= 42(9) + 0.
\end{aligned}
$$

That was quick! Since the last non-zero remainder is 42, gcd(420, 378) $= 42$, as we already computed by factoring.

(b) Again from Example 13.4, what is $\gcd(858, 1092)$? Applying our algorithm:

$$
\begin{aligned}
1092 &= 858(1) + 234 \\
858 &= 234(3) + 156 \\
234 &= 156(1) + 78 \\
156 &= 78(2) + 0.
\end{aligned}
$$

So that took four steps, but we get $\gcd(858, 1092) = 78$.

(c) Try a new one on your own. Use the Euclidean Algorithm to compute $\gcd(182, 420)$.

As we introduced back in Example 3.8, it is often useful to write the greatest common divisor of two positive integers $a$ and $b$ as a linear combination of them, i.e., find integers $x$ and $y$, one positive and one negative, so that $\gcd(a, b) = ax + by$. As stated in that example, an added bonus of the Euclidean Algorithm is that it provides a systematic way of finding such a pair $(x, y)$. To illustrate this method we use Part (b) of Example 13.5 above (because when we write it down with symbols it quickly becomes somewhat of a mess). The idea is, working from the bottom, we replace each remainder by its expression just above. We number the steps (omitting the last) and we solve each line for the remainder:

(1) $234 = 1092 - 858(1)$
(2) $156 = 858 - 234(3)$
(3) $78 = 234 - 156(1)$.

In line (3), replace 156 by its expression in line (2) and simplify:

$$78 = 234 - 156(1) = 234 - (858 - 234(3)) = 234(4) - 858.$$

Now in this, replace 234 by its expression in line (1) and simplify:

$$78 = 234(4) - 858 = (1092 - 858(1))(4) - 858 = 1092(4) - 858(5).$$

We have succeeded then in writing the gcd 78 as a linear combination of 1092 and 858.

We note that when running the Euclidean Algorithm to find the gcd, we do not actually make use of the specific values of the quotients $q_i$. However, in moving backwards to express the gcd as a linear combination, we definitely make use of all but the last of those values. Let's do one more example.

**Example 13.6.** (a) Let $a = 68$ and $b = 249$. We use the Euclidean Algorithm:

$$249 \ = \ 68(3) + 45$$
$$68 \ = \ 45(1) + 23$$
$$45 \ = \ 23(1) + 22$$
$$23 \ = \ 22(1) + 1$$
$$22 \ = \ 1(22) + 0$$

Thus, $\gcd(68, 249) = 1$, so 68 and 249 are relatively prime.

(b) Use the "working backwards" method shown above to find $x$ and $y$ such that $68x + 249y = 1$. You will need to do three replacements: 22, then 23, then 45.

As we move forward in our exploration of some ideas in number theory, divisibility will continue to play an important part. In particular, in our next chapter we discuss the central role played by numbers in $\mathbb{Z}^+$ which can only be divided by 1 and themselves. These numbers, called *prime numbers*, become the building blocks of the multiplicative structure of $\mathbb{Z}$.

**Exercises**

**13.1.** Let $m$ and $k$ be arbitrary positive integers. Label each statement as true or false and support your choice.
(a) $5 | 635$
(b) $-5 | 635$
(c) $48 | 124$
(d) $341 | 32871$
(e) $5 | (15m - 10)$
(f) $m | (-3m)$
(g) $(k + m) | (7k + 14m)$
(h) $k | (-6k^2 - k)$

**13.2.** (a) Prove Part (iii) of Lemma 13.1. (Note: Having proved this, note that Parts (i) and (ii) are special cases of Part (iii).)
(b) Prove Part (iv) of Lemma 13.1.

**13.3.** Find $\gcd(35, 180)$ (a) using factorization, and (b) using the Euclidean Algorithm.

**13.4.** Find $\gcd(44, 111)$ (a) using factorization, and (b) using the Euclidean Algorithm.

**13.5.** (a) Find $\gcd(224, 468)$.
(b) Find $(x, y)$ such that $\gcd(224, 468) = 224x + 468y$.

**13.6.** (a) Find $\gcd(381, 3837)$.
(b) Find $(x, y)$ such that $\gcd(381, 3837) = 381x + 3837y$.

**13.7.** Complete the proof of the Euclidean Algorithm (Theorem 13.3) by making use of Lemma 13.1 Part (iii):
(a) Prove that $r_n$ is a common divisor of $a$ and $b$, (Hint: Start at the bottom. By the last step $r_n$ clearly divides $r_{n-1}$. Moving up, since $r_n$ divides both itself and $r_{n-1}$, it must $\cdots$.)
(b) Prove that $r_n$ is the *greatest* common divisor of $a$ and $b$. (Hint: Start at the top now and suppose d is some common divisor of $a$ and $b$. By the first step, we must have $d|r_1$. Now move down a line $\cdots$.)

**13.8.** The sequence of numbers $1, 1, 2, 3, 5, 8, 13, 21, \ldots$ is known as the sequence of **Fibonacci numbers**. After the first two values, a given number is obtained as the sum of the previous two numbers. We denote this sequence of positive integers by $F_1, F_2, F_3, \ldots$, in honor of Fibonacci who first wrote about these numbers in his book "Liber Abaci", which was published in 1202. (We first introduced this sequence in Exercise 4.9.)
(a) Above we have written $F_1$ through $F_8$. Write down $F_9$ through $F_{12}$.
(b) Prove that for any positive integer $k \geq 1$, $\gcd(F_k, F_{k+1}) = 1$, i.e., prove that any two consecutive Fibonacci numbers are relatively prime. (Hint: Suppose $d$ divides both $F_{k+1}$ and $F_k$, then by Lemma 13.1 Part (ii) (or (iii)) it divides their difference, which is $\cdots$. Continue downward.)

**13.9.** Show that, for any positive integer $n$, 3 divides $4^n - 1$.

**13.10.** Concerning Fermat's Last Theorem (Example 13.1), we claimed that there are infinitely many Pythagorean triples.
(a) Verify this by showing that if $s$ and $t$ are odd positive integers with $s > t$, then

$$(st)^2 + \left(\frac{s^2 - t^2}{2}\right)^2 = \left(\frac{s^2 + t^2}{2}\right)^2,$$

i.e., $(st, (s^2 - t^2)/2, (s^2 + t^2)/2)$ form a Pythagorean triple.
(b) Write down the Pythagorean triples generated by $(s = 3, t = 1)$, $(s = 5, t = 1)$, $(s = 7, t = 1)$ and $(s = 7, t = 3)$.

**13.11.** Is $n^3 - n$ divisible by 6 for each positive integer $n$? If so, show it, and if not, find an example where it fails.

**13.12.** For a positive integer $a$, what are the possibilities for the quantity $\gcd(a + 3, a)$? Find specific examples to demonstrate each possibility. Now prove your conjecture. (Hint: Suppose $d$ divides both $a$ and $a + 3$; then by Lemma 13.1 Part (ii)$\cdots$.)

**13.13.** Concerning the Twin Primes Conjecture (Example 13.2):
(a) Starting with 3-5, there are eight twin prime pairs below 100. Write them down.
(b) If you have access to some mathematics software, find out how many twin prime pairs there are between 1000 and 1100. What is the smallest of these?

**13.14.** How many steps must the Euclidean Algorithm take to find the gcd? We have seen through examples and exercises that it can vary, but what is the "worst case?" It is not too hard to show that if $a$ is the first divisor, then the total number of steps can be no more than 7 times the number of decimal digits of $a$ (see, for example, [9], Exercise 5.3).

(a) Suppose the first divisor $a$ is between a billion and 9 billion (and the dividend $b$ is larger). What is the maximum number of steps to discover $\gcd(a, b)$ by the Euclidean Algorithm? (Note: You may not want to do this by hand, but a computer can do it extremely quickly.)

(b) Returning to the Fibonacci numbers (Exercise 13.8), we saw that $F_{11} = 89$ and $F_{12} = 144$. According to the above information, what is the maximum number of steps needed to compute $\gcd(F_{11}, F_{12})$ using the Euclidean Algorithm? Now do the actual computation and check the number of steps. (Note: This part illustrates that computing the gcd of two adjacent Fibonacci numbers using the Euclidean Algorithm goes about as slowly as possible.)

# CHAPTER 14

# Primes and Unique Factorization

$$4312 = 2^3 \cdot 7^2 \cdot 11$$

In this chapter we continue our focus on the integers $\mathbb{Z}$, zeroing in on prime numbers, which are, as we said at the end of Chapter 13, the building blocks of the multiplicative structure of $\mathbb{Z}$. The set of prime numbers has fascinated mathematicians throughout history, and there continue to be many unsolved mysteries about this set, some of which we shall discuss later in the chapter.

A positive integer $p$ is **prime** if $p$ has exactly two distinct positive divisors, namely 1 and $p$ itself. (If an integer $n$ is not prime, it is said to be **composite**.) Thus $2, 3, 5, 7, 11$ are the five smallest primes. We note that 2 stands apart from the other primes as the only *even* prime. We note also that with this definition the positive integer 1 is *not* prime since it has only one divisor (itself). Early mathematicians may have viewed 1 as being prime, but once we discuss the theory of unique factorization of integers, we will see why it is important not to count 1 as being prime.

A first question we might ask about prime numbers is: How can one test whether a given positive integer $n$ is prime? This, in general, turns out not to be an easy task, especially when $n$ is large.

**Example 14.1.** (a) Are 4997 and 4999 prime? Our only approach would seem to be to try to find relatively small divisors of them to establish that they are *not* prime. It turns out that 4997 is divisible by 19 and hence is not prime, but 4999 is indeed prime.

(b) Show that 5357 is not prime by finding a relatively small divisor of it. (Note: See Exercise 14.5 for an easy way to solve this particular case.)

A second question we can ask is: How can we list all of the prime numbers up to some positive value $n \geq 2$? A method to do this is known as the **Sieve of Eratosthenes**, named in honor of Eratosthenes (276BC – 194BC), who appears to be the first to make use of this process. The process is quite efficient as long as $n$ isn't *too* large. Here it is:

We begin by listing all of the numbers from 2 to $n$. Then since 2 is prime, we leave it in the list and delete all multiples of 2 (except 2 itself) up to and including $n$. That knocks out all the even numbers in our list larger than 2. We then leave 3 and delete all larger multiples of 3. The next value not already deleted is 5, so we leave it and delete all multiples of 5. We continue this process with 7 which is yet to be deleted, then 11, etc. The numbers remaining in the list give all primes up to $n$. A question you might have is: When can we stop this process? You are asked in Exercise 14.1 to show that *we need only process primes which are less than or equal to the square root of* $n$.

**Example 14.2.** We illustrate the Sieve of Eratosthenes by finding all primes up to $n = 50$. We begin by listing all of the positive integers from 2 through 50. By what we just stated, we need only process 2, 3, 5 and 7 since $11 > \sqrt{50}$.

$$
\begin{array}{cccccccccc}
 & 2 & 3 & 4 & 5 & 6 & 7 & 8 & 9 & 10 \\
11 & 12 & 13 & 14 & 15 & 16 & 17 & 18 & 19 & 20 \\
21 & 22 & 23 & 24 & 25 & 26 & 27 & 28 & 29 & 30 \\
31 & 32 & 33 & 34 & 35 & 36 & 37 & 38 & 39 & 40 \\
41 & 42 & 43 & 44 & 45 & 46 & 47 & 48 & 49 & 50
\end{array}
$$

We boldface the number 2 as the first item in this list (since we know it is prime), and then cross out each multiple of 2 that is greater than 2. It is important to note that no actual arithmetic must be done here! We simply start at 2, skip by the amount of 2 (which gets us to the number 4), cross out the 4, then skip by another 2 to get to 6, cross out the 6, and so on. This stage of the process is quite straightforward. This now leaves us with the following table.

$$
\begin{array}{cccccccccc}
 & \mathbf{2} & 3 & \cancel{4} & 5 & \cancel{6} & 7 & \cancel{8} & 9 & \cancel{10} \\
11 & \cancel{12} & 13 & \cancel{14} & 15 & \cancel{16} & 17 & \cancel{18} & 19 & \cancel{20} \\
21 & \cancel{22} & 23 & \cancel{24} & 25 & \cancel{26} & 27 & \cancel{28} & 29 & \cancel{30} \\
31 & \cancel{32} & 33 & \cancel{34} & 35 & \cancel{36} & 37 & \cancel{38} & 39 & \cancel{40} \\
41 & \cancel{42} & 43 & \cancel{44} & 45 & \cancel{46} & 47 & \cancel{48} & 49 & \cancel{50}
\end{array}
$$

Once we have traversed the entire list, we then return to the beginning of the list and look for the first number that has not been selected as a prime already and that has not been crossed out. At this stage, that number is 3. We are guaranteed that this is a prime, so we boldface that number and then cross out all multiples of 3 in the list (again by simply skipping by 3 each time and crossing out the corresponding numbers). That leaves us with the following:

|    | 2  | 3  | 4  | 5  | 6  | 7  | 8  | 9  | 10 |
|----|----|----|----|----|----|----|----|----|----|
| 11 | 12 | 13 | 14 | 15 | 16 | 17 | 18 | 19 | 20 |
| 21 | 22 | 23 | 24 | 25 | 26 | 27 | 28 | 29 | 30 |
| 31 | 32 | 33 | 34 | 35 | 36 | 37 | 38 | 39 | 40 |
| 41 | 42 | 43 | 44 | 45 | 46 | 47 | 48 | 49 | 50 |

Returning to the beginning of the list, we see that 5 is the first number which is neither boldfaced nor crossed out. We boldface it and then cross out its multiples.

|    | 2  | 3  | 4  | 5  | 6  | 7  | 8  | 9  | 10 |
|----|----|----|----|----|----|----|----|----|----|
| 11 | 12 | 13 | 14 | 15 | 16 | 17 | 18 | 19 | 20 |
| 21 | 22 | 23 | 24 | 25 | 26 | 27 | 28 | 29 | 30 |
| 31 | 32 | 33 | 34 | 35 | 36 | 37 | 38 | 39 | 40 |
| 41 | 42 | 43 | 44 | 45 | 46 | 47 | 48 | 49 | 50 |

As stated at the outset, we need now only process 7 to finish the job. Our list then looks like this:

|    | 2  | 3  | 4  | 5  | 6  | 7  | 8  | 9  | 10 |
|----|----|----|----|----|----|----|----|----|----|
| 11 | 12 | 13 | 14 | 15 | 16 | 17 | 18 | 19 | 20 |
| 21 | 22 | 23 | 24 | 25 | 26 | 27 | 28 | 29 | 30 |
| 31 | 32 | 33 | 34 | 35 | 36 | 37 | 38 | 39 | 40 |
| 41 | 42 | 43 | 44 | 45 | 46 | 47 | 48 | 49 | 50 |

Thus we find that the list of all primes up to 50 is given by

$$2, 3, 5, 7, 11, 13, 17, 19, 23, 29, 31, 37, 41, 43, 47.$$

In Exercise 14.2 below you will be asked to list all of the primes up to 100.

We have stated at the close of the previous chapter and the beginning of this chapter that the prime numbers are the "building blocks" of the multiplicative structure of the integers. We now develop that idea, and to do so we must first prove some additional results on divisibility. The first is the following lemma:

**Lemma 14.1.** *Let $a, b, c$ be positive integers with $a$ and $b$ relatively prime.*
  (i) *If $a|bc$ then $a|c$;*
  (ii) *If $a|c$ and $b|c$, then $ab|c$.*

Proof.
(i) Since $a$ and $b$ are relatively prime (i.e., $\gcd(a, b) = 1$), we know from Chapter 13 that the Euclidean Algorithm gives us a way to find integers $r$ and $s$ so that $1 = ar + bs$. Multiplying through by $c$, we get $c = car + cbs$. Since $a$ clearly divides $car$ and divides $cbs$ by hypothesis, it follows from Lemma 13.1 that $a$ divides the sum on the right. Hence $a$ divides $c$.
(ii) Since $a$ divides $c$, $ab$ divides $cbs$. Moreover, since $b$ divides $c$, $ab$ divides $car$. Thus $ab$ divides the sum, which is $c$. ∎

Our next result is often attributed to Euclid (and is referred to by many as "Euclid's Lemma"). It follows from our lemma above and provides the foundation for the truly important role that the primes play in the algebraic structure of the integers.

**Theorem 14.2. (Euclid's Lemma)** *If $p$ is a prime and $p$ divides $ab$, then $p$ divides $a$ or $p$ divides $b$.*

Proof. Since $p$ is a prime, we know $\gcd(p, a)$ is either 1 or $p$. In the latter case $p$ divides $a$. If $\gcd(p, a) = 1$ then Part (i) of Lemma 14.1 implies that $p$ divides $b$. ∎

This says that if $p$ is a prime and it divides the product of two integers, it must divide one or the other (or both). Note that this is not necessarily true of divisors which are not prime. For example, note that 4 divides $2 \cdot 6 = 12$ , but 4 divides neither 2 nor 6.

The following corollary generalizes Euclid's Lemma. You will be asked to prove this corollary in Exercise 14.10.

**Corollary 14.3.** *If $p$ is a prime and $p$ divides the product $a_1 \cdots a_r$, then $p$ must divide $a_i$ for some $i = 1, \ldots, r$.*

We now come to the central result of this chapter, often referred to as the "Fundamental Theorem of Arithmetic." It states that any positive integer can be written as a product of (not necessarily distinct) primes, and that with a possible re-ordering of those primes, this product is unique. We remark here that this result is one of many results in mathematics which establish *existence and uniqueness* of some object (in this case, of course, that object being a product of primes equaling a given positive integer).

**Theorem 14.4. (Unique Factorization of Integers)**
  (i) (*Existence*) *Every positive integer $n \geq 2$ may be written as a product of (not necessarily distinct) prime numbers; i.e., $n$ may be written in the form*

$$n = p_1 \cdots p_r$$

*where each integer $p_i, i = 1, \ldots, r$, is a prime.*
  (ii) (*Uniqueness*) *Moreover, this factorization is unique except for the order of the primes; i.e., if we also have $n = q_1 \cdots q_s$ where each $q_i$ is a prime, then $r = s$ and (if necessary) upon re-ordering, $p_i = q_i, i = 1, \ldots, r$.*

Proof. (i) (Existence) We proved this in Example 4.6, using strong induction. (Look back at that now!)
(ii) (Uniqueness) Here we proceed by induction in its more usual ("weak") form. We induct on $r$ the number of primes in the factorization of $n$ in terms of the primes $p_i$. For the case $r = 1$, we have that $n = p_1$ (so $n$ is prime) and we also have that $n = q_1 \cdots q_s$. If $s \geq 2$, then the prime $n$ would have at least the three

90

distinct divisors 1, $q_1$ and $q_1 q_2$, but this contradicts the fact that $n$ is a prime. Hence $s = 1$, and so $n = q_1$, i.e., $p_1 = q_1$.

We now use the induction hypothesis that any positive integer greater than 2 which has a factorization into $r - 1$ primes has a unique factorization in the above sense. Assume further that

$$n = p_1 \cdots p_r = q_1 \cdots q_s$$

are two prime factorizations of the positive integer $n$. The prime $p_1$ divides $n$, so by Corollary 14.3 it must divide one of the primes $q_1, \ldots, q_s$. By re-ordering the primes $q_i$ if necessary, we may assume that $p_1$ divides $q_1$. But since both $p_1$ and $q_1$ are primes, we must have that $p_1 = q_1$. We can now divide both sides of the right-hand equation above by $p_1$ to obtain

$$p_2 \cdots p_r = q_2 \cdots q_s.$$

The integer on the left side is thus a product of $r - 1$ primes, so by the induction hypothesis, we have that $r - 1 = s - 1$ and hence $r = s$. Upon re-ordering if necessary, we have $p_i = q_i, i = 2, \ldots, r$. Since $p_1 = q_1$, we have that $p_i = q_i, i = 1, \ldots, r$ and the proof is complete. ∎

We note in the statement of the Unique Factorization Theorem that $n \geq 2$. This theorem provides a good reason why 1 is not considered to be a prime, for if 1 were a prime, then for example $6 = 2 \cdot 3 = 1 \cdot 2 \cdot 3$ would be two different prime factorizations of 6.

It was stated in the Unique Factorization Theorem that the primes in the representation $n = p_1 \cdots p_r$ are not necessarily distinct. We can, however, collect like primes in the factorization of an integer $n$ and thus write $n = p_1^{a_1} \cdots p_t^{a_t}$ where each $p_i, i = 1, \ldots, t$, is a prime with $p_i \neq p_j$ if $i \neq j$ and each exponent $a_i \geq 1$. This form is often called the **canonical factorization** of the positive integer $n$.

**Example 14.3.** (a) The canonical factorization of the integer $n = 1,000$ is given by $n = 2^3 5^3$, while the canonical factorization of the integer $n = 3,500$ is $2^2 5^3 7$.
(b) Write down the canonical factorization of $n = 45,000$.

Now comes a very important point about primes and factorization: Given a positive integer $n$, the Unique Factorization Theorem tells us about the guaranteed existence and uniqueness of a prime factorization of $n$, but it tell us *nothing about how to actually find that factorization*. In fact, if $n$ is a very large number, this is a formidable problem, even with the use of a fast modern computer. We will see in Chapter 20, for example, that when the RSA cryptographic system is employed, the security of the system is based upon the difficulty of factoring a very large number.

**Example 14.4.** Why are the numbers in Example 14.3 above relatively easy to factor? The reason is that they contain some small primes which are easy to check, so we can "whittle away" at the factorization. But what if there are no small primes around to get us started? For example, find the prime factorization of $n = 4307$.

In Chapter 13 we introduced a very efficient procedure for computing the gcd of two positive integers, namely the Euclidean Algorithm. This procedure is probably not, however, how you computed a gcd in the past; rather you probably used factorization. We now formalize this simple method under the assumption that we have been able to find the prime factorizations of two positive integers $a$ and $b$. We wish to emphasize that for relatively small $a$ and $b$ the process of first factoring and then applying this method generally works well; but as $a$ and $b$ get larger, the Euclidean Algorithm is *much* more efficient.

**Theorem 14.5.** *Let* $a = p_1^{a_1} p_2^{a_2} \cdots p_r^{a_r}$ *and* $b = p_1^{b_1} p_2^{b_2} \cdots p_r^{b_r}$ *be the prime factorizations of $a$ and $b$, respectively, (where perhaps some of the exponents are zero in order to allow a common list of primes to be used). Here, $p_1, p_2, \ldots, p_r$ are distinct primes, $a_1, a_2, \ldots, a_r \geq 0$ and $b_1, b_2, \ldots, b_r \geq 0$. Then*

$$gcd(a,b) = p_1^{min(a_1,b_1)} p_2^{min(a_2,b_2)} \cdots p_r^{min(a_r,b_r)}$$

*where $min(x,y)$ is the smaller of the two values $x$ and $y$.*

**Example 14.5.** (a) Let us use Theorem 14.5 to calculate $gcd(350, 450)$. The canonical factorizations of 350 and 450 are $2^1 5^2 7^1$ and $2^1 3^2 5^2$ respectively. In order to apply the theorem we include each of the primes $2, 3, 5$, and 7 in both of our prime factorizations, hence rewriting the factorizations as $350 = 2^1 3^0 5^2 7^1$ and $450 = 2^1 3^2 5^2 7^0$. Now the theorem tells us that $gcd(350, 450) = 2^1 3^0 5^2 7^0$, i.e., $gcd(350, 450) = 50$.
(b) Try to apply this method to find $gcd(589, 899)$. You may find yourself getting frustrated pretty soon.
(c) Use the Euclidean Algorithm to compute $gcd(589, 899)$.

We finish this chapter now by discussing some of the very many interesting properties and questions involving prime numbers. For example, we asked previously if there are infinitely many prime numbers and found out that Euclid proved long ago that there are. You should look back at that beautiful proof now (Example 3.4). When reading it, note that we actually used there two results we proved subsequently: Example 4.5 (that every integer is divisible by a prime) and Lemma 13.1 (that if an integer divides two others, then it divides their difference). We have also observed (in Example 13.2) that whether there are or are not infinitely many *twin prime pairs* is an unsolved problem.

What follows then is a sampler of prime number questions and ideas. We suppose that $p_1, p_2, p_3, \ldots$ are the primes in the natural order $2, 3, 5, \ldots$ .

• Can you determine a formula for the $n$th prime $p_n$ (i.e., a function which takes in $n$ and returns $p_n$)? Currently no such formula has been discovered.

• Given the $n$th prime $p_n$, what is the next prime $p_{n+1}$ in the list? Currently there is no known way to answer this question other than to consecutively test each positive integer after $p_n$ to see if it is a prime.

• Let $F_k = 2^{2^k} + 1$ for $k = 0, 1, 2, \ldots$. Note that these numbers get really big very quickly. You should check here that $F_k$ is a prime for $k = 0, 1, 2, 3$. Moreover, $F_4 = 65,357$ is also prime, but it turns out that $F(5) = 4,294,967,297$ is divisible by 641 and hence is not prime. Currently it is not known if any other $F_k$ is a prime for any value of $k \geq 6$, though it is known that some of the values $F_k$ for $k \geq 6$ are not prime; for example $F_6 = 18,446,744,073,709,551,616$ is also not prime. A number $F_k$ is called a **Fermat number** and those Fermat numbers which are primes are called **Fermat primes**, named after the same mathematician Fermat we introduced in Example 13.1.

• Given an even integer $n \geq 4$, can $n$ be written as $n = q_1 + q_2$ where $q_1$ and $q_2$ are both primes? Looking at some small examples, $4 = 2 + 2$, $6 = 3 + 3$, $8 = 3 + 5$, $10 = 5 + 5 = 3 + 7$, and so on. Note that starting with 10 there may well be more than one way to write $n$ as a sum of two primes. For example, 100 can be so written in 6 different ways (try writing them down). The conjecture that *all* even positive integers $n \geq 4$ can be written in at least one way as a sum of two primes is known as **Goldbach's Conjecture** (first stated by Goldbach in 1642). Currently, this conjecture has been verified by computer for very large even numbers, but no one has proved as yet that it holds for *all* even numbers.

• Around 1644 the mathematician Mersenne conjectured that if $p$ is a prime, then the number $M_p = 2^p - 1$ (called a **Mersenne number**) is itself prime. Checking some small examples, $M_2 = 3$, $M_3 = 7$, $M_5 = 31$ and $M_7 = 127$ (so far so good). However, he should have looked more carefully at $M_{11} = 2047$ since it factors into 23 times 89. Nonetheless, we can still conjecture that since there are infinitely many Mersenne *numbers*, perhaps there are infinitely many Mersenne *primes*, i.e., Mersenne numbers which are prime. This conjecture is called the **Mersenne prime conjecture**, and it remains unsolved. As of the writing of this text, only 51 Mersenne primes have been identified by computers, the largest being $2^{82,589,933} - 1$, which has $24,862,048$ digits! You should check the web (try googling "largest known Mersenne prime") to see if some new ones have since been found.

• The sequence of primes is known to be very irregularly spaced within the positive integers. On the one hand, we know that except for 2 and 3, primes can be as close together as having a gap of one composite number between them, and we have already discussed such primes, known as twin primes. Again, it is unknown whether there are infinitely many twin prime pairs (Example 13.2). Currently, the largest *known* twin prime pair is

$$2,996,863,034,895 \times 2^{1,290,000} \pm 1,$$

each of which has 388,342 decimal digits.

On the other hand, it turns out that there can be arbitrarily large gaps between adjacent primes. In Exercise 14.12, you will be asked to show that if $n$ is any positive integer, there is a gap in the sequence of primes of length at least $n$ composite numbers. For example, if we wish to find two adjacent primes which have a gap of a million composite numbers between them, we can find such a pair. Hence the smallest possible gap (after 2 and 3, which have a gap of 0 between them) is of length 1; but there does not exist a largest gap.

• We finish this sampler with a fact about primes which *has* been proved (late in the 19th century) and is called the ***Prime Number Theorem***. It can be stated in a few different ways; here are two of them:
(a) If $N$ is any positive integer, then the *approximate* number of primes less than $N$ is $N/\ln(N)$, i.e., is $N$ divided by the natural log of $N$. For example, then, if $N = 1,000$ this says that there are approximately $1,000/\ln(1,000) \approx 145$ primes below 1,000 (the actual count is 168). The theorem states that as $N$ goes to infinity, the ratio of $N/\ln(N)$ to the actual count approaches 1.
(b) Another nice way to express what this theorem says is that "in the vicinity" of the number $N$, the *density* of primes there is about one out of every $\ln(N)$ integers. So, for example, near 1,000, about one out of every $\ln(1,000) \approx 7$ integers is prime. We shall ask you to use calculus to explore the relationship of these two statements of the theorem in Exercise 14.13.

We hope this chapter has given you an appreciation of the central role played by the prime numbers in the algebraic structure of the integers $\mathbb{Z}$ and also an appreciation of the many mysteries surrounding this fascinating subset of $\mathbb{Z}$.

## Exercises

**14.1.** Suppose that $n$ is a composite integer and $p$ is a prime which divides $n$. Prove that if $p > \sqrt{n}$, then $n$ must be divisible by another prime $q$ which satisfies that $q < \sqrt{n}$. Conclude that in using the Sieve of Eratosthenes to identify all the primes up to $n$, we only need to sieve by primes which are less than or equal to $\sqrt{n}$.

**14.2.** (a) According to the previous exercise, what primes must you sieve by when using the Sieve of Eratosthenes in order to find all of the primes less than 100?

(b) Now use sieving to list all the primes below 100. (Note: You should find 25 of them.)

(c) What is the largest prime by which you need to sieve to discover all the primes below 1,000?

**14.3.** Find the canonical factorizations of (a) 384, (b) 1,155.

**14.4.** Find the canonical factorization of 9,360.

**14.5.** Some small primes have simple criteria for divisibility by them in terms of the digits of the number being factored. The primes 2 and 5 have obvious criteria, but 3 and 11 also have less obvious but simple criteria:

*An integer $n$ is divisible by 3 if and only if the sum of its decimal digits is also divisible by 3; $n$ is divisible by 11 if and only if the alternating sum (i.e., plus, minus, plus, minus, etc.) of its decimal digits is also divisible by 11.*

(a) Use these criteria to determine if the following numbers are divisible by 3 only, 11 only, both 3 and 11, or neither 3 nor 11:

$$5,412 \quad 5,421 \quad 5,071 \quad 5,701.$$

(b) Suppose $w$, $x$, $y$ and $z$ are the decimal digits of the integer $n$, i.e., $n = 10^3w + 10^2x + 10y + z$. Use the fact that $10 = 3^2 + 1$ to establish the criterion for divisibility by 3. You will make use of the Binomial Theorem and Lemma 13.1.

(c) Suppose again that $n = 10^3w + 10^2x + 10y + z$. Use the fact that $10 = 11 - 1$ to establish the criterion for divisibility by 11.

**14.6.** We know that there are many twin prime pairs, but show that 3, 5, and 7 form the only "triple" of consecutive primes (i.e., three consecutive primes with gaps of length 1 between them).

**14.7.** Using Theorem 14.5, find the canonical factorization of $\gcd(2^3 3^4 5^{10}, 3^4 5^5 7^3)$.

**14.8.** Using Theorem 14.5, find the canonical factorization of $\gcd(2^5 3^8 5, 3^3 5^5 7^6)$.

**14.9.** A positive integer $m$ is a **square** if $m$ can be written as $m = n^2$ for some integer $n$. Prove that if the canonical factorization of $m$ is $p_1^{a_1} \cdots p_r^{a_r}$, then $m$ is a square if and only if each $a_i$ is even.

**14.10.** Use induction on the number $r$ of factors $a_i$ to prove Corollary 14.3, which generalizes Euclid's Lemma (Theorem 14.2, in which $r = 2$).

**14.11.** Concerning Mersenne numbers, we defined them above to be of the form $2^p - 1$ where $p$ is prime. Here is why: Prove that if $2^n - 1$ is a prime, then the exponent $n$ must be a prime. (Hint: A contrapositive proof should work here. Suppose that $n$ is *not* prime and see what that says about $2^n - 1$. Go back to algebra and think about factoring $x^{ab} - 1 = (x^a)^b - 1$ where both $a$ and $b$ are greater than 1.)

**14.12.** (a) Concerning gaps between consecutive primes, show that for any positive integer $n$ there exist $n$ consecutive composite (i.e., not prime) integers. (Hint: Recall that for any positive integer $m \geq 1$, $m! = m(m-1)\cdots(2)(1)$. Now start with the number $(n+1)! + 2$, which is divisible by 2. Continue.)

(b) Using the method of Part (a), set $n = 5$ and write down the resulting 5 consecutive composite numbers.

(c) Looking at the primes below 100 (see Exercise 14.2), where is the first gap of at least 5 composite integers between consecutive primes? Note that this is *considerably* sooner than your answer to Part (b).

**14.13.** Concerning the two statements of the Prime Number Theorem (look back at that now), let's use calculus to explore the relation between them. By the latter statement about density, we know that if we "add up" the densities over the range from 2 to $N$, we should get approximately the total number of primes in that range. But "adding up" is simply *integrating*. Changing from an integer variable $n$ to a real variable $x$, we see that the number of primes in the interval $[2, N]$ is approximately

$$\int_2^N \frac{1}{\ln(x)}\, dx.$$

(a) Using integration by parts, show that

$$\int_2^N \frac{1}{\ln(x)}\, dx = \frac{N}{\ln(N)} - \frac{2}{\ln(2)} + \int_2^N \frac{1}{(\ln(x))^2}\, dx.$$

(b) Observe that the first term on the right above is exactly the estimate given in the first statement of the theorem. The second term is essentially negligible, but what about the integral in the third term? Now use l'Hopital's Rule to show that

$$\lim_{N \to \infty} \frac{\int_2^N \frac{1}{(\ln(x))^2}\, dx}{\int_2^N \frac{1}{\ln(x)}\, dx} = 0.$$

Hence as $N$ goes to infinity, that integral on the right also becomes negligible in comparison to the original integral. We can conclude then that the number of primes below $N$ is approximately

$$\int_2^N \frac{1}{\ln(x)}\, dx \approx N/\ln(N),$$

i.e., the second statement of the theorem implies the first statement.

# CHAPTER 15

# Congruences and the Finite Sets $\mathbb{Z}_n$

# $59 \equiv 4 \pmod{11}$

Up to this point nearly all of the sets on which we have concentrated have been infinite sets ($\mathbb{Z}$, $\mathbb{R}$, and so on), the sole exception being the set $S_n$ of all permutations of a set of $n$ objects, which, you will recall, has $n!$ elements. It turns out, however, that we can form *finite* sets from the integers which themselves have interesting algebraic structures. The key tool for forming these sets is an important idea defined on the integers which was originally developed by Carl Friedrich Gauss in his book *Disquisitiones Arithmeticae*, published in 1801. This idea, as we shall see in subsequent chapters, also plays an important role in computer science in general and in cryptography in particular.

Let $n \geq 2$ be a fixed integer. We define two integers $a$ and $b$ to be **congruent modulo** $n$ if $n$ divides the difference $a - b$. We will denote this by writing $a \equiv b$ (mod $n$). We call the integer $n$ the **modulus** of the congruence. We note that by the definition of "divides," $a \equiv b \pmod{n}$ means that $a - b = nk$ for some integer $k$. In Exercise 15.4 you are asked to show that congruence mod $n$ is an equivalence relation; that is, it is reflexive, symmetric, and transitive (see Chapter 7). Finally, we require $n$ to be greater than 1 since if $n = 1$ then every integer is equivalent to every other integer.

Probably without realizing it, you have already encountered congruences in everyday life. For example, the US clock system works modulo 12 whereas the military clock system works modulo 24. Days of the week are determined modulo 7 because if a given day is Monday, then seven days later we have another Monday. Similarly, except for leap years, our yearly calendars work modulo 365. Let's look at some examples.

**Example 15.1.** (a) We know $27 \equiv 5 \pmod{11}$ since $27 - 5 = 22 = 11(2)$. Note that 27 is also congruent to 5 modulo 2 since, again, $27 - 5 = 22 = 2(11)$. Is the following statement true or false and why: $43 \equiv 17 \pmod{13}$?
(b) One has to be a bit more careful with negative numbers, but the idea is the same. For example, $4 \equiv -21 \pmod{5}$ since $4 - (-21) = 4 + 21 = 25 = 5(5)$. True or false and why: $-16 \equiv 5 \pmod{11}$?

An alternative way to determine if two integers $a$ and $b$ are congruent modulo $n$ is to use the Division Algorithm to divide each integer by the modulus $n$ and check to see if the two remainders are the same. In Example 15.1 we noted that $27 \equiv 5 \pmod{11}$. Dividing 27 by 11 we obtain a remainder of 5, and when dividing 5 by 11, we also obtain the same remainder of 5. Because these two remainders are the same, we can conclude that $27 \equiv 5 \pmod{11}$.

**Example 15.2.** Consider the positive integers 235 and 147 with modulus $n = 11$. Dividing 235 by 11 we obtain a remainder of 4; similarly dividing 147 by 11 we also obtain a remainder of 4. So 235 and 147 are congruent modulo 11. As a check we can also calculate $235 - 147 = 88 = 11(8)$ so that $235 \equiv 147 \pmod{11}$.

This idea of finding the two remainders upon division by the modulus $n$ leads us to an important point which follows directly from the Division Algorithm (Theorem 13.2): *Every integer is congruent modulo $n$ to exactly one of $n$'s possible remainders.* For each $n > 1$, this finite set of remainders, i.e., the set $\{0, 1, \ldots, n - 1\}$, turns out to be very important because *it has algebraic structure provided that we do its arithmetic modulo $n$.* This set is called the *integers mod $n$* and is denoted $\mathbb{Z}_n$. To emphasize, we repeat:

$$\mathbb{Z}_n = \{0, 1, \ldots, n - 1\} \text{ with arithmetic done modulo } n.$$

If $a$ is any integer, we shall use the notation $a \pmod{n}$ to denote the unique remainder of $a$ divided by $n$, which is of course an element of $\mathbb{Z}_n$. This remainder is also referred to as the *least non-negative residue* of $a$ modulo $n$. We shall refer to this operation as *reduction mod $n$*. Note that "$a \pmod{n}$" is an object; "$a \equiv b \pmod{n}$" is a statement.

**Example 15.3.** The least non-negative residue of 27 modulo 5 (which we are denoting as $27 \pmod{5}$) is 2 because, by the Division Algorithm, $27 = (5)(5) + 2$. Similarly, the least non-negative residue of -27 modulo 5 (i.e., $-27 \pmod{5}$) is 3 since $-27 = (-6)(5) + 3$. (See Exercise 15.2 for lots of practice on this operation.)

**Example 15.4.** (a) Here are the addition and multiplication tables for $\mathbb{Z}_5$. Remember that any time an operation results in an answer which is greater than or equal to the modulus, we "reduce" that answer by dividing by the modulus and taking the remainder.

| + | 0 | 1 | 2 | 3 | 4 |
|---|---|---|---|---|---|
| 0 | 0 | 1 | 2 | 3 | 4 |
| 1 | 1 | 2 | 3 | 4 | 0 |
| 2 | 2 | 3 | 4 | 0 | 1 |
| 3 | 3 | 4 | 0 | 1 | 2 |
| 4 | 4 | 0 | 1 | 2 | 3 |

| · | 0 | 1 | 2 | 3 | 4 |
|---|---|---|---|---|---|
| 0 | 0 | 0 | 0 | 0 | 0 |
| 1 | 0 | 1 | 2 | 3 | 4 |
| 2 | 0 | 2 | 4 | 1 | 3 |
| 3 | 0 | 3 | 1 | 4 | 2 |
| 4 | 0 | 4 | 3 | 2 | 1 |

Note that in both tables (except Row 1 and Column 1 in the multiplication table) every element appears exactly once in each row and column.

(b) Here are the tables for $\mathbb{Z}_6$. Note that the addition table looks similar to that of $\mathbb{Z}_5$, but the multiplication table definitely does not. Can you guess why such a difference?

| + | 0 | 1 | 2 | 3 | 4 | 5 |
|---|---|---|---|---|---|---|
| 0 | 0 | 1 | 2 | 3 | 4 | 5 |
| 1 | 1 | 2 | 3 | 4 | 5 | 0 |
| 2 | 2 | 3 | 4 | 5 | 0 | 1 |
| 3 | 3 | 4 | 5 | 0 | 1 | 2 |
| 4 | 4 | 5 | 0 | 1 | 2 | 3 |
| 5 | 5 | 0 | 1 | 2 | 3 | 4 |

| · | 0 | 1 | 2 | 3 | 4 | 5 |
|---|---|---|---|---|---|---|
| 0 | 0 | 0 | 0 | 0 | 0 | 0 |
| 1 | 0 | 1 | 2 | 3 | 4 | 5 |
| 2 | 0 | 2 | 4 | 0 | 2 | 4 |
| 3 | 0 | 3 | 0 | 3 | 0 | 3 |
| 4 | 0 | 4 | 2 | 0 | 4 | 2 |
| 5 | 0 | 5 | 4 | 3 | 2 | 1 |

(c) Write down the two tables for $\mathbb{Z}_4$. Does the multiplication table more resemble that of $\mathbb{Z}_5$ or $\mathbb{Z}_6$? Does this make sense to you?

Returning now to congruences, we state in the following lemma some of their properties which will be needed as we move forward. We prove several of the properties and leave the remaining proofs for you to do in Exercise 15.3. The basic proof technique is to translate a statement about congruences into a corresponding equation in the integers and proceed from there.

**Lemma 15.1.** *Let $n \geq 2$ be a fixed integer. Assume that $a \equiv b \pmod{n}$ and that $c \equiv d \pmod{n}$. Then*
(i) $a + c \equiv b + d \pmod{n}$;
(ii) $a - c \equiv b - d \pmod{n}$;
(iii) $ac \equiv bd \pmod{n}$;
(iv) *If $m$ is an integer, then $ma \equiv mb \pmod{n}$;*
(v) *If $d$ is a divisor of $n$, then $a \equiv b \pmod{d}$.*

Proof. We prove Parts (i), (iii), and (v), leaving proofs of the remaining parts to you. From the assumptions of the lemma, we have that $a - b = nk$ and $c - d = nj$ for some integers $k$ and $j$. To prove Part (i), we calculate

$$(a + c) - (b + d) = (a - b) + (c - d) = nk + nj = n(k + j).$$

Since $k + l$ is an integer, we can conclude that $a + c \equiv b + d \pmod{n}$.

For Part (iii) we have that

$$ac = (b + nk)(d + nj) = bd + bnj + dnk + n^2kl = bd + n(bj + dk + nkj).$$

Therefore $ac - bd = n(bj + dk + nkj)$ where $bj + dk + nk$ is an integer, and so $ac \equiv bd \pmod{n}$.

Finally for Part (v) we have that $n = de$ for some integer $e$. Hence $a - b = nk = dek = d(ek)$ so that $a \equiv b \pmod{d}$, as desired. ∎

We observe that Parts (i), (ii) and (iii) of Lemma 15.1 tell us that congruences and reduction mod $n$ are "compatible" with integer addition, subtraction, and multiplication; that is, one can add, subtract and multiply congruences and the new congruence will remain true, with the same modulus. So, now we must ask, what about *division*? It turns out that, in the theory of congruences and in the sets $\mathbb{Z}_n$, sometimes one can divide and sometimes one can't! We note the similarity here to the situation in the integers $\mathbb{Z}$, where division may or may not be possible, as opposed to the rational numbers $\mathbb{Q}$, the real numbers $\mathbb{R}$ and the complex numbers $\mathbb{C}$, where division (except by 0) can always be done. We again mention that in the study of abstract algebra, these latter three sets are called *fields*, whereas $\mathbb{Z}$ and now $\mathbb{Z}_n$ are called *rings* (see, for example, Exercise 12.14).

**Example 15.5.** (a) The statement $40 \equiv 30 \pmod{5}$ is true. If we divide both sides of the congruence by 2 and leave the modulus the same, we get the statement $20 \equiv 15 \pmod{5}$, which is also true, so in this case division by 2 is okay. However, if we instead divide both sides of $40 \equiv 30 \pmod{5}$ by 10, we obtain the statement $4 \equiv 3 \pmod{5}$, which is, of course, not true. So when can one divide both sides of a congruence and arrive at a new (correct) congruence while maintaining the same modulus? This question is answered in the following lemma.

**Lemma 15.2.** *If $ac \equiv bc \pmod{n}$ and $c$ and $n$ are relatively prime, then $a \equiv b$* (mod n).

Proof. By the definition of congruence, we have that $ac - bc = (a - b)c = nk$ for some integer $k$. Thus $n$ divides $(a - b)c$. Since $c$ and $n$ are relatively prime, by Part (i) of Lemma 14.1, $n$ must divide $a - b$. Thus $a \equiv b \pmod{n}$. ∎

Hence in Example 15.5 it "worked" to divide by 2 since 2 and 5 are relatively prime, but it did not work to divide by 10. See Exercise 15.12 for a look at the possibility of simplifying a congruence by dividing out a common factor of both sides and the modulus.

To end this chapter, let's take a closer look at division in the sets $\mathbb{Z}_n$. In the sets $\mathbb{Q}$, $\mathbb{R}$ and $\mathbb{C}$ we know we can always divide by any non-zero element. Another way to say this is that every non-zero element of these sets possesses a (unique) *multiplicative inverse* (see the discussion between Examples 12.5 and 12.6, Example 12.6 Part (b), and Exercise 12.8). We note that dividing by an element is the same thing as *multiplying by its multiplicative inverse*, for example, in $\mathbb{Q}$, 7 divided by $5 = 7(1/5) = 7/5$. Thinking of division in this way will help us understand what happens in our new sets $\mathbb{Z}_n$. Recall that in the integers $\mathbb{Z}$, the only elements possessing multiplicative inverses are 1 and $-1$, so they are the only numbers by which you can always divide in $\mathbb{Z}$.

We ask then, which elements of $\mathbb{Z}_n$ possess multiplicative inverses and which do not? Now Lemma 15.2 tells us the answer in slightly different language: You can always divide by $c$ in $\mathbb{Z}_n$ provided that $c$ and $n$ are relatively prime. Put in our alternative way, $c$ will possess a multiplicative inverse in $\mathbb{Z}_n$ if and only if $c$ and $n$ are relatively prime. We observe that in $\mathbb{Q}$, $\mathbb{R}$ and $\mathbb{C}$ it's easy to write down the inverse of an element $x$ (it's just $1/x$), but in $\mathbb{Z}_n$ it's not so obvious. We need to look at some examples.

**Example 15.6.** (a) Look back now at the multiplication table of $\mathbb{Z}_6$ in Example 15.4 Part (b). Recall that the multiplicative inverse $a^{-1}$ of an element $a$ in any set with an associative multiplication and multiplicative identity 1 is the unique element for which $(a^{-1})(a) = 1$. So, in that table, we see that $(1)(1) = 1$ and $(5)(5) = 1$ (i.e., both are their own multiplicative inverses, just like 1 and $-1$ in $\mathbb{Z}$), but 2, 3 and 4 do not possess multiplicative inverses. This of course makes sense since 1 and 5 are relatively prime to 6, but 2, 3 and 4 are not.
(b) Without writing out the whole table, figure out the multiplicative inverses of all elements of $\mathbb{Z}_8$ which can possess them.
(c) Now look at the multiplication table for $\mathbb{Z}_5$ in Example 15.4 Part (a). It should be no surprise that since 5 is prime, every non-zero element possesses a multiplicative inverse (what are they specifically?) So in $\mathbb{Z}_5$, you can always divide (except, of course, by 0).

Let us formalize what we observed in the examples of $\mathbb{Z}_5$ and $\mathbb{Z}_6$.

**Lemma 15.3.** *If $a$ is an element of $\mathbb{Z}_n$, then $a$ possesses a multiplicative inverse $a^{-1}$ in $\mathbb{Z}_n$ if and only if $a$ and $n$ are relatively prime.*

Proof. First suppose that $a$ is relatively prime to $n$. By the Euclidean Algorithm (Theorem 13.3) we know that in $\mathbb{Z}$ we can find integers $x$ and $y$ such that $ax + ny = 1$. In $\mathbb{Z}_n$ then, we have $a(x \pmod{n}) = 1$, i.e., $a^{-1} = x \pmod{n}$.

On the other hand, suppose that $a$ is *not* relatively prime to $n$ and suppose that their gcd is $d > 1$ with $dk = n$ and $jd = a$ (so, in particular, $k > 0$). Just suppose $d$ has a multiplicative inverse $d^{-1}$ in $\mathbb{Z}_n$; then, doing arithmetic in $\mathbb{Z}_n$, we have

$$k = (1)k = (d^{-1}d)k = d^{-1}(dk) = d^{-1}0 = 0,$$

which is a contradiction. Thus $d$ does not have a multiplicative inverse in $\mathbb{Z}_n$.
Finally then, just suppose that $a$ has a multiplicative inverse $a^{-1}$ in $\mathbb{Z}_n$; then $1 = a^{-1}a = a^{-1}jd$, so $(a^{-1}j) \pmod{n}$ is a multiplicative inverse of $d$.

This again is a contradiction, so we must conclude that $a$ has no multiplicative inverse, and we are done. ∎

We see then that if $n$ is composite, $\mathbb{Z}_n$ will have at least one element by which you cannot always divide, and so $\mathbb{Z}_n$ (like $\mathbb{Z}$) will be a *ring* but not a field (again, see Exercise 12.14). However, if $p$ is prime, then $\mathbb{Z}_p$ has division by all non-zero elements, and so $\mathbb{Z}_p$ (like $\mathbb{Q}$, $\mathbb{R}$ and $\mathbb{C}$) is indeed a *field*.

A final question we should ask about multiplicative inverses in $\mathbb{Z}_n$ is whether there is an efficient method for computing them. The answer is yes and is supplied by the Euclidean Algorithm (see Theorem 13.3 together with the discussion proceeding Example 13.6). Given relatively prime integers $a$ and $n$, we showed how to compute integers $x$ and $y$ such that $ax + ny = 1$. This leads us directly to the following useful lemma:

**Lemma 15.4.** *If $a$ and $n$ are relatively prime and if the Euclidean Algorithm gives is that $ax + ny = 1$ for integers $x$ and $y$, then the multiplicative inverse $a^{-1}$ of $a$ in $\mathbb{Z}_n$ is $x$ (mod $n$).*

Proof. The integer equation $ax + ny = 1$ tells us that $ax \equiv 1$ (mod $n$). Hence in $\mathbb{Z}_n$, $a^{-1} = x$ (mod $n$). ∎

**Example 15.7.** Let us use the Euclidean Algorithm to compute the multiplicative inverse of 24 in $\mathbb{Z}_{37}$. Running the algorithm forward, we have

$$
\begin{aligned}
37 &= 24 + 13 \\
24 &= 13 + 11 \\
13 &= 11 + 2 \\
11 &= 5(2) + 1 \\
2 &= 2(1) + 0.
\end{aligned}
$$

Now working backwards from the next to last line:

$$
\begin{aligned}
1 &= 11 - 5(2) \\
1 &= 11 - 5(13 - 11) = 6(11) - 5(13) \\
1 &= 6(24 - 13) - 5(13) = 6(24) - 11(13) \\
1 &= 6(24) - 11(37 - 24) = 17(24) - 11(37).
\end{aligned}
$$

By Lemma 15.4, we conclude that the multiplicative inverse of 24 in $\mathbb{Z}_{37}$ is 17.

In the following chapters we take a closer look at congruences and the sets $\mathbb{Z}_n$, aiming toward their key roles in modern cryptography.

**Exercises**

**15.1.** Which of the following congruences are true?
  (a) $17 \equiv 5$ (mod 9)

(b) $33 \equiv 0 \pmod{11}$

(c) $55 \equiv -9 \pmod{16}$

(d) $283 \equiv 177 \pmod 5$

(e) $220 \equiv 34 \pmod 6$

(f) $17 \equiv -35 \pmod 9$

(g) $5m + 1 \equiv 2m - 1 \pmod m$ for any integer $m \geq 2$

(h) $3m + 3 \equiv m^2 - 4m + 3 \pmod m$ for any integer $m \geq 2$

**15.2.** For each of the following congruences, fill in the blank with the least non-negative residue (i.e., with the element $a \pmod n$ in $\mathbb{Z}_n$):

(a) $17 \equiv$ _____ $\pmod 9$

(b) $21 \equiv$ _____ $\pmod 7$

(c) $0 \equiv$ _____ $\pmod 9$

(d) $-25 \equiv$ _____ $\pmod 6$

(e) $334 \equiv$ _____ $\pmod{55}$

(f) $220 \equiv$ _____ $\pmod 6$

(g) $-221 \equiv$ _____ $\pmod{33}$

(h) $5m - 1 \equiv$ _____ $\pmod m$ for any integer $m \geq 2$

**15.3.** (a) Prove Part (ii) of Lemma 15.1.

(b) Prove Part (iv) of Lemma 15.1.

**15.4.** Fix an integer $n \geq 2$. Show that the relation on the integers defined by saying that two integers $a$ and $b$ are related if $a \equiv b \pmod n$ is an equivalence relation. (Note: See Chapter 7 if you need a reminder of what you need to show.)

**15.5.** (a) Suppose in the preceding exercise $n = 7$. List eight elements of $\mathbb{Z}$, five positive and three negative, which are in the equivalence class of 2 under congruence mod 7.

(b) More generally, suppose $x$ is an element of $\mathbb{Z}_n$. Describe the equivalence class of $x$ in $\mathbb{Z}$ for the equivalence relation congruence mod $n$.

**15.6.** (a) Write down the 8 by 8 multiplication table of the non-zero elements of $\mathbb{Z}_9$. (Note: Since multiplication mod 9 is commutative, your table will be symmetric with respect to the main diagonal.)

(b) Using your table, for the six elements of $\mathbb{Z}_9$ which possess multiplicative inverses, write down what they are (e.g., $2^{-1} =?$, etc.).

**15.7.** In $\mathbb{Z}_{15}$ there are eight elements which are relatively prime to 15 and hence possess multiplicative inverses. (Note: We shall see in Chapter 19 an easy way to count such elements.) For each of these, determine its multiplicative inverse.

**15.8.** (a) For any positive integer $k$, find the least non-negative residue of $6^k$ modulo 10.

(b) Do any other numbers $a$ besides 6 in the set $\{1, 2, \ldots, 9\}$ have this property that the numbers $a^k$ have the same least non-negative residue modulo 10 for all $k$? If so, which ones?

103

**15.9.** Show that if $a \equiv b \pmod{n}$, then $a^k \equiv b^k \pmod{n}$ for any positive integer $k$.

**15.10.** It's important to remember that many times there are multiple routes to a correct proof of a statement. Let's try three different approaches here:
(a) Using induction, prove that $8^k - 1$ is divisible by 7 for all positive integers $k$.
(b) Using the fact that $8 = 7 + 1$ and the Binomial Theorem, prove that $8^k - 1$ is divisible by 7 for all positive integers $k$.
(c) An integer $n$ divides another integer $a$ if and only if $a \equiv 0 \pmod{n}$. Using this and Exercise 15.9 above, prove (very quickly) that $8^k - 1$ is divisible by 7 for all positive integers $k$. (Hint: What is 8 (mod 7)?)

**15.11.** Using congruences and reduction mod 5, prove that 5 divides $18^k - 13^k$ for all positive integers $k$. (See Exercise 15.10 Part (c).)

**15.12.** (a) Suppose that $a \equiv b \pmod{n}$ and suppose that $d$ is a common divisor of $a$, $b$, and $n$. Show that $a/d \equiv b/d \pmod{n/d}$.
(b) Use Part (a) to "simplify" the congruence $65 \equiv 50 \pmod{15}$ to one with a smaller modulus.

**15.13.** Use the Euclidean Algorithm to compute the multiplicative inverse of 16 in $\mathbb{Z}_{23}$. (See Lemma 15.4 and Exercise 15.7.)

# CHAPTER 16

## Solving Congruences

$$5x \equiv 6 \pmod 8$$

We now turn to the matter of attempting to solve congruences which contain an unknown. We focus in this chapter on *linear* congruences; i.e., congruences of the form $ax \equiv b \pmod n$, where $a$ and $b$ are fixed integers, $n$ is a fixed integer greater than 1, and $x$ is an unknown integer. In fact, *we shall seek solutions $x$ which lie in $\mathbb{Z}_n$* (i.e., which lie in the set $\{0, 1, 2, \ldots, n-1\}$), so the set of possible solutions is finite to start with. We shall see that in this set a given linear congruence may have no solutions, a unique solution, or multiple solutions. Let us look at some examples.

**Example 16.1.** (a) Let's solve the congruence $5x \equiv 6 \pmod 8$. Because there are only 8 numbers which are candidates as solutions, we can just check each one. Doing arithmetic modulo 8 (i.e., working in $\mathbb{Z}_8$), we get $5(0) = 0$, $5(1) = 5$, $5(2) = 2$, $5(3) = 7$, $5(4) = 4$, $5(5) = 1$, $5(6) = 6$ and $5(7) = 3$. Hence we have a unique solution $x = 6$.
(b) Work through the congruence $6x \equiv 5 \pmod 8$ yourself "by hand". If you do it right, you should discover *no* solutions.
(c) Do the same with the congruence $6x \equiv 4 \pmod 8$. Here you should discover 2 solutions.

What's going on here? It turns out, as Theorem 16.1 will soon prove, that the key to the existence and uniqueness of solutions to the linear congruence $ax \equiv b \pmod n$ is the greatest common divisor $d$ of $a$ and $n$ and then $d$'s relationship to $b$. Looking back at Example 16.1, in Part (a) $\gcd(5, 8) = 1$, 1 divides 6, and there is exactly one solution; in Part (b) $\gcd(6, 8) = 2$, but 2 does not divide 5, and there are no solutions; and finally in Part (c) $\gcd(6, 8) = 2$, 2 *does* divide 4, and we get exactly two solutions. Let's settle this matter in general.

**Theorem 16.1.** (i) *The linear congruence $ax \equiv b$ (mod $n$) has solutions if and only if the greatest common divisor $d$ of $a$ and $n$ divides $b$.*
(ii) *If $d$ divides $b$, then there are $d$ distinct solutions modulo $n$. More specifically, if $c$ is any one of the solutions, then the set of all $d$ solutions is*

$$\{c, c + n/d, c + 2(n/d), \ldots, c + (d-1)(n/d)\},$$

*with all numbers reduced modulo $n$.*

Proof. For Part (i), we must prove the implications in both directions. First, suppose the congruence does have a solution $x = c$; then $ac - b = nr$ for some integer $r$, and hence $b = ac - nr$ must be divisible by $d$ since $d$ divides the right-hand side (by its definition and by Lemma 13.1 Part (ii)).

On the other hand, suppose $d$ divides $b$, so that $b = de$ for some integer $e$. Since $d$ is the greatest common divisor of $a$ and $n$, by the Euclidean Algorithm we have $d = ak + nj$ for some integers $k$ and $j$. This gives us $b = ake + nje$, so $a(ke) - b = n(-je)$. Finally, this equation in $\mathbb{Z}$ can be rewritten as the congruence

$$a(ke) \equiv b \pmod{n}.$$

This shows that $ke$ (mod $n$) is a solution of our congruence $ax \equiv b$ (mod $n$), and the proof of Part (i) is complete.

For Part (ii), we must again prove two things: first that the numbers in the given list are indeed solutions; second that any solution $c_1$ other than $c$ must be in our list. So, first, we know that $ac - b = nr$ (for some integer $r$). Hence for $1 \leq t \leq d-1$ (and using the fact that both $a/d$ and $n/d$ are integers), we have

$$a(c + t(n/d)) - b = (ac - b) + at(n/d) = nr + at(n/d) = n(r + (a/d)t).$$

The left-hand and right-hand ends here say that $a(c + t(n/d)) \equiv b$ (mod $n$), as desired.

On the other hand, suppose that $c_1$ is another element in $\mathbb{Z}_n$ (besides $c$) which is a solution to our congruence . We have then that $n$ divides both $ac - b$ and $ac_1 - b$, so $n$ divides $a(c - c_1)$, i.e., $a(c - c_1) = ns$ for some integer $s$. Dividing both sides by $d$, we obtain $(a/d)(c - c_1) = (n/d)s$, i.e., $n/d$ divides $(a/d)(c - c_1)$. But now by the definition of $d$, $a/d$ and $n/d$ are relatively prime, so by Lemma 14.1 Part (i), $n/d$ must divide $c - c_1$. Hence $c_1 = c + t(n/d)$ for some integer $t$, and reducing modulo $n$ if necessary, we see that $c_1$ is in our given list. Hence our congruence has exactly $d$ solutions in $\mathbb{Z}_n$, and we have identified all of them. Whew! ∎

This result together with other results preceding it gives us a procedure for solving a given linear congruence. However, if one or more solutions exist, there will always be some work to do to find a first solution. A procedure for solving $ax \equiv b$ (mod $n$) using as small numbers as possible is as follows:

(1) If either $a$ or $b$ is not in $\mathbb{Z}_n$ (i.e., is not in the set $\{0, 1, 2, \ldots, n-1\}$, reduce it modulo $n$. We'll continue to label these possibly reduced values in $\mathbb{Z}_n$ as $a$ and $b$.

(2) Compute $d = \gcd(a, n)$. If $d$ does not divide $b$, there are no solutions and we are done.

(3) If $d$ *does* divide $b$, as Exercise 15.12 showed us, we can divide our entire congruence through by $d$, resulting in a new (if $d > 1$) reduced congruence

$$(a/d)x \equiv b/d \pmod{n/d}.$$

(4) This new congruence has a unique solution $c$ in $\mathbb{Z}_{n/d}$ (since $a/d$ and $n/d$ are relatively prime) which we must find by inspection or by using the Euclidean Algorithm to compute the multiplicative inverse $(a/d)^{-1}$ of $a/d$ in $\mathbb{Z}_{n/d}$ (see Lemma 15.4) and then multiplying both sides of the congruence by $(a/d)^{-1}$. That is, we have

$$(a/d)^{-1}(a/d)x \equiv (a/d)^{-1}(b/d) \pmod{n/d},$$

i.e., $x \equiv (a/d)^{-1}(b/d)$ (mod $n/d$), and so our unique solution $c$ is

$$(a/d)^{-1}(b/d) \pmod{n/d}.$$

(See Example 16.5 for an illustration of this "multiplicative inverse" method.)
(5) Finally, we know from Part (ii) of Theorem 16.1 that the full set of $d$ solutions in $\mathbb{Z}_n$ to our original congruence is

$$\{c, c + n/d, c + 2(n/d), \ldots, c + (d-1)(n/d)\},$$

and again we are done.

It is definitely time for some more examples.

**Example 16.2.** Consider the congruence $12x \equiv 22$ (mod 8). We first reduce both 12 and 22 by 8, obtaining the simpler (but equivalent) congruence $4x \equiv 6$ (mod 8). Since $\gcd(4, 8) = 4$ and since 4 does not divide 6, we have no solutions. (Feel free to check this conclusion "by hand".)

**Example 16.3.** Consider the congruence $33x \equiv 15$ (mod 12). We first reduce 33 and 15 modulo 12, obtaining $9x \equiv 3$ (mod 12). Since $\gcd(9, 12) = 3$ and 3 divides itself, we can divide our congruence through by 3, obtaining

$$3x \equiv 1 \pmod{4}.$$

Now we seek the unique solution $c$ to this new congruence "by hand", since we only have 4 possibilities, and we find that $c = 3$. Finally, the full set of three solutions is $\{3, 3+4, 3+8\} = \{3, 7, 11\}$. You should verify these solutions in the original congruence!

**Example 16.4.** Trying out our procedure from Theorem 16.1, solve $16x \equiv 52 \pmod{20}$. Check your solution(s) (if any) just to be sure.

**Example 16.5.** If now you are thinking that using this procedure turns solving every congruence into an easy task, consider the following example:

$$24x \equiv 5 \pmod{37}.$$

Since 24 and 37 are relatively prime, we know there is a unique solution, but no reductions are possible, so we can either work though the numbers 1 through 36 until we find the one that works, or we can apply the Euclidean Algorithm to find the multiplicative inverse of 24 in $\mathbb{Z}_{37}$. Using this latter method (which will be by far the most efficient if the modulus is large), we solved this exact problem in Example 15.7, getting that $(17)(24) + (-11)(37) = 1$ (check this), and so $(17)(24) \equiv 1 \pmod{37}$, i.e., 17 is the multiplicative inverse of 24 in $\mathbb{Z}_{37}$. Multiplying our congruence through by 17, we obtain $x \equiv (17)(5) \pmod{37}$, and so our unique solution is $(17)(5) \pmod{37} = 11$. Check this in our given congruence.

Let us now consider the possibility of trying to solve a system of two or more *simultaneous linear congruences*. Any methods to be discovered here have numerous applications, one of which will appear in Chapter 19 when we learn how to calculate using what's called "Euler's function." Let's look at an example.

**Example 16.6.** Is there a positive integer $x$ below, say, 80 which has the property that upon division by 7 we get a remainder of 5 and upon division by 11 we get a remainder of 4? If so, is $x$ unique in that range?

How can we go about seeking $x$? One way is to write down all the numbers between 1 and 80 which are congruent to 4 modulo 11 and then check each of them to see if they are congruent to 5 modulo 7 (we choose to use 11 first since the list will be shorter). Here is that list: $\{4, 15, 26, 37, 48, 59, 70\}$. Now let's form the corresponding list we get by reducing each of these numbers by 7: $\{4, 1, 5, 2, 6, 3, 0\}$. So the answer to our question is that $x = 26$ and it is unique. What's most interesting here, however, is that in the list of remainders upon division by 7, every possible remainder value appears exactly once.

Was what happened in this example a coincidence? It's likely that at this point you suspect it is not a coincidence and that it may have to do with the fact that 7 and 11 are prime, or at least that they are relatively prime to each other. This is correct, as our next important result proves. This result appears to have been first published by the Chinese mathematician Sun Tzu, who lived sometime between the third and fifth centuries A.D.

**Theorem 16.2. (Chinese Remainder Theorem)** *Let $m \geq 2$ and $n \geq 2$ be integers which are relatively prime. Let $a$ and $b$ be integers. Then there is a simultaneous solution to the pair of congruences*

$$x \equiv a \pmod{m}, \ x \equiv b \pmod{n}.$$

*Moreover this solution is unique modulo $mn$, i.e., there is only one solution $x$ with $0 \leq x < mn$.*

Proof. Existence: Since $m$ and $n$ are relatively prime, the Euclidean Algorithm tells us that there are integers $r$ and $s$ so that $mr + ns = 1$. (Notice how often we use this idea!) We claim that $c = bmr + ans$ is a simultaneous solution to the pair of congruences. We know that

$$c \equiv ans \pmod{m} \text{ and } ns \equiv 1 \pmod{m},$$

and thus $c \equiv a(1) \pmod{m}$. The proof that $c$ is also a solution of the second congruence is similar (write it down now).

Uniqueness: Assume that $c$ and $d$ are both solutions; then $c \equiv a \pmod{m}$ and $d \equiv a \pmod{m}$. Thus $c - d \equiv 0 \pmod{m}$ and similarly $c - d \equiv 0 \pmod{n}$. Thus $c - d$ is divisible by both $m$ and $n$, and since $m$ and $n$ are relatively prime, $c - d$ is divisible by the product $mn$ (Lemma 14.1, Part (ii)). Hence $c \equiv d \pmod{mn}$. ∎

We note that the relative primeness of the moduli is crucial to both parts of the proof. We also note, however, that the theorem gives us little information about how to actually find the simultaneous solution. In Example 16.6 we found the solution by listing all the candidates with respect to one modulus and then testing those with respect to the other. Another approach, which we now illustrate, is to use algebra.

**Example 16.7.** Consider the two congruences

$$x \equiv 3 \pmod{4}$$
$$x \equiv 4 \pmod{5}.$$

Our technique will be to use the definition of congruence to write down an equation (in $\mathbb{Z}$) for $x$ in terms of one modulus and substitute that into the second congruence. So, starting with the larger modulus 5, we have $x - 4 = 5k$, so that $x = 5k + 4$ for some integer $k$. Substituting this into the other congruence we obtain $5k + 4 \equiv 3 \pmod{4}$, and after reducing the coefficients modulo 4 we have $k \equiv -1 \equiv 3 \pmod{4}$. We choose $k = 3$ since it is in $\mathbb{Z}_4$. Then $x = 5(3) + 4 = 19$. Note that $0 \le 19 < 4(5) = 20$ so our solution $x$ lies in the correct range. You should now check that 19 really is the desired solution.

Note that in finding the above solution, in analogy to solving a pair of simultaneous linear equations in the variables $x$ and $y$ over the real numbers $\mathbb{R}$, we first solve for $x$ in one of the congruences (to save arithmetic, it's best to first do this in the congruence with the larger modulus). We then substitute this expression for $x$ in the other congruence and solve this second congruence. The arithmetic in the second congruence is easier since its modulus is smaller.

**Example 16.8.** Let's redo the congruences in Exercise 16.6 using this algebraic technique.

$$x \equiv 5 \pmod{7}$$
$$x \equiv 4 \pmod{11}.$$

From the congruence with the larger modulus we have $x - 4 = 11k$ for some integer $k$, so $x = 11k + 4$. Substituting this into the other congruence we obtain $11k + 4 \equiv 5 \pmod{7}$. After reducing modulo 7 we have that $4k \equiv 1 \pmod{7}$, and by testing cases we see that $k = 2$ is a solution to this congruence (we want to select for $k$ an element of $\mathbb{Z}_7$). Hence $x = 11k + 4 = 11(2) + 4 = 26$ is the simultaneous solution to our system of linear congruences, as we already discovered using the "make a list" technique.

It should be no surprise that the conclusions of the Chinese Remainder Theorem do not hold if the hypothesis of relative primeness of the moduli is removed. We illustrate this in the following example, attempting to use the "make a list" technique.

**Example 16.9.**

$$x \equiv 3 \pmod{6}$$
$$x \equiv 2 \pmod{8}.$$

We make a list of integers up to $(6)(8) = 48$ which are congruent to 2 modulo 8: $\{2, 10, 18, 26, 34, 42\}$. Reducing this list modulo 6, we obtain $\{2, 4, 0, 2, 4, 0\}$. Hence there are no simultaneous solutions to this pair of congruences. Note that if the first congruence had been, say, $x \equiv 4 \pmod{6}$, then there would have been solutions (10 and 34), but they obviously would not have been unique in the range $0 \leq x < 48$.

Finally, the conclusions of the Chinese Remainder Theorem continue to hold for three or more congruences provided, of course, that all of the moduli are pairwise relatively prime. It is simply a matter of solving two of the congruences, then combining that information with a third congruence, and so on. We illustrate by solving a set of three simultaneous congruences, the latter two being from Examples 16.6 and 16.8.

**Example 16.10.** Consider the simultaneous congruences

$$x \equiv 2 \pmod{5}$$
$$x \equiv 5 \pmod{7}$$
$$x \equiv 4 \pmod{11}.$$

We know already that for the latter two $x \equiv 26 \pmod{77}$, so $x = 77j + 26$ for some integer $j$. Substituting this into the modulo 5 congruence, we get $77j + 26 \equiv 2 \pmod{5}$, and reducing and simplifying we have
$2j + 1 \equiv 2 \pmod{5}$, i.e., $2j \equiv 1 \pmod{5}$. By inspection then, the least non-negative value for $j$ is 3, and we get $x = (77)(3) + 26 = 257$. This value does lie below $(5)(7)(11) = 385$, and you should check that it satisfies all three congruences.

Had we instead used the "make a list" method, the latter list would be $\{26, 103, 180, 257, 334\}$, and reducing modulo 5 we get $\{1, 3, 0, 2, 4\}$, so our answer is 257.

At this point you should feel reasonably comfortable about solving linear congruences, whether single ones or sets of simultaneous ones. You do usually need to do some work beyond reduction (as Example 16.5 tried to illustrate), but otherwise the methods are straightforward. You might be asking now, what about *non-linear* congruences, i.e., what about when exponents are present? We tackle some cases of this problem in our next chapter, again with an eye toward an important application to cryptography in Chapter 18. See Exercise 16.13 for a preview.

To finish this chapter, we would like to update a chart on sets with algebraic structure which we first displayed at the beginning of Chapter 12. We have now added the finite sets $\mathbb{Z}_n$ (with the special cases $\mathbb{Z}_p$ for $p$ prime) and have also added the primary algebraic types of these sets. Take a little time now to review this chart and be sure it makes sense to you, with the understanding that we have not as yet carefully defined groups, rings and fields nor studied some of their basic properties. We shall do this starting with Chapter 21.

| Set(s) | Binary operation(s) | Type |
|---|---|---|
| $\mathbb{Z}^+$ | add, multiply | |
| $\mathbb{Z}$ | add, subtract, multiply | ring |
| $\mathbb{Q}^+, \mathbb{R}^+$ | add, multiply, divide* | group $(\cdot)$ |
| $\mathbb{Q}, \mathbb{R}, \mathbb{C}$ | add, subtract, multiply, divide* | field |
| permutations ($S_n$, etc.) | composition | group |
| $n \times n$ matrices | add, subtract, multiply | ring |
| $\mathbb{Z}_n$ ($n$ composite) | add, subtract, multiply | ring |
| $\mathbb{Z}_p$ ($p$ prime) | add, subtract, multiply, divide* | field |

*No division by 0.

## Exercises

**16.1.** Determine whether the following congruences have solutions. If they do, determine the number of non-negative solutions smaller than the modulus $n$ (i.e., lying in $\mathbb{Z}_n$) and then find all such solutions.
  (a) $5x \equiv 6 \pmod{11}$
  (b) $5x \equiv 12 \pmod{20}$
  (c) $8x \equiv 12 \pmod{20}$
  (d) $32x \equiv 48 \pmod{18}$ (Reduce first!)

**16.2.** Determine whether the following congruences have solutions. If they do, determine the number of non-negative solutions smaller than the modulus $n$ (i.e., lying in $\mathbb{Z}_n$) and then find all such solutions.
  (a) $8x \equiv 14 \pmod{36}$
  (b) $8x \equiv 7 \pmod{13}$
  (c) $60x \equiv 24 \pmod{144}$
  (d) $-6x \equiv 48 \pmod{15}$ (Reduce first!)

**16.3.** Find all the elements $x$ (if any) of $\mathbb{Z}_{25}$ which have the property that 10 times them is congruent to 5 modulo 25.

**16.4.** Find all the elements $x$ (if any) of $\mathbb{Z}_{14}$ which have the property that 20 times them is congruent to 10 modulo 14.

**16.5.** By the Chinese Remainder Theorem, there is a unique non-negative solution below 77 to the following pair of simultaneous linear congruences. Find it.

$$x \equiv 4 \pmod 7$$
$$x \equiv 5 \pmod{11}.$$

**16.6.** By the Chinese Remainder Theorem, there is a unique non-negative solution below 187 to the following pair of simultaneous linear congruences. Find it.

$$x \equiv 5 \pmod{11}$$
$$x \equiv 6 \pmod{17}.$$

**16.7.** Find the smallest positive solution to the pair of simultaneous linear congruences

$$x \equiv 2 \pmod 6$$
$$2x \equiv 1 \pmod 7.$$

(Hint: Consider multiplying both sides of the second congruence by 4 and then proceeding. Why do we multiply by the value 4?)

**16.8.** Find the unique non-negative solution below $792 = (8)(9)(11)$ to the simultaneous system of linear congruences

$$x \equiv 5 \pmod 8$$
$$x \equiv 4 \pmod 9$$
$$x \equiv 5 \pmod{11}.$$

**16.9.** The following exercise is Problem 26 in Volume 3 of "Sun Tzu's Mathematical Manual" (circa 300-400 AD): "We have a number of things, but we do not know exactly how many. If we count them by threes, we have two left over. If we count them by fives, we have three left over. If we count them by sevens, we have two left over. How many things are there?" (Note: He no doubt was asking "what is the smallest possibility for the number of things?" Answer this question.)

**16.10.** Find the smallest positive integer whose remainder when divided by 11 is 8, which has the last digit of 4, and is divisible by 27. (Note: This one takes some fortitude. Hint on for the answer: Year of D-Day.)

**16.11.** Use the "make a list" method to show that there are no common solutions below 96 to the simultaneous congruences

$$x \equiv 4 \pmod 8$$
$$x \equiv 2 \pmod{12}.$$

**16.12.** Find the smallest positive solution to the simultaneous system of linear congruences

$$2x \equiv 1 \pmod 5$$
$$5x \equiv 2 \pmod 6$$
$$x \equiv 2 \pmod 7.$$

(Note: See the hint in Exercise 16.7.)

**16.13.** Here is a preview of some of what we will soon learn about congruences containing exponents:
(a) Compute $2^4 \pmod 5$, $3^4 \pmod 5$ and $4^4 \pmod 5$.
(b) Compute $2^6 \pmod 7$, $3^6 \pmod 7$ and $4^6 \pmod 7$.
(c) Compute $2^{10} \pmod{11}$ and $2^{12} \pmod{13}$.
(d) Form a conjecture:
  "If $p$ is prime and if $a$ and $p$ are relatively prime, then $\cdots$."

# CHAPTER 17

## Fermat's Theorem

$$3^{52} \equiv 3^4 \equiv 4 \pmod{7}$$

Up to now we have concentrated on congruences which do not contain exponents, so here (and in Chapter 19) we shall study how to deal with congruences which *do* contain exponents greater than 1. A first immediate question is: Can we reduce an exponent which is larger than a modulus $n$ in the same way we can reduce numbers in the base, that is, by reducing modulo $n$? This turns out to be *not true*, and here is a simple *counterexample*:

**Example 17.1.** *Question*: What is $2^9$ modulo 5?
*Correct solution method*: $2^9 = 512$, and 512 (mod 5) = 2.
*Incorrect solution method*: Reducing the exponent 9 modulo 5 gives a new exponent of 4. $2^4 = 16$ and 16 (mod 5) = 1, which evidently is *wrong*!

Hence we see that in general *we cannot reduce exponents by the modulus* the way can do with numbers in the base. This is an issue we need to deal with since we could be asked to compute, say, $2^{90}$ modulo 5. Surely there is a better and faster way to obtain this least non-negative residue than to multiply out the value $2^{90}$ (which, by the way is a number with 28 decimal digits). Fortunately, there is indeed a reduction method for exponents, which we asked you in Exercise 16.13 to form a conjecture about by looking at data. If you did not do that exercise already, please turn back and do it now!

**Example 17.2.** Let us take a closer look at one of the pieces of data in Exercise 16.13, $2^6$ (mod 7), and use it to illustrate the proof technique below. Consider the set $\{1, 2, 3, 4, 5, 6\}$ of all non-zero elements of $\mathbb{Z}_7$. If we now multiply each of these elements by the base 2 and then reduce modulo 7, we obtain

$$\{2, 4, 6, 8, 10, 12\} \equiv \{2, 4, 6, 1, 3, 5\} \quad (\text{mod } 7).$$

Note that we get the same set back, but with the numbers rearranged. This says that modulo 7 the products of all the elements of each set will be equal; that is,

$$(1 \cdot 2)(2 \cdot 2)(3 \cdot 2)(4 \cdot 2)(5 \cdot 2)(6 \cdot 2) \equiv (1)(2)(3)(4)(5)(6) \quad (\text{mod } 7),$$

or, gathering the six factors of 2 on the left,

$$(2^6)(1)(2)(3)(4)(5)(6) \equiv (1)(2)(3)(4)(5)(6) \quad (\text{mod } 7)$$

But finally all the numbers $\{1, 2, 3, 4, 5, 6\}$ in the product are relatively prime to 7, so by Lemma 15.2 we may divide them out, leaving us with $2^6 \equiv 1$ (mod 7).

We arrive then at the following powerful theorem which was first discovered by the French lawyer, government official, and amateur mathematician Pierre de Fermat (1601 – 1665). The only catch with this result is that *the modulus must be prime*, but we shall be able to remove this restriction in Chapter 19. We note that this result is not to be confused with Fermat's *Last* Theorem (as discussed in Example 13.1).

**Theorem 17.1. (Fermat)** *If $p$ is a prime and $gcd(a, p) = 1$, then*

$$a^{p-1} \equiv 1 \quad (\text{mod } p).$$

Proof. Consider the product of all the non-zero elements of $\mathbb{Z}_p$, i.e., the product $1(2)(3) \cdots (p-1)$. Also consider the product of each of these elements multiplied by the base element $a$, i.e., the product

$$(1a)(2a)(3a) \cdots ((p-1)a).$$

We claim that all these multiples of $a$ are distinct modulo $p$, for if $i$ and $j$ are in $\mathbb{Z}_p$ and $ia \equiv ja$ (mod $p$), then by Lemma 15.2, using the fact that $a$ is relatively prime to $p$, we can divide by $a$ to obtain $i \equiv j$ (mod $p$); that is, in $\mathbb{Z}_p$, $i = j$. Thus this new set of integers $\{1a, 2a, \ldots, (p-1)a\}$ must be the same modulo $p$, except for the order, as the set $\{1, 2, \ldots, p-1\}$. It follows that

$$(1a)(2a)(3a) \cdots ((p-1)a) \equiv (1)(2)(3) \cdots (p-1) \quad (\text{mod } p).$$

Notice now that we have the common factors $1, 2, \ldots, p-1$ on both sides of the congruence. Each of the values is relatively prime to $p$. so we can divide each one from both sides of the congruence (again applying Lemma 15.2) to obtain

$$\underbrace{(a)(a) \cdots (a)}_{p-1 \text{ times}} \equiv \underbrace{(1)(1) \cdots (1)}_{p-1 \text{ times}} \equiv 1 \quad (\text{mod } p).$$

We have then that $a^{p-1} \equiv 1 \pmod{p}$ and the proof is complete. ∎

Fermat's Theorem has a nice corollary:

**Corollary 17.2.** *For any integer $a$ and any prime $p$, $a^p \equiv a \pmod{p}$.*

Proof. If $a$ is not divisible by $p$ then Fermat's Theorem implies that $a^{p-1} \equiv 1 \pmod{p}$ and thus by multiplying both sides by $a$, we have $a^p \equiv a \pmod{p}$. If $p$ divides $a$, then $a^p \equiv 0 \pmod{p}$ and $a \equiv 0 \pmod{p}$, so $a^p \equiv a \pmod{p}$. ∎

Let us look at some examples using Fermat's Theorem.

**Example 17.3.** Because 5 is prime and $\gcd(3,5) = 1$, Fermat's Theorem tells us immediately that $3^4 \equiv 1 \pmod 5$. Similarly we know that $12^{16} \equiv 1 \pmod{17}$, $54^{96} \equiv 1 \pmod{97}$, and so on.

The power of the theorem really kicks in, however, whenever the exponent is larger than the modulus $p$, for then we can reduce the exponent modulo $p - 1$. So, for example, what is $3^{100}$ modulo 5? By our theorem, we have

$$3^{100} \equiv 3^{4(25)} \equiv (3^4)^{25} \equiv 1^{25} \equiv 1 \pmod 5.$$

And what if our exponent is not divisible by $p - 1$? Consider the exponent 103 rather than 100. By the Division Algorithm $103 = 4(25) + 3$, and so

$$3^{103} \equiv 3^{4(25)+3} \equiv 3^{4(25)}3^3 \equiv (3^4)^{25}3^3 \equiv 1^{25}3^3 \equiv 3^3 \equiv 2 \pmod 5.$$

Fermat's Theorem tells us then that in any congruence with a prime modulus $p$ and a base $a$ which is relatively prime to $p$, we can reduce any *exponent* modulo $p - 1$ (and of course, we know we can reduce the base $a$ modulo $p$). In trying to do computations with relatively small numbers and in as few steps as possible, this theorem is a great help, but it may not be enough help if the modulus is not small. Let us look at an example now in which we illustrate a method called *fast exponentiation.*

**Example 17.4.** Suppose that we are asked to compute $3^{99} \pmod{101}$. Fermat's theorem only allows us to reduce the exponent 99 by 100, so we gain nothing in this example. Whatever multiplications we do need to do, we should always remember to *immediately reduce the answer modulo $p$*, which will guarantee that all intermediate numbers we arrive at will be smaller than $p$. But to get $3^{99}$, do we need to do 98 (or so) multiplications? The answer is: far fewer using fast exponentiation! The idea is to do *repeated squaring*, as follows:

(1) Write 99 as a sum of powers of 2: $99 = 64 + 32 + 2 + 1$.
(2) Square 3 and reduce modulo 101: $3^2 \pmod{101} = 9$.
(3) Square $3^2 = 9$ and reduce modulo 101: $3^4 \equiv 9^2 \pmod{101} = 81$.
(4) Square $3^4 = 81$ and reduce modulo 101: $3^8 \equiv 81^2 \pmod{101} = 97$.
(5) Square $3^8 \equiv 97$ and reduce modulo 101: $3^{16} \equiv 97^2 \pmod{101} = 16$.
(5) Square $3^{16} \equiv 16$ and reduce modulo 101: $3^{32} \equiv 16^2 \pmod{101} = 54$.
(6) Square $3^{32} \equiv 54$ and reduce modulo 101: $3^{64} \equiv 54^2 \pmod{101} = 88$.
(7) Finally, by Step 1, $3^{99} = 3^{64}3^{32}3^23^1 \equiv 88 \cdot 54 \cdot 9 \cdot 3 \equiv 34 \pmod{101}$.

Hence in order to do this computation, we required only 7 steps (or, if you consider Step 7 as being three multiplications followed by reductions, 9 steps). This is very efficient, and the beauty of this method is that the larger the modulus, the greater (relatively speaking) the efficiency. For example, suppose we are asked to compute $14^{904,237}$ (mod 1,564,379) (this last number is prime, but that is not required to apply this method). Well, since the highest power of 2 below 904,237 is $2^{19} = 524,288$, we'll be able to do this calculation in, say, 20 to 30 steps, and every number we arrive at along the way (after reduction) will be below $1,564,539$. Okay, so you certainly don't want to do this with your hand calculator, but for a properly programmed computer it's a complete snap. The answer is 601,202, and *Mathematica*© can knock it off in less than a tenth of a second. We'll ask you in Exercises **17.9** and **17.10** to work through two relatively small examples "by hand."

Finally for this brief chapter, a natural question to ask at this point is the following: What if our modulus is *not* prime? For example, how do we calculate something like $5^{29}$ (mod 12)? We shall be able to settle this question completely in Chapter 19, once we learn about an important function called Euler's Function. We invite you to begin exploring this "composite modulus question" in Exercise **17.14** below.

In the meantime, we shall begin exploring some ideas from modern cryptography, many of which depend on:
(1) Reduction by the modulus (on base numbers) and reduction using Fermat's Theorem or its generalizations (on exponents) in order to keep computed numbers from being too large,
(2) fast exponentiation to compute powers after the exponent has been reduced below the modulus, and
(3) the Euclidean Algorithm to compute greatest common divisors.
Using these tools, very efficient cryptographic algorithms can be employed, as we shall see in Chapters 18 and 20.

## Exercises

**17.1.** In each of the following, fill in the blank with the least non-negative residue.

    (a) $2^7 \equiv$ _____ (mod 5)
    (b) $3^{14} \equiv$ _____ (mod 7)
    (c) $16^{22} \equiv$ _____ (mod 19)
    (d) $3^{144} \equiv$ _____ (mod 17)

**17.2.** In each of the following, fill in the blank with the least non-negative residue.

    (a) $2^{10} \equiv$ _____ (mod 7)
    (b) $3^{13} \equiv$ _____ (mod 11)
    (c) $12^{22} \equiv$ _____ (mod 19)
    (d) $5^{131} \equiv$ _____ (mod 17)

**17.3.** Compute $2^{80}$ (mod 29).

**17.4.** Compute $3^{125}$ (mod 31).

**17.5.** Compute $55^{147}$ (mod 13).

**17.6.** Compute $104^{564}$ (mod 11).

**17.7.** Find the least non-negative residue of $2^{14}4^{15}$ modulo 7.

**17.8.** Calculate $3^{30}7^{20}$ (mod 5).

**17.9.** Using fast exponentiation by hand and showing all the steps, compute $3^{18}$ (mod 50). (See Example 17.4.)

**17.10.** Using fast exponentiation by hand and showing all the steps, compute $4^{19}$ (mod 25). (Again, see Example 17.4.)

**17.11.** Note that the number 100 is not a prime, so Fermat's Theorem does not apply modulo 100. Nevertheless, find the least non-negative residue of $5^8$ modulo 100.

**17.12.** Concerning fast exponentiation, one can get a pretty good estimate of the maximum number of steps required by counting the number of decimal digits in the modulus and then multiplying that by twice the base 2 log of 10, which value is a little less than 7. According to this, what is the maximum number of steps required if the modulus is (a) 2,376 (b) 5,768,105,357,451?

**17.13.** (a) Is the number 499 is prime? (Recall: Since $\sqrt{499}$ is about 22, you must check all primes below that to be sure.)
(b) If the answer to Part (a) is yes, find the least non-negative residue of $3^{503}$ modulo 499.

**17.14.** Let's explore the question of how to reduce an exponent when the modulus $n$ is possibly composite.
(a) Compute $7^2$ (mod 10), $7^3$ (mod 10) and $7^4$ (mod 10).
(b) List the elements of $\mathbb{Z}_{10}$ which are relatively prime to 10. How many are there?
(c) Compute $2^2$ (mod 9), $2^3$ (mod 9), $2^4$ (mod 9), $2^5$ (mod 9) and $2^6$ (mod 9).
(d) List the elements of $\mathbb{Z}_9$ which are relatively prime to 9. How many are there?
(e) How many elements of $\mathbb{Z}_{22}$ are relatively prime to 22? Call this number $k$. Compute $3^k$ (mod 22). (Note: Employing fast exponentiation works nicely here.)
(f) Make a conjecture about a congruence modulo $n$ (an arbitrary integer greater than 1) and a base element $a$ which is relatively prime to $n$: "If $k$ is the number of elements of $\mathbb{Z}_n$ which are relatively prime to $n$ and if $a$ is relatively prime to $n$, then $\cdots$."

# CHAPTER 18

# Diffie-Hellman Key Exchange

Now, and in Chapter 20, we introduce some ideas from modern cryptography which make use of congruences and Fermat's Theorem as well as (the soon to be studied) Euler's Function and Euler's Theorem. Let's start by discussing the basic idea of encryption and the need for one or more **keys** to implement that encryption. Here is an example of a relatively simple encryption technique.

**Example 18.1.** Suppose I want to send the message "SELL" to my stock broker but I want the message to be encrypted so that only she knows what I want. A simple technique is called *linear encryption*, which in this case would involve using a modulus of 26 and two keys, a multiplier $m$ (which must be relatively prime to 26) and an adder $b$. Then if $x$ is a numerical representation of a letter (say A = 0, B = 1, and so on), then my encryption of each letter $x$ will be $y = mx + b$ (mod 26) and her decryption would then be $x = m^{-1}(y - b)$ (mod 26). What she and I need to do before any communication can occur is *agree on our two private keys $m$ and $b$*, and, using the Euclidean Algorithm, we can each compute the multiplicative inverse $m^{-1}$ of $m$ in $\mathbb{Z}_{26}$. So, suppose we agree on $m = 5$ and $b = 12$. Since S = 18, E = 4 and L = 11, I now do the three computations modulo 26: $5(18) + 12 \equiv 24$, $5(4) + 12 \equiv 6$, and $5(11) + 12 \equiv 15$, and send over to her $\{24, 6, 15, 15\}$ (that is, "YGPP"). Since $m^{-1} = 21$, she now computes (again modulo 26) $21(24 - 12) \equiv 18$, $21(6 - 12) \equiv 4$, and $21(15 - 12) \equiv 15$; that is, she decrypts the message to $\{18, 4, 11\}$, which is, of course, "SELL."

The primary point for us in this example is that my broker and I need to share a private key (or keys) to do secure communication, but how can we *securely* agree on the shared keys? The point is that we need shared secure keys to communicate, but, ironically, we need other keys to communicate our desired keys, and so on. What was traditionally used was a "trusted carrier," but who/what is that? An answer to this seemingly unsolvable situation would be some method by which a key or keys can be agreed upon with an unsecured communication exchange which does not reveal what the keys are to anyone but the two communicators, even if the process is somehow hacked by an outsider. This is what Diffie and Hellman devised in the 1970's (see [3]). Their system, which we describe below, was for example used during the Cold War when the United States and the Soviet Union wanted to establish their "hotline," and it continues to be used in setting up extremely fast and secure modern computer communication channels.

Before describing the Diffie-Hellman method, we need to investigate the multiplicative structure of $\mathbb{Z}_p$ where $p$ is prime. Fermat's Theorem tells us that if $a$ is a non-zero element of $\mathbb{Z}_p$, then $a^{p-1} \equiv 1 \pmod{p}$. A question is: Is $p-1$ the *smallest* exponent on $a$ for which the reduced answer is 1? Let's look at a couple of examples.

**Example 18.2.** (a) For each non-zero element $a$ of $\mathbb{Z}_{11}$, the following chart shows the smallest power $k$ for which $a^k \equiv 1 \pmod{11}$

| $a$ | 1 | 2 | 3 | 4 | 5 | 6 | 7 | 8 | 9 | 10 |
|---|---|---|---|---|---|---|---|---|---|---|
| $k$ | 1 | 10 | 5 | 5 | 5 | 10 | 10 | 10 | 5 | 2 |

We see then that four of the ten non-zero elements (2, 6, 7 and 8) of $\mathbb{Z}_{11}$ need to be raised all the way to the 10th power to get to 1. Such an element will be called a *primitive root* of $\mathbb{Z}_{11}$ (see the general definition below). We observe that each $k$ above divides 10; we shall be able to prove this once we become familiar with *groups* a bit later on. We also remark that we shall have a way of counting how many primitive roots there are in $\mathbb{Z}_p$, but not necessarily a way to identify which elements they are.

(b) Do the same experiment for $p = 7$. You should find that there are two primitive roots, i.e., two elements which must be raised all the way to the 6th power modulo 7 to get an answer of 1.

We make then the following definition: Suppose $a$ is a non-zero element of $\mathbb{Z}_p$ where $p$ is prime and suppose that $k$ is the smallest exponent such that $a^k \equiv 1 \pmod{p}$. If $k = p - 1$, then $a$ is called a ***primitive root*** of $\mathbb{Z}_p$.

We are now in a position to describe Diffie's and Hellman's idea. Let us suppose that Alice and Bob wish to communicate securely by setting up a common key only they will know. Here are the steps they follow:

(1) They agree upon a large prime $p$ to act as the modulus (in practice, $p$ may have 100 decimal digits or more!), and they agree upon a primitive root $g$ of $\mathbb{Z}_p$. (We leave aside for the time being the difficulties involved in identifying such a large prime and such a primitive root.)

(2) Alice chooses a number $a$ and Bob likewise chooses a number $b$, both satisfying that $2 \le a, b \le p - 2$. For security reasons, it is essential that Alice keeps her number $a$ to herself and that Bob also keeps his number $b$ to himself.

(3) Now Alice computes $g^a$ (mod $p$) and sends this number $c$ to Bob. Bob then computes $c^b$ (mod $p$). Notice that the net effect is that Bob (without knowing $a$) has then the value

$$c^b \equiv (g^a)^b \equiv g^{ab} \pmod{p}.$$

(4) In the same way, Bob calculates $g^b$ (mod $p$) and sends this number $d$ to Alice. She then computes $d^a$ (mod $p$), but again notice that she (without knowing $b$) has the value

$$d^a \equiv (g^b)^a \equiv g^{ba} \pmod{p}.$$

(5) But now $g^{ab}$ (mod $p$) $= g^{ba}$ (mod $p$), and we have arrived at a *secret common key*, namely the number $g^{ab}$ (mod $p$), which Alice and Bob can now use for secure communication. Note that an evil eavesdropper Eve cannot obtain this key from the information exchange since she will only have seen the numbers $c$ and $d$. Even if Eve knows $p$ and $g$ (which Alice and Bob had to agree on to get started), she cannot compute the common key since she does not know $a$ and $b$. After a couple of examples we shall discuss this point further.

**Example 18.3.** We first illustrate the Diffie-Hellman key-exchange method using numbers which of course cannot be used in practice since they are *much too small* to be secure.

(1) Alice and Bob agree that their modulus $p$ will be 11 and that their primitive element $g$ will be 2. (We saw in Example 18.2 that 2 is a primitive root in $\mathbb{Z}_{11}$.)

(2) Alice chooses her secret number $a$ to be 8 and Bob chooses his secret number $b$ to be 9.

(3) Alice computes $2^8$ (mod 11) $= 3$ (check this and the other calculations in this example) and sends it to Bob. Upon receiving the value 3 from Alice, Bob calculates $3^9$ (mod 11) $= 4$. Bob now knows the secret common key is 4.

(4) Meanwhile Bob computes $2^9$ (mod 11) $= 6$ and sends it to Alice. Upon receipt, Alice computes $6^8$ (mod 11) $= 4$. She now also knows that the common key is 4.

(5) Note that the eavesdropper Eve has only seen the 3 and the 6, from which she has no way of finding the 4 (thanks to the secrecy of the 8 and 9, even though she may know the modulus 11 and the primitive root 2).

Let's look now at another example which is somewhat larger but still not close to large enough to be secure, showing how Alice and Bob might use a mathematical software program like *Mathematica*© to do the necessary calculations.

**Example 18.4.** (1) Alice and Bob decide to use a modulus $p$ which lies between a million and ten million. According to the Prime Number Theorem (see the last bullet at the end of Chapter 14), about one out of every $\ln(5,000,000) \approx 15$ numbers in this range is prime, so it won't take too long to happen on one. They discover that $p = 2,134,879$ will work; for example, in *Mathematica*, they see

$$\text{PrimeQ}[2134879] = \text{True}.$$

Their software also can quickly identify a primitive root $g$; specifically, they see

$$\text{PrimitiveRoot}[2134879] = 6.$$

(2) Alice picks $a = 8,254$ for her secret number and Bob picks $b = 47,235$ for his.

(3) Alice's software will use fast exponentiation to compute any base (say $g$) raised to any power (say $a$) modulo any modulus (say $p$). So she computes

$$\text{PowerMod}[6,8254,2134879] = 808247,$$

which is the value $c$ which she sends over to Bob. Bob now computes

$$\text{PowerMod}[808247,47235,2134879] = 1205481.$$

Hence he now knows that the shared secure key is 1,205,481.

(4) Likewise, Bob computes

$$\text{PowerMod}[6,47235,2134879] = 1981047$$

and sends this value $d$ to Alice, who then computes

$$\text{PowerMod}[1981047,8254,2134879] = 1205481.$$

Now she also knows the shared key to be 1,205,481, but the eavesdropper Eve does not, and we are done.

You may be wondering how secure the Diffie-Hellman scheme really is. The numbers $c = g^a \in \mathbb{Z}_p$ and $d = g^b \in \mathbb{Z}_p$ are sent across the communication line, so Eve definitely has access to the values $c$ and $d$. As we have noted, she may also know the prime modulus $p$ and the primitive element $g$ (since Alice and Bob had to somehow agree on them). Hence the real problem here, known as the *Discrete Logarithm Problem*, is the following:

***Discrete Logarithm Problem.*** In the equation $g^a \pmod{p} = c$, if we know the base $g$, the modulus $p$ where $p$ is prime, and the answer $c$, can we discover the exponent $a \in \mathbb{Z}_p$?

This problem is considered by theoretical computer scientists to be very difficult to solve for large primes $p$. It is this difficulty which keeps an eavesdropper from being able to work out the secret key even if she intercepts $p$, $g$, $c$ and $d$.

Being a little more formal, we make the following definition: Let $g$ be a primitive root in $\mathbb{Z}_p^*$ where $p$ is prime and where by $\mathbb{Z}_p^*$ we denote the non-zero elements of $\mathbb{Z}_p$. The **Discrete Logarithm Function** is the function

$$\log_g : \mathbb{Z}_p^* \to \{0, 1, \ldots, p-2\}$$

which makes the equation

$$c = g^{\log_g(c)}$$

hold for every $c \in \mathbb{Z}_p^*$.

As a quick example, If $p = 7$ and $g = 3$ (check that 3 is a primitive root), then since $3^3 \pmod 7 = 6$ (check!), we have $\log_3(6) = 3$.

Again, the function $\log_g$ is thought to be very difficult to compute for large $p$, although its inverse $a \to g^a \pmod p$ is very easy to compute. If an efficient method for computing discrete logarithms were discovered, the Diffie-Hellman Key Exchange would no longer be secure.

In [7] the following formula for the Discrete Logarithm Function is given:

**Theorem 18.1.** *Let $p$ be a prime and let $g$ be a primitive root of $\mathbb{Z}_p^*$. Then*

$$\log_g(c) = -1 + \sum_{j=1}^{p-2} c^j ((g^{-1})^j - 1)^{-1} \pmod p.$$

**Example 18.5.** Let's use this formula to compute $\log_3(6)$ in $\mathbb{Z}_7$; that is, we wish to compute

$$-1 + \sum_{j=1}^{5} 6^j ((3^{-1})^j - 1)^{-1} \pmod 7.$$

We know the answer is 3 from above. Even for a little prime like 7, this does not look like fun to do by hand, but here goes. We need the following values, all reduced modulo 7:

$$6^1 = 6, \quad 6^2 = 1, \quad 6^3 = 6, \quad 6^4 = 1, \quad 6^5 = 6.$$

Since in $\mathbb{Z}_7$, $3^{-1} = 5$, again reducing modulo 7, we have:

$$5^1 = 5, \quad 5^2 = 4, \quad 5^3 = 6, \quad 5^4 = 2, \quad 5^5 = 3.$$

Subtracting 1 from each of these values and computing the inverse in $\mathbb{Z}_p$:

$$4^{-1} = 2, \quad 3^{-1} = 5, \quad 5^{-1} = 3, \quad 1^{-1} = 1, \quad 2^{-1} = 4.$$

So, our summation is

$$-1 + 6(2) + 1(5) + 6(3) + 1(1) + 6(4) \equiv -1 + 5 + 5 + 4 + 1 + 3 \equiv 3 \pmod 7.$$

So why can't Eve, or anyone else, use this formula to compute $\log_g(c)$ modulo a very large prime $p$? The obvious answer is that the summation has $p - 2$ terms, and there are no apparent shortcuts, so using the formula in that setting is simply computationally infeasible, even with today's fastest computers. So, for the time being the Diffie-Hellman method remains secure.

We shall return to our relatively brief look at modern cryptography in Chapter 20, but in the meantime we need to learn about the ultra-important function known as *Euler's Function* and then the lovely generalization of Fermat's Theorem known as *Euler's Theorem*.

## Exercises

**18.1.** Assume that $p = 13$. First, find a primitive root modulo 13. Now assume that Alice and Bob are using the Diffie-Hellman key exchange system to create a common secure key. If Alice chooses her secret number to be $a = 3$ and Bob chooses his secret number to be $b = 5$, determine the common key.

**18.2.** Repeat Exercise 18.1 with $p = 17$ with the same values for the secret numbers $a$ and $b$.

**18.3.** If you have access to some good mathematical software, repeat Exercise 18.1 with $p = 577$, $a = 52$ and $b = 34$. (Note: If you have trouble finding a primitive root modulo 577, see the solutions for this exercise.)

**18.4.** Using the exact same linear encryption scheme as in Example 18.1, my broker sends me an encrypted reply to my "SELL" message. It translates as "EKMC." By decrypting, what is her message?

**18.5.** (a) In Example 18.2 Part (b) you should have discovered that there are two primitive roots in $\mathbb{Z}_7$. Now, how many elements of $\mathbb{Z}_7$ are relatively prime to $6 (= 7 - 1)$? Did the two counts come out the same?
(b) If $p = 13$, determine all primitive roots in $\mathbb{Z}_{13}$. How many are there? Now count the number of elements of $\mathbb{Z}_{13}$ which are relatively prime to $12 (= 13 - 1)$. Did the two counts come out the same?
(c) Conjecture as to how many primitive roots there are in $\mathbb{Z}_{17}$.
(d) Make a general conjecture: "If $p$ is prime, then the number of primitive roots in $\mathbb{Z}_p$ is $\cdots$."

**18.6.** Concerning computing the discrete $\log_g(c)$ modulo $p$, we currently have two methods at our disposal: (1) trial and error, i.e., raise $g$ to various powers $a$ until we get $c$ for an answer, or (2) apply Theorem 18.1. If $p = 11$, we know from Example 18.2 Part (a) that $g = 2$ is a primitive root in $\mathbb{Z}_{11}$. Using *both* of our methods, calculate $\log_2(7)$.

**18.7.** Since 2 is a primitive root modulo 13, calculate the base-2 discrete logarithm of each non-zero element in $\mathbb{Z}_{13}$. Because you are calculating discrete logarithms for all the non-zero elements in $\mathbb{Z}_{13}$, it should be pretty obvious which of our two methods (see the previous exercise) is easiest.

**18.8.** If $g$ is a primitive root modulo $p$, consider the function

$$\exp_g : \{0, 1, 2, \ldots, p-2\} \to \mathbb{Z}_p^*$$

given by $\exp_g(a) = g^a$. Prove that $\exp_g$ is one-to-one. Conclude that this function is a bijection (since its domain and co-domain are finite and have the same cardinality) and so its inverse function $\log_g$ is also a bijection. (Note: Again, by $\mathbb{Z}_p^*$ we mean the non-zero elements of $\mathbb{Z}_p$.)

**18.9.** Question: If $g$ is a primitive root modulo $p$, is $p - g$ also a primitive root modulo $p$? Let's examine some data, assuming $p > 2$.
(a) We have seen that 2 is a primitive root modulo 11. Is it true or false that $11 - 2 = 9$ is also a primitive root modulo 11?
(b) We have also seen that 2 is a primitive root modulo 13. Is it true or false that $13 - 2 = 11$ is also a primitive root modulo 13?
(c) See if you can formulate a conjecture about what must be true of $p - 1$ in order for $g$ being a primitive root modulo $p$ to imply that $p - g$ is also a primitive root modulo $p$. (Hint: If $g$ is primitive, think of $p - g$ as $-g$. When will $g^{(p-1)/2}$ be the same as $(-g)^{(p-1)/2}$?)

**18.10.** Looking toward our next three chapters, let's try to generalize our results to moduli of the form $pq$ where $p$ and $q$ are prime.
(a) Compute $3^{10}$ (mod 22) and $7^{10}$ (mod 22).
(b) Compute $5^{32}$ (mod 51).
(Note: Fast exponentiation (see Example 17.4) is definitely helpful here.)
(c) Form a conjecture: If $p$ and $q$ are primes and if $c$ is an integer which is relatively prime to $pq$, then $\cdots$.

---

# CHAPTER 19

Euler's Formula and Euler's Theorem

---

# $a^{\phi(n)} \equiv 1 \pmod{n}$

Fermat's Theorem (Theorem 17.1) gives us very useful information about congruences involving exponents when the modulus is a prime number $p$, but one can immediately wonder about how to generalize this result to a possibly composite modulus $n$. In fact, the mathematician Leonard Euler (1707 − 1783) developed such a generalization, called Euler's Theorem, but the statement of the theorem required a new function which he also developed, called Euler's Function. We shall focus on first the function and then the theorem in this chapter.

We have seen repeatedly during our exploration of some of the main ideas of number theory that the concept of *relative primeness* plays a central role. For example, the Chinese Remainder Theorem (Theorem 16.2), which we shall use in this chapter, depends entirely on the moduli being relatively prime. Hence it makes perfect sense that we would like to be able to *count* the number of elements of $\mathbb{Z}_n$ which are relatively prime to $n$. As an example, you conjectured in Exercise 18.5 that the number of primitive roots in $\mathbb{Z}_p^*$ ($p$ prime) is equal to the number of elements of $\mathbb{Z}_p^*$ which are relatively prime to $p - 1$.

Euler then introduced a function which takes in a positive integer $n$ and returns the number of elements in the range $\{1, 2, \ldots, n\}$ which are relatively prime to $n$. He denoted this function using the Greek letter phi ($\phi$, pronounced "phee"), so it is often referred to as "the Euler $\phi$ Function" as well as "Euler's Function."

**Example 19.1.** (a) It is easy to check that $\phi(2) = 1$, $\phi(3) = 2$, $\phi(4) = 2$, $\phi(5) = 4$, $\phi(6) = 2$, $\phi(7) = 6$, $\phi(8) = 4$, $\phi(9) = 6$, $\phi(10) = 4$, and so on. Some of this data is "obvious", but some is not. Why is $\phi(9) = 6$? Why is $\phi(10) = 4$?
(b) By hand, compute $\phi(20)$. Is it equal to $\phi(2)\phi(10)$? Is it equal to $\phi(4)\phi(5)$? Why do you think one of these "works" and the other doesn't?
(c) Would you like to compute $\phi(120)$ by hand? Probably not! We need some theorems on how to break down a "$\phi$-calculation" into smaller parts.

The following three theorems tell us how to calculate Euler's function $\phi(n)$ for any positive integer $n$. The idea is to develop a formula for $\phi(p^k)$ for all primes $p$ and all positive integer exponents $k$, and then show that $\phi$ "is compatible with relative primeness," by which we mean that if $\gcd(a, b) = 1$, then $\phi(ab) = \phi(a)\phi(b)$. Given these results we can compute $\phi(n)$ for any positive integer $n$ provided that we can write $n$ in its canonical factorization $p_1^{k_1} p_2^{k_2} \cdots p_r^{k_r}$ since all these factors are obviously relatively prime in pairs. We begin with a result whose proof is very short.

**Theorem 19.1.** *If $p$ is a prime, then $\phi(p) = p - 1$.*

Proof. Each element of the set $\{1, 2, \ldots, p - 1\}$ is relatively prime to $p$. ∎

We remark that using our $\phi$ notation, Fermat's Theorem (Theorem 17.1) can be stated in the form "If $p$ is prime and if $\gcd(a, p) = 1$, then $a^{\phi(p)} \equiv 1 \pmod{p}$."

A more interesting result is when the prime $p$ is raised to an arbitrary positive power $k$.

**Theorem 19.2.** *If $p$ is a prime and $k$ is a positive integer, then*

$$\phi(p^k) = p^k - p^{k-1}.$$

Proof. The "back door counting technique" introduced in Example 3.10 works very well here. There are $p^k$ positive integers $a$ with $1 \leq a \leq p^k$. Of these values, the ones which are *not* relatively prime to $p^k$ are all the multiples of $p$, namely

$$1p, 2p, \ldots, pp, (p+1)p, \ldots, p^{k-1}p.$$

There are $p^{k-1}$ values in this list. Hence the remaining values in our set are all relatively prime to $p^k$, and so $\phi(p^k) = p^k - p^{k-1}$. ∎

Note that Theorem 19.1 follows from Theorem 19.2 by setting $k = 1$.

We now state and prove our third theorem, which tells us that if we can compute the $\phi$-function for prime powers, we can then piece those values together to get $\phi(n)$ for any composite positive integer $n$ (provided that we can write $n$ in its canonical prime factorization).

**Theorem 19.3.** *If $a$ and $b$ are relatively prime positive integers, then*

$$\phi(ab) = \phi(a)\phi(b).$$

Before embarking on the proof, let's illustrate the idea of the proof via an example.

**Example 19.2.** Suppose $a = 4$ and $b = 5$, then $ab = 20$ and note that 4 and 5 are relatively prime. In Example 19.1 Part (b) you were asked to write down the numbers in the set $\{1,,2,\ldots,20\}$ which are relatively prime to 20. Define $U_{20}$ to be this set: $U_{20} = \{1,3,7,9,11,13,17,19\}$. By definition, $U_{20}$ has $\phi(20)$ elements. In the same manner, define $U_4 = \{1,3\}$ and $U_5 = \{1,2,3,4\}$. In what follows, recall that by $\mathbb{Z}_n^*$ we mean the non-zero elements of $\mathbb{Z}_n$.

Let us now define a function $f : U_{20} \to \mathbb{Z}_4^* \times \mathbb{Z}_5^*$ given by

$$f(c) = (c \ (\text{mod } 4), c \ (\text{mod } 5)).$$

You should check (!) that the function $f$ does the following:
$f(1) = (1,1)$, $f(3) = (3,3)$, $f(7) = (3,2)$, $f(9) = (1,4)$,
$f(11) = (3,1)$, $f(13) = (1,3)$, $f(17) = (1,2)$, and $f(19) = (3,4)$.

Look closely at these values and observe that:
(a) The image of $f$ is exactly the Cartesian product of $U_4$ and $U_5$ (as defined above); that is, elements of $\mathbb{Z}_{20}^*$ which are relatively prime to 20 get sent by $f$ to pairs in $\mathbb{Z}_4^*$ and $\mathbb{Z}_5^*$ which are relatively prime to 4 and 5 respectively, and
(b) the function $f$ is a bijection of $U_{20}$ to $U_4 \times U_5$, so that we must have $\phi(20) = \phi(4)\phi(5)$.

The key ingredients of the general proof will then be:
(1) Define a function like $f$ in Example 19.2 from $U_{ab}$ to $\mathbb{Z}_a^* \times \mathbb{Z}_b^*$ and show that the image of $f$ is actually contained in $U_a \times U_b$.
(2) Use the Chinese Remainder Theorem (Theorem 16.2; look back at that theorem now!) to prove that $f$ is a bijection between $U_{ab}$ and $U_a \times U_b$, hence implying our desired conclusion since the number of ordered pairs in a Cartesian product is the product of the number of elements in each of its component sets.

Proof of Theorem 19.3. Suppose that the positive integers $a$ and $b$ are relatively prime. Let $U_{ab}$ be the set of elements of $\mathbb{Z}_{ab}^*$ which are relatively prime to $ab$, so by definition $U_{ab}$ has $\phi(ab)$ elements. In the same way define $U_a = \{k \in \mathbb{Z}_a^* | \gcd(k,a) = 1\}$, so $U_a$ has $\phi(a)$ elements, and $U_b = \{j \in \mathbb{Z}_b^* | \gcd(j,b) = 1\}$, so $U_b$ has $\phi(b)$ elements. We note then that $U_a \times U_b$ has $\phi(a)\phi(b)$ elements.

Now define a function $f : U_{ab} \to \mathbb{Z}_a^* \times \mathbb{Z}_b^*$ given by

$$f(c) = (c \ (\text{mod } a), c \ (\text{mod } b)).$$

We claim that the image of $f$ is actually contained in $U_a \times U_b$; that is, if $c \in U_{ab}$, then $c \pmod{a}$ is relatively prime to $a$ and $c \pmod{b}$ is relatively prime to $b$. Let $k = c \pmod{a}$, then $c = sa + k$ for some integer $s$. If $d = \gcd(k, a)$ were greater than 1, then $d|(sa+k)$, i.e., $d|c$, so $c$ and $a$ would not be relatively prime, and so $c$ and $ab$ would not be relatively prime, which contradicts the choice of $c$. We conclude that $k$ is in $U_a$, and a similar proof shows that $c \pmod{b}$ lies in $U_b$. Hence the image of $f$ is contained in $U_a \times U_b$.

We can now apply the Chinese Remainder Theorem. If $(k, j) \in U_a \times U_b$, we know from that theorem that there exists a *unique* $c \in \mathbb{Z}_{ab}^*$ such that $f(c) = (k, j)$, so we need only show now that $c$ actually lies in $U_{ab}$, i.e., that $c$ is relatively prime to $ab$. As before we have $c = sa + k$ for some $s$, and if $d = \gcd(c, a)$ were greater than 1, then $d|(c - sa)$, so $d|k$, and so $k$ and $a$ would not be relatively prime, which contradicts the choice of $k$. Hence $c$ is relatively prime to $a$, and a similar argument shows that $c$ is relatively prime to $b$. Thus $c$ is relatively prime to $ab$, i.e., $c \in U_{ab}$, and now by the existence and uniqueness of $c$, we have verified that $f$ is a bijection of $U_{ab}$ to $U_a \times U_b$. We conclude that $\phi(ab) = \phi(a)\phi(b)$, and we are done. ∎

We can summarize our theorems in the following corollary:

**Corollary 19.4.** *If $n$ is a positive integer divisible by $r$ distinct primes and if its canonical factorization is $p_1^{k_1} p_2^{k_2} \cdots p_r^{k_r}$, then*

$$\phi(n) = (p_1^{k_1} - p_1^{k_1-1})(p_2^{k_2} - p_2^{k_2-1}) \cdots (p_r^{k_r} - p_r^{k_r-1}).$$

Proof. Each factor in our equation above follows from Theorem 19.2, and because all the primes $p_i$ are distinct, multiplying the factors to get the answer follows from Theorem 19.3. ∎

It's certainly time for some examples.

**Example 19.3.** (a) $\phi(81) = \phi(3^4) = 3^4 - 3^3 = 81 - 27 = 54$.

(b) $\phi(490) = \phi(2^1 5^1 7^2) = (2 - 1)(5 - 1)(7^2 - 7^1) = (1)(4)(42) = 168$.

(c) You compute $\phi(363)$.

It is important here to emphasize that our ability to use our theorems to compute Euler's Function $\phi(n)$ depend first of all on our ability to write $n$ in its canonical factorization. For relatively small $n$ this is usually not too hard to do, but as we get into large numbers, factoring can turn out to be very difficult if not impossible even using today's powerful computers.

**Example 19.4.** (a) To compute $\phi(3713)$, we must factor 3713. This is not easy "by hand", but it turns out that $3713 = (79)(47)$, both of which are prime, so by our theorems we get $\phi(3713) = \phi(79)\phi(47) = (78)(46) = 3588$.

(b) To compute

$$\phi(1454111047978735887325875101489423882404399543603007),$$

we must factor it. As in Part (a), it turns out that this number is a product of two primes, each with 25 decimal digits, and it took *Mathematica* 13.6 seconds of CPU on a PC to find these two factors, 5443833602771186901735043 and 267111589751479499392266349. Had our number been a product of two primes with 30 digits each, *Mathematica* would have had great difficulty factoring it. Had they been 50-digit primes, factorization would have been impossible.

You may well wonder why anyone would want to find the canonical factorization and/or compute the $\phi$-function of a 50 or 60 or 100 digit number. It turns out that the great difficulty (or impossibility) of accomplishing these calculations on very large numbers is a cornerstone of the so-called RSA encryption system, which is widely used to ensure secure digital communications. A basic idea of this system is: "Multiplying is easy; factoring is hard." We shall study the RSA system in the following chapter, but now it is time to state Euler's Theorem, a beautiful and important generalization of Fermat's Theorem which turns out to be, among other things, another cornerstone of the RSA system.

**Theorem 19.5. (Euler's Theorem)** *If $a$ and $n$ are integers with $n > 1$ and with $\gcd(a, n) = 1$, then*

$$a^{\phi(n)} \equiv 1 \pmod{n}.$$

Proof. Because it is a very direct translation of the proof of Fermat's Theorem (Theorem 17.1), we leave this proof to you the student. Start by taking a very close look at that proof. You now replace the set $\mathbb{Z}_p^*$ (which has $p - 1$ elements) with the set $U_n$ as used in the proof of Theorem 19.3 above; that is, $U_n = \{a \in \mathbb{Z}_n^* \mid \gcd(a, n = 1)\}$, and note that the number of elements in $U_n$ is $\phi(n)$. Now proceed exactly as before, writing down all the details yourself. ∎

Because $\phi(p) = p - 1$ when $p$ is prime, we see that Euler's Theorem is indeed a direct generalization of Fermat's Theorem.

Among other things, Euler's Theorem allows us to reduce exponents in a congruence involving a composite modulus in the same way Fermat's Theorem did for a prime modulus. Let's look at a couple of examples.

**Example 19.5.** (a) Let us calculate $2^{41} \pmod{75}$. Since

$$\phi(75) = \phi(25)\phi(3) = (20)(2) = 40,$$

we have by Euler's Theorem

$$2^{41} = (2^{40})(2^1) \equiv (1)(2) = 2 \pmod{75}.$$

(b) Why can we not use Euler's Theorem to compute $5^{41} \pmod{75}$? If you tried to use it, you would get 5 as your answer, but the correct answer is 50.
(c) Use Euler's Theorem and the Division Algorithm to compute $3^{99} \pmod{56}$.

Equipped with Euler's Function and Euler's Theorem, we can now be introduced to one of the most commonly used "public key" encryption systems in modern secure digital communication, the RSA system.

**Exercises**

**19.1.** Calculate (a) $\phi(27)$, (b) $\phi(28)$, (c) $\phi(29)$ and (d) $\phi(30)$.

**19.2.** Calculate (a) $\phi(64)$, (b) $\phi(65)$, (c) $\phi(66)$ and (d) $\phi(67)$.

**19.3.** (a) Calculate $\phi(18)$.
(b) Write down the $\phi(18)$ elements of $\mathbb{Z}_{18}^*$ which are being counted by $\phi$.

**19.4.** (a) Calculate $\phi(24)$.
(b) Write down the $\phi(24)$ elements of $\mathbb{Z}_{24}^*$ which are being counted by $\phi$.

**19.5.** Calculate $\phi(525)$.

**19.6.** Calculate $\phi(594)$.

**19.7.** Calculate $\phi(10^8)$.

**19.8.** Calculate $\phi(2407)$. (Hint: This number is the product of two primes, one of which is less than 30. This illustrates that even relatively small numbers can be hard to factor.)

**19.9.** Use Euler's Theorem and the Division Algorithm to compute $2^{187}$ (mod 77).

**19.10.** Use Euler's Theorem and the Division Algorithm to compute $2^{152}$ (mod 135).

**19.11.** (a) Prove that if $n = 2^k$ for some positive integer $k$, then $\phi(n) = n/2$.
(b) Describe the elements of $\mathbb{Z}_n^*$ which are being counted by $\phi(n)$ in this case.

**19.12.** (a) Prove that if $n$ is greater than 2, then $\phi(n)$ must be even.
(b) Describe all integers $n \geq 5$ which have the property that 4 does *not* divide $\phi(n)$. (Note: This requires some careful thought.)

**19.13.** In Exercise 18.5 Part (d) you were asked to conjecture about the number of primitive roots in $\mathbb{Z}_p$ where $p$ is prime. It turns out to be true that this number is exactly $\phi(p-1)$. Given this, how many primitive roots are there in $\mathbb{Z}_{101}$?

**19.14.** (a) Prove that if $\{p_1, p_2, \ldots, p_r\}$ is the set of distinct primes dividing $n$, then

$$\phi(n) = n(1 - \frac{1}{p_1})(1 - \frac{1}{p_2}) \cdots (1 - \frac{1}{p_r}).$$

(b) Use Part(a) to compute $\phi(3600)$. (Note: Isn't this a nice formula?)

**19.15.** The corollary of Fermat's Theorem (Corollary 17.2) says that if $p$ is prime and $a$ is *any* integer, then $a^p \equiv a$ (mod $p$). One might conjecture then that an analogous corollary of Euler's Theorem might be true; that is, perhaps for every modulus $n > 1$ and every integer $a$ (whether or not $a$ is relatively prime to $n$), $a^{\phi(n)+1} \equiv a$ (mod $n$). Find a single counter-example to this conjecture, rendering it false.

# CHAPTER 20

## RSA Cryptographic System

# Private: $p, q$; Public: $n=pq$

In Chapter 18 we introduced the idea of *secure digital communication*, which is very widely needed in today's world. There we discussed the Diffie-Hellman method of creating a *shared private key* to be used for message exchanges. For example, the communicators could use two shared private keys to act as the multiplier and adder in a linear encryption scheme, as illustrated in Example 18.1. Recall that the Diffie-Hellman key exchange method depends upon Fermat's Theorem (Theorem 17.1) and on the existence of primitive roots in the set $\mathbb{Z}_p^*$ where $p$ is prime. The cryptographic system we shall introduce here, called the RSA system, depends instead on Euler's Function and Euler's Theorem (Theorem 19.5), as introduced in the previous chapter. The system is named after its founders R. Rivest, A. Shamir, and L. Adleman; see [8]. Though introduced in the late 1970's, this system remains in wide use today for digital communications of all sorts, including in particular financial transactions such as on-line payments with a credit card.

An important new idea in the RSA system is that it involves *public keys*. This conceptual breakthrough showed that it is possible to avoid the dependence on private keys which themselves require secure exchange. As designed in RSA, the public keys are made possible by the fact that *factorization of integers is hard*, especially when the primes involved in the factorization are large.

So let us suppose that Alice (as in Chapter 18) would like *anyone* to be able to send her a secure encrypted message which she and only she can decrypt. Here is her procedure:

(1) She selects two large primes $p$ and $q$ of the same approximate size and carefully keeps these choices private! In practice these primes need to have at least 100 decimal digits to guarantee security. She then computes her modulus $n = pq$.

(2) Now Euler comes in. She computes $\phi(n) = \phi(pq) = (p-1)(q-1)$ and keeps this value private! Now she selects a number $e$, called her *encrypting exponent*, which by using the Euclidean Algorithm she carefully checks is relatively prime to $\phi(n)$. (If it is not relatively prime to $\phi(n)$, she picks another $e$ and checks it, etc.) As we saw in Lemma 15.3, this calculation can be run backwards to discover the multiplicative inverse $d$ (called the *decrypting exponent*) of $e$ modulo $\phi(n)$. That is, for some positive integer $k$, we have $ed - k\phi(n) = 1$, so that $ed = 1 + k\phi(n)$. Again, *she keeps the value of $d$ very secret*.

(3) She publishes, for all the world (including the eavesdropper Eve) to see, her modulus $n$ and her encrypting exponent $e$. This is why RSA is an example of *public key cryptography*. She keeps the values of $p$ and $q$ a secret, so in addition the values of $\phi(n)$ and in particular the decrypting exponent $d$ are unknown to the outside world. She is now ready to receive messages encrypted via her public $n$ and $e$ which she alone can decrypt.

Now Bob would like to send a message $m$ to Alice which will be encrypted to ensure its security. The message $m$ needs to be *digitized* in some standard agreed-upon way (that is, if the message contains symbols other than just digits, it must be converted to all digits), and $m$ must be a number which is positive but less than Alice's public modulus $n$. If $m$ is larger than $n$, Bob breaks $m$ into smaller "packets" and transmits each separately. (We'll assume here that $m$ is small enough to send in full.) In addition, we shall assume that $m$ and $n$ are relatively prime in order to apply Euler's Theorem. (When $n = pq$ is large, the chances that $n$ and $m$ are *not* relatively prime are infinitesimal.) Bob now sends the following encrypted digitized message to Alice:

$$c = m^e \pmod{n}.$$

Upon receiving this message, Alice uses Euler's Theorem, which says in this case that $m^{\phi(n)} \equiv 1 \pmod{n}$. She applies her decrypting exponent $d$ to $c$ and reduces modulo $n$. Here is what happens, because of Euler's Theorem and the fact that $ed = 1 + k\phi(n)$ (for some $k$, as given in Step (2) above):

$$c^d \equiv m^{ed} = m^{1+k\phi(n)} = (m)(m^{\phi(n)})^k \equiv (m)(1^k) = m \pmod{n}.$$

We see then that Alice, using her secret $d$, has successfully decrypted Bob's message, which he encrypted using *her* public $n$ and $e$. Isn't RSA a truly elegant system? Again, it depends on two basic things: the difficulty of factoring large numbers, and Euler's Theorem. Eve the Eavesdropper would love to know the value of $d$, but she cannot compute it unless she can figure out $p$ and $q$, which involves factoring the huge number $n$.

Let's look at a couple of examples, the first using small numbers to illustrate the various calculations and the second with larger numbers (but not as large as needed in "real life" to guarantee security).

**Example 20.1.** First, Alice selects $p = 11$ and $q = 17$ for her two (secret) primes, so that her modulus is $n - (11)(17) = 187$. Next, she computes $\phi(n) = (11 - 1)(17 - 1) = 160$ and selects her encrypting exponent $e = 13$. Using the Euclidean Algorithm she verifies that 13 and 160 are relatively prime, and in fact she computes that $(13)(37) - 3(160) = 1$, which says that 37 is the multiplicative inverse of 13 modulo 160. Thus Alice's very secret decrypting exponent is $d = 37$. She now publishes her public modulus $n = 187$ and her public encrypting exponent $e = 13$.

Now Bob wants to secretly meet Alice at 8 tonight, so he wishes to send her the encrypted message $m = 8$. Using fast exponentiation (see Example 17.4), he computes and sends to her

$$c = m^e \ (\text{mod } n) = 8^{13} \ (\text{mod } 187) = 94.$$

Upon receiving Bob's encrypted message $c = 94$, she uses fast exponentiation to decrypt it:

$$c^d \ (\text{mod } n) = 94^{37} \ (\text{mod } 187) = 8,$$

so now she knows what time to meet Bob, but no one else does.

As mentioned above, all messages must be digitized, and in some standardized way. In practice what's often used is the so-called ASCII codes, which are a standardized conversion of letters and symbols into base-10 numbers between 0 and 127, but here let us just use a slightly simpler system as follows: "A" will be denoted by 01, "B" will be denoted by 02, ..., "Z" will be denoted by 26. We will likely want to have spaces, commas, etc., so a space will be denoted by 00, a comma by 27, a period by 28, and an exclamation point by 29.

Let's now do an example with more realistic numbers for security, but still not large enough to be used in practice. We show the functions which would be employed if we were doing the calculations in *Mathematica*.

**Example 20.2.** This time Alice selects as her primes ones with about 25 digits, making use of a *Mathematica* function which identifies prime numbers in the given range using advanced algorithms (not factorization!):

$$p = \text{RandomPrime}[\{10^{24}, 10^{25}\}] = 5401712780740566847737779,$$

$$q = \text{RandomPrime}[\{10^{24}, 10^{25}\}] = 8599036489738585051862237,$$

so her public modulus is

$$n = pq = 46449525308675415080549048730750651874434608351623.$$

She computes $\phi(n) = (p-1)(q-1)$:

$$\phi(n) = 4644952530867541508054903473000138139528270875160 8.$$

Now she randomly selects her encrypting exponent $e$ to be 137 and uses the Euclidean Algorithm to verify that it is relatively prime to $\phi(n)$ and to identify the multiplicative inverse $d$ of $e$ modulo $\phi(n)$ (the *Mathematica* function used here is ExtendedGCD) :

$$d = 4746666819864640957136397709635177660831809653449.$$

Finally, she publishes $n$ and $e$.

Now Bob wants to send Alice the encrypted message "MATH IS FUN!" (Bob is a math nerd). He uses the scheme above to digitize his message: 130120080009190006211429. Since he knows that $n$ has about 50 digits whereas his message has 24, he can send it all in one packet. Using Alice's public $n$ and $e$, Bob uses fast exponentiation to send over

$$\text{PowerMod}[130120080009190006211429,e,n]$$

$$=1492183157298840308043406047727939115124952139663 5.$$

Upon receiving the message, Alice uses her decrypting exponent $d$ to get

$$\text{PowerMod}[149218315729884030804340604772793911512495213966 35,d,n]$$

$$= 130120080009190006211429,$$

which she can easily translate to "MATH IS FUN!" (At which point she rolls her eyes, wondering why he bothered to encrypt such a message.)

We finish this chapter by discussing the need for *authenticity* as well as security in modern digital communication. In Example 20.1 Alice decrypts Bob's message to meet him at 8, *but* how can she be sure that Bob sent the message? Perhaps it was the evil Eve who actually sent it and plans to trick her into giving up her decrypting exponent $d$ when they meet. Alice would like to know that the message from Bob is authentic. Well, it turns out that RSA can also be used to establish authenticity as well as guarantee security, via what's called a *digital signature*. This can be done by having as the last packet in a message a "signature" which, unlike the main part of the message, is encoded using *the sender's public modulus and private decoding exponent.*

Here's how it works. Let us now denote Alice's public keys by $n_A$ and $e_A$ and her private decrypting key by $d_A$. Of course Bob can also have public and private keys which we shall denote by $n_B$, $e_B$ and $d_B$. Now Bob wants to send the message $m$ and his signature $s$ to Alice. As before he uses her public modulus and encrypting exponent on the $m$, but for the signature part he uses his own public modulus and his private decrypting exponent. Hence Alice receives two numbers $c$ and $t$, say, which are

$$c = m^{e_A} \pmod{n_A}; \; t = s^{d_B} \pmod{n_B}.$$

Now upon receipt of the pair $(c, t)$, she can decrypt both as follows:

$$m = c^{d_A} \pmod{n_A}; \; s = t^{e_B} \pmod{n_B}.$$

If the message really was from Bob, the resulting digitized signature $s$, when "undigitized," should make sense. On the other hand, if the signature was actually from anyone besides Bob, what would come out of her computation for $s$, when undigitized, would be gobbledygook.

**Example 20.3.** In Exercise 20.1 Alice assumed that the message "8" about when to meet had come from Bob but couldn't be sure. Let us assume now that a general agreement between them is that when sending a message to the other, they will include a signature at the end. Remembering that these numbers are far too small to be secure, let's assume that Alice's keys are as in that exercise but are now labeled $p_A = 11$, $q_A = 17$, $n_A = 187$, $e_A = 13$ and $d_A = 37$. Without going into the details, let's assume that Bob's private primes are $p_B = 17$ and $q_B = 23$, so his public modulus is $n_B = 391$. He selects for his public encrypting exponent $e_B = 15$ and then computes that his private decrypting exponent is $d_B = 47$.

Now all the pieces are in place, and Bob can send his message $m = $ "8" to Alice, following that with his signature $s$, say "2" since "B" is second in the alphabet. So he sends over

$$c = m^{e_A} \pmod{n_A} = 8^{13} \pmod{187} = 94, \text{ and}$$

$$t = s^{d_B} \pmod{n_B} = 2^{47} \pmod{391} = 77.$$

Alice, upon receipt of the pair $(c = 94, t = 77)$, decrypts as follows:

$$c^{d_A} \pmod{n_A} = 94^{37} \pmod{187} = 8, \text{ and}$$

$$t^{e_B} \pmod{n_B} = 77^{15} \pmod{391} = 2.$$

Thus Alice knows that the message is both secure and authentic.

However, let's now suppose that Eve the Eavesdropper, whom Alice does not know, actually is the one who sends the message. Eve knows that Alice will be looking for Bob's signature, but she can't encrypt that because *she does not know his private $d_B$*, so she randomly picks a value, say 101, and sends the pair $(c = 94, t = 101)$ to Alice. But when Alice decrypts the signature, hoping the message is from Bob, she gets

$$t^{e_B} \pmod{n_B} = 101^{15} \pmod{391} = 305,$$

and she now knows that this message is definitely not from Bob. In fact Eve had a one out of 391 chance of picking the right number to sent to Alice in order for the decryption of the signature to yield a "2", so Eve is foiled by RSA again!

Once the idea of public key encryption emerged, numerous other public key systems have been developed, some of which (for example "elliptic curve encryption") are more efficient than RSA in that they are secure using smaller keys. Nonetheless, RSA, the simplest and most elegant of these systems, remains in significant use. When you use your computer today to do any secure communication, RSA may well be working for you in the background.

This brings us now to an end of our relatively brief look at number theory and its very modern applications. If you are interested in pursuing any of these topics further, there are a number of excellent texts, including [9].

### Exercises

**20.1.** Suppose you wish to use the RSA system to receive secure messages. You pick as your private primes $p = 17$ and $q = 23$.
(a) Compute your public modulus $n$ and your private $\phi(n)$.
(b) Verify that $e = 5$ is a possible public encrypting exponent by using the Euclidean Algorithm applied to $\phi(n)$ and $e$, and find the decrypting exponent $d$ by running the algorithm backwards. (Note: It may be helpful to review Theorem 13.3 and Examples 13.5 and 13.6 before proceeding.)

**20.2.** Repeat Exercise 20.1 with the private primes $p = 19$ and $q = 29$, and with the public encrypting exponent $e = 5$.

**20.3.** In the RSA system in Exercise 20.1, how many different possible encrypting exponents $e$ are there with $1 \le e < \phi(n)$?

**20.4.** (a) In the RSA system in Exercise 20.2, how many different possible encrypting exponents $e$ are there with $1 \le e < \phi(n)$?
(b) More generally, if $n = pq$ is *any* RSA modulus, how many possible encrypting exponents $e$ are there as a function of $n$?

**20.5.** Your friend Chuck wants to send you the secret message 4. He uses your public modulus $n$ and encrypting exponent $e$ from Exercise 20.1. What encrypted message will you receive from Chuck?

**20.6.** Your friend Sally wants to send you the secret message 6. She uses your public modulus $n$ and encrypting exponent $e$ from Exercise 20.2. What encrypted message will you receive from Sally?

**20.7.** Referring to Exercises 20.1 and 20.5, suppose you want to send a message back to Chuck to which you wish to append your digital signature, a number between 1 and 10. Chuck receives the number 9 and decrypts it. What was your signature?

**20.8.** Referring to Exercises 20.2 and 20.6, suppose you want to send a message back to Sally to which you wish to append your digital signature, a number between 1 and 10. Sally receives the number 62 and decrypts it. What was your signature?

**20.9.** If you and a friend have access to good mathematical software, set up an RSA-enabled secure communication channel as follows:

(1) You each privately pick two primes $p$ and $q$ with $5200 < p, q < 10000$. Compute your modulus $n$, your encrypting exponent $e$, and your decrypting exponent $d$. Share your values of $n$ and $e$ with each other.

(2) Using the digitizing scheme described prior the Example 20.2, send to each other a digitized encrypted message followed by an encrypted digital signature. Each of these must consist of between 1 and 4 characters (so the digitized message and signature each has at most 8 digits).

(3) Decrypt the incoming messages and signatures, and compare notes. Good luck!

**20.10.** Concerning the difficulty of factoring larger numbers, especially "by hand," it can help to have some special information about the factors. For example, without the aid of mathematical software see if you can factor $n = 246,973$ given the following information: $n$ is the product of two primes $p$ and $q$ which differ by 12. (Hint: What is $\sqrt{n}$?)

# CHAPTER 21

# Groups - Definition and Examples

$$a * a^{-1} = a^{-1} * a = e$$

In Chapter 12 we introduced the concept of *sets with algebraic structure* and displayed a chart of examples of such sets. We then displayed a slightly expanded chart at the end of Chapter 16 and promised to go into details about the terms "group", "ring" and "field" in subsequent chapters. The time has now come for us to go into those details, starting with groups. Before that, let us once again display our chart, with one more class of sets which we encountered in Chapter 19 added at the end:

| Set(s) | Binary operation(s) | Classification |
|---|---|---|
| $\mathbb{Z}^+$ | add, multiply | |
| $\mathbb{Z}$ | add, subtract, multiply | ring |
| $\mathbb{Q}^+, \mathbb{R}^+$ | add, multiply, divide* | group under $\cdot$ |
| $\mathbb{Q}, \mathbb{R}, \mathbb{C}$ | add, subtract, multiply, divide* | field |
| permutations ($S_n$, etc.) | composition | group |
| $n \times n$ matrices | add, subtract, multiply | ring |
| $\mathbb{Z}_n$ ($n$ composite) | add, subtract, multiply | ring |
| $\mathbb{Z}_p$ ($p$ prime) | add, subtract, multiply, divide* | field |
| $U_n(\{a \in \mathbb{Z}_n \mid \gcd(a, n) = 1\})$ | multiply, divide | group under $\cdot$ |

*No division by 0.

In our chart, you can see that three types of sets are classified as groups. However, in subsequent chapters we will see that every ring (see Chapter 26) contains the structure of a group within it, and every field (see Chapter 29) contains the structure of a group in two different ways! Vector spaces (see Chapter 30), to which you may already have been introduced if you took a course in Linear Algebra, contain the structure of a group as well. Hence you can see that groups are a central concept in the study of sets with algebraic structure.

In this chapter we shall define what a group is and give numerous examples of groups, including the ones in our chart above. In the three following chapters, we shall learn about basic properties of groups, be introduced to the concepts of *subgroups* (which are subsets of a group which themselves have group structure) and *cosets*, and finally learn about the fundamentally important Lagrange Theorem on finite groups. So here we go.

**Definition 21.1.** A *group* is a non-empty set $G$ with a binary operation $*$ such that the following properties hold:

1. Closure: $a * b \in G$ for each $a, b \in G$;

2. Associativity: $(a * b) * c = a * (b * c)$ for all $a, b, c \in G$;

3. Identity: There is an element $e \in G$, called the **identity element** of $G$, such that $e * a = a * e = a$ for all $a \in G$;

4. Inverses: For each $a \in G$ there is an element $a^{-1} \in G$, called the **inverse** of $a$, such that $a * a^{-1} = a^{-1} * a = e$.

We remark that it is quite common to suppress the notation $*$ and simply write $ab$ instead of $a * b$. We also remark that we introduced the ideas of binary operation, commutativity, associativity, identity element and inverse element in Chapter 12; it may be worth looking back at that now (see in particular Examples 12.1, 12.5 and 12.6). Finally, it is common to denote a group by $(G, *)$ (for example, $(\mathbb{Z}, +)$ for the integers under addition) to emphasize that a group is defined both by its underlying set and by its associative binary operation.

It is important to note that in the definition we do *not* require that for all $a$ and $b$, $a * b = b * a$. If this condition *does* hold in a group $G$, we say that $G$ is **Abelian** or **commutative**. As we will see, some groups are Abelian, or commutative, and some are not. The term Abelian is used in honor of Niels Abel (1802-1829) who was an early pioneer in the study of groups. Not surprisingly, a group G in which this condition does not hold (i.e., in which there exist elements $a$ and $b$ for which $a * b \neq b * a$) is called **non-Abelian** or **non-commutative**. (Math nerd joke: Question: What's purple and commutative? Answer: An Abelian grape.)

Finally, for our terminology, we say that $G$ is a **finite** group if $G$ contains a finite number of distinct elements; otherwise $G$ is an **infinite** group. Using this terminology, it is helpful to think of groups as falling into four basic categories:

| finite Abelian | finite non-Abelian |
|---|---|
| infinite Abelian | infinite non-Abelian |

As we move forward we will want to distinguish between working with an abstract group $G$ and working with an *example* of a group. Here are three areas for us to be clear on the distinction:

(1) When working with an abstract group $G$, we shall use the notation $*$ to represent a "generic" binary operation. However, as we have seen (for example in our chart), in an example of a group, the binary operation may be addition (of numbers or matrices), or multiplication (of numbers or matrices), or composition of functions, or any number of other possibilities.

(2) In an abstract group $G$, we use the generic notation $e$ for the identity element of $G$. In an example, the identity could be 0 (or the zero matrix) if the binary operation is addition, could be 1 (or the identity matrix $I_n$ - see Exercise 12.10) if the operation is multiplication, and so on.

(3) Finally, in an abstract group $G$ we use the generic notation $a^{-1}$ for the inverse of $a$. In an example, the inverse of $a$ (or the matrix $A$) could be $-a$ (or the matrix $-A$) if the binary operation is addition, could be $1/a$ if the binary operation is multiplication on $\mathbb{R}$, $\mathbb{Q}$ or $\mathbb{C}$, could be the multiplicative inverse of $a$ in $\mathbb{Z}_n$ if it exists), could be the multiplicative inverse of the matrix $A$ in the set of all $n \times n$ matrices over, say, $\mathbb{R}$ (if it exists) and so on.

As you get more familiar with Abstract Algebra, you will get used to the transition from working with an abstract group $G$ to applying what we learn to its many examples. For now though, let's look at a number of those examples.

**Example 21.1.** The simplest group is the set $G = \{e\}$ which consists of just one element $e$. We define the binary operation $*$ by $e * e = e$. You should check that $G$ forms a finite Abelian group, known as the *trivial group*.

**Example 21.2.** Under the operation of addition, the sets of all integers $\mathbb{Z}$, all rational numbers $\mathbb{Q}$, all real numbers $\mathbb{R}$, and all complex numbers $\mathbb{C}$ form infinite Abelian groups. Each set is clearly closed under additon, addition of numbers is associative, the identity element is 0, and the inverse of any element $a$ is $-a$.

**Example 21.3.** Under the operation of multiplication, none of the sets $\mathbb{Z}$, $\mathbb{Q}$, $\mathbb{R}$ and $\mathbb{C}$ form groups because the number 0 does not have a multiplicative inverse (remember that in a group, *every* element must have an inverse in that group). Moreover, the only numbers in $\mathbb{Z}$ which have multiplicative inverses *in* $\mathbb{Z}$ (we must have closure!) are 1 and $-1$, so the integers $\mathbb{Z}$ do *not* form a group under multiplication. However, if we remove 0 from the sets $\mathbb{Q}$, $\mathbb{R}$ and $\mathbb{C}$, forming the sets we denote by $\mathbb{R}^*$, $\mathbb{Q}^*$ and $\mathbb{C}^*$, we do indeed get infinite Abelian groups under multiplication. The identity element in each case is 1, and the multiplicative inverse of $a$ in each case is $1/a$. It is also the case (as we have in our chart) that if we restrict the rational or real numbers to only those which are positive, the sets $\mathbb{Q}^+$ and $\mathbb{R}^+$ do form infinite Abelian groups under multiplication.

**Example 21.4.** Again from our chart, you should check that the set $\mathbb{Z}_n$ (i.e., the set $\{0, 1, 2, \ldots, n-1\}$ under addition modulo $n$) forms a finite Abelian group with $n$ elements. Modular addition is associative, the identity element is 0, and the additive inverse of $a$ is $-a \pmod{n}$.

**Example 21.5.** What about the sets $\mathbb{Z}_n$ under the operation of multiplication? Well, we know we need to remove 0, but recall from Lemma 15.3 that $a \in \mathbb{Z}_n$ possesses a multiplicative inverse $a^{-1}$ if and only if $a$ and $n$ are relatively prime. We introduced this subset of $\mathbb{Z}_n$ in the proof of Theorem 19.3 and labeled it $U_n$, that is, $U_n = \{a \in \mathbb{Z}_n \,|\, \gcd(a, n) = 1\}$. You should now check that for all $n \geq 2$, $U_n$ is a finite Abelian group with $\phi(n)$ elements, where $\phi$ is the Euler function introduced in Chapter 19. For example, we have closure since if $a$ and $b$ are relatively prime to $n$, then so is their product modulo $n$ (check this). We note that if $n = p$ with $p$ a prime, then $U_p = \mathbb{Z}_p^*$; i.e., we need only remove 0 from $\mathbb{Z}_p$ in order to form a multiplicative group. This will be relevant when we study fields.

**Example 21.6.** For any finite group we can form a *group table* which shows in the $a$th row and $b$th column the result of $a * b$. As an example, here is the group table for the group $U_8 = \{1, 3, 5, 7\}$ in which the operation is multiplication modulo 8.

| $\times$ | 1 | 3 | 5 | 7 |
|---|---|---|---|---|
| **1** | 1 | 3 | 5 | 7 |
| **3** | 3 | 1 | 7 | 5 |
| **5** | 5 | 7 | 1 | 3 |
| **7** | 7 | 5 | 3 | 1 |

**Example 21.7.** So far all our examples of group have been Abelian (finite and infinite). However, in Chapter 10 we learned about permutations of the set $\{1, 2, \ldots, n\}$ for some positive integer $n$, and in particular we saw that composition of permutations is non-commutative (this is a special case of the general fact that composition of functions is non-commutative). Since we have an identity permutation which fixes every element and since every permutation has an inverse permutation (because permutations are bijections of a set to itself), we see that we have the makings of a group. Specifically, the set $S_n$ of all permutations of the set $\{1, 2, \ldots, n\}$ forms a finite non-Abelian group containing $n!$ elements (as we observed in Example 10.1). The group $S_n$, called the *symmetric group* on $n$ objects turns out to be a very important example of a group.

**Example 21.8.** Viewing permutations in a different context, we now consider the set of all possible rotations and reflections (i.e., "rigid motions") of an equilateral triangle. This set turns out to form a finite non-Abelian group with six elements and is commonly labeled $D_3$.

Consider the triangles which appear below. We can rotate the triangle counter-clockwise about its mid-point by an angle of $2\pi/3$ radians (120 degrees), and we will denote this operation by $\rho$. We will denote by $R$ the operation of reflecting the triangle about a vertical line from the top of the triangle to the mid-point of the base. Since there are six different orientations in which the triangle can end up, there must be six different operations which can be accomplished by $e, \rho, \rho^2, R, \rho R$, and $\rho^2 R$, the identity (i.e., no movement) being denoted by $e$. Here are three of the six:

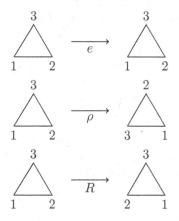

You should now draw pictures of the other three possible operations and convince yourself that they can be accomplished by applying $\rho^2$, $\rho R$ and $\rho^2 R$. These six operations form a non-Abelian group as described in the following group table:

|        | $e$      | $\rho$   | $\rho^2$ | $R$      | $\rho R$ | $\rho^2 R$ |
|--------|----------|----------|----------|----------|----------|------------|
| $e$      | $e$      | $\rho$   | $\rho^2$ | $R$      | $\rho R$ | $\rho^2 R$ |
| $\rho$   | $\rho$   | $\rho^2$ | $e$      | $\rho R$ | $\rho^2 R$ | $R$      |
| $\rho^2$ | $\rho^2$ | $e$      | $\rho$   | $\rho^2 R$ | $R$    | $\rho R$   |
| $R$      | $R$      | $\rho^2 R$ | $\rho R$ | $e$    | $\rho^2$ | $\rho$     |
| $\rho R$ | $\rho R$ | $R$      | $\rho^2 R$ | $\rho$ | $e$      | $\rho^2$   |
| $\rho^2 R$ | $\rho^2 R$ | $\rho R$ | $R$   | $\rho^2$ | $\rho$ | $e$        |

We note the group table reveals that this group is non-Abelian since the group table is not symmetric about the main diagonal. For example, note that $R\rho \neq \rho R$ since $R\rho = \rho^2 R$ which is not the same as $\rho R$.

Instead of just considering triangles, one could also consider the rigid motion of a square, or even more generally of a regular $n$-sided polygon. These operations always form finite non-Abelian groups containing $2n$ elements and are referred to as the **dihedral groups**, denoted $D_n$.

**Example 21.9.** Consider the set $G$ of all $n \times n$ matrices whose entries are real numbers. In Chapter 12 we defined addition of matrices and observed that matrix addition on $n \times n$ matrices is closed (i.e., the sum of two $n \times n$ matrices is an $n \times n$ matrix), associative and commutative. The identity is the zero matrix (i.e., all entries are 0), and the inverse of a matrix $A = (a_{ij})$ is the matrix $-A = (-a_{ij})$. Hence $G$ under matrix addition forms an infinite Abelian group.

**Example 21.10.** What about $n \times n$ matrices under matrix multiplication as defined in Chapter 12? We observed there that matrix multiplication on this set is closed (i.e., the product of two $n \times n$ matrices is an $n \times n$ matrix) and associative but is *not* commutative. There is an identity, specifically the identity matrix $I_n$ which has all 1's down the main diagonal and all 0's elsewhere. However, not all $n \times n$ matrices possess multiplicative inverses. As discussed near the end of Chapter 12, an $n \times n$ matrix will possess a multiplicative inverse if and only if it is *non-singular*, meaning that its rows are linearly independent, or put another way, if its determinant is non-zero. Hence the set of non-singular $n \times n$ matrices over $\mathbb{R}$ form an infinite non-Abelian group. For each positive $n$, it is called the **general linear group of dimension $n$** and is commonly denoted $GL(n, \mathbb{R})$. We note that these are the first examples we have seen of groups which are infinite and non-Abelian.

**Example 21.11.** Though polynomials were a familiar topic in your algebra, pre-calculus and calculus courses, we have thus far not concentrated much on them in this text (but see Chapters 33, 34 and 35). In the current context, however, we note that groups can exist which have polynomials as their elements. For example, consider the set $G$ of all polynomials in a single variable $x$ with real coefficients under the operation of polynomial (i.e., "term-by-term") addition. This operation is clearly closed on $G$, associative and commutative. The identity element is the constant polynomial $f(x) = 0$ and the inverse of a polynomial $f(x) = \sum_{i=0}^{m} c_i x^i$ is the polynomial $-f(x) = \sum_{i=0}^{m} -c_i x^i$. Thus $G$ forms an infinite Abelian group. In a similar way, we could let $G_m$ be the set of all polynomials of degree less than or equal to a positive integer $m$ under the same operation. Then for each $m$, $G_m$ is another infinite Abelian group. Finally, we remark that all of this works in the same way if the coefficients are taken from any of our other number sets $\mathbb{Z}$, $\mathbb{Q}$, $\mathbb{C}$ or even $\mathbb{Z}_n$; the only difference being that over $\mathbb{Z}_n$, the group $G$ remains infinite but the group $G_m$ is *finite*.

**Example 21.12.** We introduced the concept of Cartesian products in Examples 6.8 and 6.9. Here we note that if $G_1$ and $G_2$ are any two groups, then their Cartesian product $G_1 \times G_2$ is a group under the "component-wise" operation. That is, in the ordered pairs, the operation in the first component is $G_1$'s operation and the operation in the second component is $G_2$'s operation. For example, in the Cartesian product $\mathbb{Z}_8 \times \mathbb{Z}_5$ where both operations are modular addition, then $(3, 3) + (4, 4) = (7, 2)$.

Hopefully all of the above examples will seem to you to involve natural sets of numbers, permutations, matrices and polynomials, together with natural operations on those sets. Let us finish this list with an example of a group which may seem to be rather unnatural. The point here is to demonstrate that whether seeming natural or unnatural, we need only establish that the given set and the given operation satisfy the requirements (closure, associativity, identity, inverses) to verify that we have a group.

**Example 21.13.** Let $G = \mathbb{Z}$ denote the set of integers. For two integers $a$ and $b$, we define a binary operation $*$ by $a * b = a + b + 2$. Does this strange looking operation on $\mathbb{Z}$ provide another example of a group? Let's check.
(1) The operation $*$ is closed since if $a$ and $b$ are integers, so is $a + b + 2$.
(2) Given integers $a, b, c$, we have

$$(a * b) * c = (a + b + 2) * c = (a + b + 2) + c + 2 = a + b + c + 4.$$

Similarly, we have

$$a * (b * c) = a * (b + c + 2) = a + (b + c + 2) + 2 = a + b + c + 4.$$

Hence the operation $*$ is associative.
(3) The identity $e$ must satisfy that $a * e = a$ for all integers $a$, so we must have $a + e + 2 = a$, i.e., $e = -2$. Likewise then, $e * a = -2 + a + 2 = a$, as required.
(4) Finally for inverses, we are required to have $a * a^{-1} = e$, i.e., $a + a^{-1} + 2 = -2$, so we get that $a^{-1} = -a - 4$. Likewise then, $a^{-1} * a = (-a - 4) + a + 2 = -2$; as required.

Thus, the integers with this rather unusual operation do indeed form an infinite group which is clearly Abelian since $a + b + 2 = b + a + 2$.

Finally, in Exercise 21.14, we ask if there was anything special about 2 in this example.

This list of examples of different groups is just "scratching the surface," but we hope it's given you an idea of how extensively this algebraic structure appears in mathematics. We shall now turn to learning about a number of the basic properties which all groups share. The beauty here is that once we show that an *abstract* group $G$ has a certain property, then every example of that type of abstract group must also have that property.

**Exercises**

**21.1.** In each case, either prove that the given set with the given operation is a group, or say exactly why it fails to be a group:
(a) The integers $\mathbb{Z}$ under the operation of subtraction.
(b) The positive real numbers $\mathbb{R}^+$ under addition.
(c) The positive real numbers $\mathbb{R}^+$ under multiplication.

**21.2.** In each case, either prove that the given set with the given operation is a group, or say exactly why it fails to be a group:
(a) The non-zero real numbers $\mathbb{R}^*$ under the operation of division.
(b) The set of all $3 \times 5$ matrices over the real numbers under matrix addition.
(c) The set of all $3 \times 5$ matrices over the real numbers under matrix multiplication.

**21.3.** Write down the group table for the set $\mathbb{Z}_4$ under the operation of addition modulo 4.

**21.4.** Write down the group table for the set $U_{12}$ (which has $\phi(12) = 4$ elements) under multiplication modulo 12 (see Example 21.5).

**21.5.** Show that in any group, the identity element $e$ must be unique; i.e., a group has only one identity element. (Hint: Suppose that $e_1$ and $e_2$ are both identity elements. Show that they are equal.)

**21.6.** Show that the inverse $a^{-1}$ of an element $a$ in a group must be unique. (See the hint in the previous exercise. Call the two inverses, say, $b_1$ and $b_2$ and look at $b_1ab_2$. Be careful; here you will need to use associativity. Note also that as mentioned earlier in this chapter, we often suppress the $*$ notation in an abstract group and simply use "juxtaposition.")

**21.7.** If $G$ is any group and $a, b \in G$, prove that $(ab)^{-1} = b^{-1}a^{-1}$.
(Note: This is simply another instance of the "socks and shoes" concept we introduced in Exercise 10.6. When you want to undo a sequence of operations, you need to reverse the order.)

**21.8.** Give an example of a group $G$ with elements $a$ and $b$ in $G$ for which $(ab)^{-1} \neq a^{-1}b^{-1}$.

**21.9.** Show that if $a, b$ are elements in any group $G$, then the equation $ax = b$ has a unique solution $x \in G$. Similarly show that $xa = b$ also has a unique solution $x \in G$.

**21.10.** (a) If $G$ is any group and $a, b \in G$, prove that if $ab = ac$, then $b = c$.
(b) Does the same conclusion as in Part (a) hold if $ab = ca$? Either prove it's true or give a counter-example.

**21.11.** We introduced the concept of equivalence relations in Chapter 7. Briefly, a relation between the elements of a given set is an *equivalence relation* if it is reflexive, symmetric and transitive.
Let $G$ be any group. For elements $a, b \in G$, define $a$ to be related to $b$ if there is an element $g \in G$ so that $b = g^{-1}ag$. Show that this relation is an equivalence relation on the group $G$.
(Note: The operation which takes $a$ to $g^{-1}ag$ for some $g$ turns out to be an important operation in group theory called "conjugation.")

**21.12.** Use mathematical induction to prove that if $G$ is a group with elements $a$ and $g$, then for all positive integers $n$,

$$(g^{-1}ag)^n = g^{-1}a^ng.$$

**21.13.** (a) Show that in Example 21.13 there was nothing special about the number 2. That is, carefully show that on the set $\mathbb{Z}$ of integers, the operation $*$ given by $a * b = a + b + k$ defines a group for each $k \in \mathbb{Z}$.
(b) In the group in Part (a) with $k = 7$, solve the equation $50 * x = 100$ for $x$.

**21.14.** Let's consider the following question: Modulo $n$, what is the sum of all the elements of the finite Abelian additive group $\mathbb{Z}_n$?
(a) Compute this sum modulo $n$ for $n = 3, 4, 5, 6, 7, 8$. (Note: It's worth looking back at what we proved in Example 4.1, but note that here we are adding up the numbers from 1 to $n - 1$.)

(b) Make a conjecture about the value of this sum modulo $n$ depending on whether $n$ is even or odd.

(c) Prove your conjecture. (Suggestion: Write $n$ in the form $2k$ if $n$ is even and $2k + 1$ if $n$ is odd. Now apply Example 4.1 in terms of $k$, and finally reduce modulo $n$ (also in terms of $k$).)

**21.15.** (a) In Example 21.12 we introduced the idea of Cartesian products of groups. Given the group $\mathbb{R}^*$ (i.e., $\mathbb{R}$ with 0 removed) under multiplication, $\mathbb{R}^* \times \mathbb{R}^*$ is called the "punctured" $xy$-plane. Compute $(3, 5.2)(1.3, -6)$ in this group.

(b) The non-zero complex numbers $\mathbb{C}^*$ can also be viewed as the same punctured plane with the real part as the horizontal coordinate and the imaginary part as the vertical coordinate. However, $\mathbb{C}^*$ under complex number multiplication is *not* the same group as the group $\mathbb{R}^* \times \mathbb{R}^*$ in Part (a). To verify this fact, compute $(3 + 5.2i)(1.3 - 6i)$ and compare with Part (a).

**21.16.** Inside the non-zero complex numbers $\mathbb{C}^*$, show that the finite set $\{1, -1, i, -i\}$ with complex number multiplication is a group.

(Note: Given a subset of an established group (as here with $\mathbb{C}^*$ under complex number multiplication), to prove that you have a group structure on that subset you need to verify closure, existence of the identity, and existence of an inverse *contained in the subset* for each element of the subset. Associativity need not be checked as it is already established. Such a subset, once verified, is called a *subgroup*.)

**21.17.** Let $G$ be a group with an operation $*$. An *automorphism* $f$ of a group $G$ is a function from $G$ to itself which is one-to-one, onto, and has the property that for all $a, b \in G$

$$f(a * b) = f(a) * f(b).$$

That is, a group automorphism is a bijection of $G$ to itself which preserves $G$'s operation. Show that the specific function $f : G \to G$ defined by $f(a) = a^{-1}$ for each $a \in G$ is a group automorphism if and only if the group is Abelian.

# CHAPTER 22

## Groups - Basic Properties

$$\mathbb{Z}_7^* = \{3^k(\bmod\ 7)|1 \le k \le 6\}$$

Having learned the formal definition of a group and seen a number of examples of different kinds of groups in the previous chapter, we shall start this chapter by summarizing some of the basic properties which all groups share. We wish to emphasize again that when working with an abstract group $G$, we usually suppress the $*$ notation and simply use juxtaposition (i.e., writing $ab$ rather that $a*b$), which means we will refer to the operation as "multiplication." Remember, however, that in any given *example* of a group, the operation may well *not* be multiplication, but may instead be addition of numbers, addition of matrices, composition of permutations, rigid motions of an object, and so on.

**Theorem 22.1. (Basic Group Properties)**
(i) *The identity element $e$ of every group $G$ is unique.*
(ii) *For each element $a$ in a group $G$, the inverse $a^{-1}$ of $a$ is unique.*
(iii) *For all $a, b$ in a group $G$, each of the equations $ax = b$ and $xa = b$ has a unique solution $x$ in $G$.*
(iv) *For all $a$ and $b$ in a group $G$, $(ab)^{-1} = b^{-1}a^{-1}$.*

We prove (ii) below and note that you were asked to establish all four of these properties in Exercises 21.5, 21.6, 21.7 and 21.9. If you did not do those exercises previously, do them now, making use of the hints and suggested solutions given there if you need them. You should note that in proving the latter three properties in this theorem, you do need to make use of the fact that the binary operation is *associative*.

Proof of (ii). Suppose that $b$ and $c$ are both inverses of an element $a$ in $G$, whose identity is denoted by $e$. Then using the definitions of identity and inverse and using the associativity of the operation, we have

$$b = be = b(ac) = (ba)c = ec = c.$$

We conclude that inverses are unique.　■

We note again that Property (iv) is an example of the "socks and shoes" principle: if we wish to undo a sequence of operations, we must reverse the order. This property in group theory consolidates a number of results from other areas of mathematics. For example, when discussing invertible matrices in linear algebra, you may recall that for $n \times n$ invertible matrices

$$(AB)^{-1} = B^{-1}A^{-1}.$$

Because invertible $n \times n$ matrices form a group under matrix multiplication, this property of matrices must follow from our theorem. Similarly when studying permutations in Chapter 10, we observed that if $\sigma$ and $\pi$ are permutations on the same set, then

$$(\sigma\pi)^{-1} = \pi^{-1}\sigma^{-1}.$$

This result for permutations must hold because, once again, permutations form groups under the operation of composition. These observations are examples of the reason that *abstraction pays dividends*, namely, *any property we prove holds for an abstract group must then hold for any example of a group!*

You might wonder if $(ab)^{-1} = a^{-1}b^{-1}$ holds in general in groups. It is clear that if a group $G$ is Abelian, then this property would follow from Part (iv) of Theorem 22.1. But what if $G$ is non-Abelian? Let's seek a counter-example in the simplest situation we can find.

**Example 22.1.** It turns out that the smallest non-Abelian group is the symmetric group $S_3$, which contains 6 permutations. Some of these 6 elements commute with each other but some do not; for example, using the cycle notation introduced following Example 10.4, (12) and (123) do not commute since $(12)(123) = (23)$ whereas $(123)(12) = (13)$. (Recall that when composing cycles, we work from right to left.) Letting $a = (12)$ and $b = (132)$, then $a^{-1} = (12)$ (i.e., $a$ is its own inverse) and $b^{-1} = (123)$, so our calculation above shows that in this example

$$(ab)^{-1} = b^{-1}a^{-1} \neq a^{-1}b^{-1}.$$

We shall now turn to defining and giving examples of arguably the simplest class of groups, namely the *cyclic groups*. These groups have the property that starting with an appropriate single element, called a *generator* of the group, we can obtain the entire group by computing the *powers* of that element. Once again, note that in an abstract group we discuss "powers" of an element because we think of our operation as being multiplication. In an example of a group for which the operation is addition, say, then we will look at *multiples* of our generator, not powers, and so on.

We first wish to observe that powers of group elements obey the same rules as powers of, say, real numbers. Specifically, if $G$ is an abstract group and $a, b$ are elements in $G$, then for any integers $r$ and $s$ we have:
  (i) $a^r a^s = a^{r+s}$;
  (ii) $(a^r)^s = a^{rs}$;
  (iii) $a^{-r} = (a^r)^{-1} = (a^{-1})^r$;
  (iv) if $ab = ba$, i.e., if the elements $a$ and $b$ commute, then $(ab)^r = a^r b^r$.

Suppose now that $G$ is an abstract group with the property that there exists an element $g \in G$ so that every element in the group can be written as a power of the element $g$, i.e., every element $a \in G$ can be written in the form $a = g^m$ for some integer $m$. Then the group is said to be a ***cyclic group***, and the element $g$ is called a ***generator*** for the group $G$. We also remark that a very common *notation* for a cyclic group $G$ generated by the element $g$ is

$$G = \langle g \rangle.$$

Let's look at some specific examples of cyclic groups.

**Example 22.2.** Consider the group $\mathbb{Z}$ under addition, so that now we will be looking at *multiples*, not powers, of a possible generator. Well, certainly every integer is a multiple of 1, so in fact $\mathbb{Z} = \langle 1 \rangle$ is an infinite cyclic group generated by 1. It is a fact, which can be made precise when you pursue abstract algebra more fully, that $\mathbb{Z}$ under addition is *the fundamental example* of all infinite cyclic groups. By this we mean a different infinite cyclic group $G$ may have a different underlying set and/or a different operation, but the *structure* of $G$ is exactly the same as that of $\mathbb{Z}$ under addition. See the next couple of examples.

**Example 22.3.** Inside the group $\mathbb{Z}$ of all integers under addition we can form the cyclic group $\langle 2 \rangle$ generated by 2, which is then all integer multiples of 2, in other words, all the even integers. This cyclic group is commonly denoted $2\mathbb{Z}$. More generally, if $n$ is any positive integer, then $n\mathbb{Z} = \langle n \rangle$ is the cyclic group consisting of all integer multiples of $n$. These of course are all infinite Abelian groups.

**Example 22.4.** Inside the group $\mathbb{R}^+$ of all positive real numbers under multiplication, the cyclic group $\langle 2 \rangle$ generated by 2 is now all *powers* of 2, i.e., it is the infinite set $\{\ldots, 1/8, 1/4, 1/2, 1, 2, 4, 8, \ldots\}$ under multiplication. We can of course start with any positive real number $r$ and form the infinite cyclic group $\langle r \rangle$ inside $\mathbb{R}^+$ consisting of all integer powers of $r$.

**Example 22.5.** So far all of our examples of cyclic groups have been infinite. Consider then the class of finite Abelian groups $\mathbb{Z}_n$ where $n$ is an integer greater than 1 and the operation is addition modulo $n$. As was the case in $\mathbb{Z}$, the group

$\mathbb{Z}_n$ is a cyclic group generated by 1; i.e., the set $\{0, 1, 2, \ldots, n-1\}$ consists of multiples of 1. We remark that here is really where the term "cyclic" makes sense, since $(n)(1) \equiv 0 \pmod{n}$ and so we "cycle" back to 0. We note also then that for every $n > 1$, we have an example of a finite Abelian cyclic group with $n$ elements.

**Example 22.6.** How about a finite *multiplicative* cyclic group? Here we can look back at Chapter 18, where we discussed the existence and uses of "primitive roots" in the set $\mathbb{Z}_p^*$ under multiplication modulo $p$, where $p$ is a prime number. We saw later in Example 21.5 that $\mathbb{Z}_p^*$ under multiplication modulo $p$ does form a finite Abelian group with $p-1$ elements. But now a primitive root $g$ has the property that the smallest positive power $k$ for which $g^k = 1$ is $k = p-1$; in other words, *a primitive root of $\mathbb{Z}_p^*$ is exactly a generator for this group!* One can prove that $\mathbb{Z}_p^*$ has exactly $\phi(p-1)$ primitive roots, and hence that many generators (where $\phi$ is the Euler function studied in Chapter 19), and so we can conclude that for any prime $p$, $\mathbb{Z}_p^*$ is a finite Abelian cyclic group with $p-1$ elements. For a specific example, since $\phi(6) = 2$, the group $\mathbb{Z}_7^*$ will have two generators, one of which turns out to be 3 (the other is 5 - check this). Working modulo 7, we get

$$3^1 \equiv 3, 3^2 \equiv 2, 3^3 \equiv 6, 3^4 \equiv 4, 3^5 \equiv 5, 3^6 \equiv 1.$$

You probably have noticed that every example of cyclic group (finite and infinite) we have seen is Abelian. Here is the easily verified reason for that.

**Theorem 22.2.** *If $G$ is a cyclic group, then $G$ is Abelian.*

Proof. If the group $G$ is cyclic, it has a generator $g \in G$. Hence if $a, b \in G$, then by definition of a cyclic group, $a = g^m$ and $b = g^n$ for some integers $m$ and $n$. Hence
$$ab = g^m g^n = g^{m+n} = g^{n+m} = g^n g^m = ba,$$
and the proof is complete. ∎

One could now ask if the converse of Theorem 22.2 is true; i.e., is every Abelian group cyclic? The answer is a definite "no."

**Example 22.7.** Consider the group $U_8 = \{1, 3, 5, 7\}$ under multiplication modulo 8. Does this little group have a generator? Well, modulo 8,

$$1^2 \equiv 3^2 \equiv 5^2 \equiv 7^2 \equiv 1,$$

i.e., no elements generate the whole group, and so $U_8$ is *not* cyclic. In fact, $U_8$ is an example of what is known as a *Klein four-group*; that is, a group consisting of four elements, each of which is its own inverse. This group structure appears in various guises (see, for example, Exercises 22.9 and 22.10) and is named for the mathematician Felix Klein (1849-1925; the famous "Klein bottle" is also named after him).

Even if it is known that a group *is* cyclic, it may be take some effort to find a generator.

**Example 22.8.** The number 4999 is prime (check this!), so we know that there are exactly $\phi(4998) = 1344$ generators of the cyclic group $\mathbb{Z}_{4999}^*$ under the operation of multiplication modulo 4999 (see Example 22.6 above). Hence about 1 out of every 4 of the elements of this group is a generator, but it's not clear how to identify even one of these. It turns out that relatively fast algorithms are available to identify generators, but we shall not pursue those methods here. By the way, the smallest generator of $\mathbb{Z}_{4999}^*$ turns out to be 3.

In numerous of the examples of groups given in this and the previous chapter, one group was sitting inside another group, with them *sharing the same operation*. These "groups within a group" are called *subgroups*, and we shall now turn our attention to them, as they play a central role in the theory of groups.

### Exercises

**22.1.** Given elements $a, b, c$ in a group $G$, solve the equation $ab^2xc = b$ for $x$.

**22.2.** Given elements $a, b, c, d$ in a group $G$, solve the equation $abxc = d$ for $x$.

**22.3.** In Example 22.1 we used the group $S_3$ to show that it is *not true* in general in groups that $(ab)^{-1} = a^{-1}b^{-1}$. More generally, show that if in a group $G$ there exist elements $a$ and $b$ such that $ab \neq ba$, then $(ab)^{-1} \neq a^{-1}b^{-1}$. (Note: It is probably easiest to prove the contrapositive statement.)

**22.4.** If $a^2 = e$ for all $a \in G$ a group with identity element $e$, show that the group $G$ is an Abelian group. (Note: The Klein four-group in an example of such a group, though the smallest such groups are the trivial group and $\mathbb{Z}_2$.)

**22.5.** Use a cardinality argument to conclude that the real numbers $\mathbb{R}$ under addition, the positive real numbers $\mathbb{R}^+$ under multiplication, and the non-zero real numbers $\mathbb{R}^*$ under multiplication are *not* cyclic groups (see Chapter 9).

**22.6.** (a) Unlike in the previous exercise, we cannot use a cardinality argument to conclude that the rational numbers $\mathbb{Q}$ under addition do not form a cyclic group (why?). Use a "spacing" argument instead, i.e., if $a/b$ *were* a generator ($a$ and $b$ in $\mathbb{Z}^*$), then the "next" element in the cyclic group would be $2a/b$, etc. (b) Show that the positive rational numbers $\mathbb{Q}^+$ under multiplication do not form a cyclic group by considering the primes which make up the factorizations of $a$ and $b$ in a potential generator $a/b$.

**22.7.** (a) How many generators does the cyclic group $\mathbb{Z}_{11}^*$ (under multiplication modulo 11) have?
(b) Find one of them.

**22.8.** (a) How many generators does the cyclic group $\mathbb{Z}_{13}^*$ (under multiplication modulo 13) have?
(b) Find one of them.

**22.9.** (a) Show that the set $U_{12}$ under multiplication modulo 12 is a Klein four-group, and hence not cyclic.
(b) Show that the set $U_{14}$ under multiplication modulo 14 is a cyclic group.

**22.10.** Recall that the group $\mathbb{Z}_n$ of $n$ elements under addition modulo $n$ is a cyclic group generated by 1.
(a) Show that the Cartesian product $\mathbb{Z}_2 \times \mathbb{Z}_2$ under component-wise addition is a Klein four-group.
(b) Show that the Cartesian product $\mathbb{Z}_2 \times \mathbb{Z}_3$ under component-wise addition is a cyclic group.
(c) Show that the Cartesian product $\mathbb{Z}_2 \times \mathbb{Z}_4$ under component-wise addition is not cyclic.
(d) On the basis of this small amount of data, make a conjecture: The Cartesian product $\mathbb{Z}_n \times \mathbb{Z}_m$ under component-wise addition is a cyclic group if and only if $\cdots$. Can you prove your conjecture?

# CHAPTER 23

Groups - Subgroups

$$(\mathbb{Z}, +) < (\mathbb{Q}, +) < (\mathbb{R}, +) < (\mathbb{C}, +)$$

Most groups besides the trivial group $\{e\}$ possess a rich "internal structure." Indeed, all non-trivial groups, whether finite or infinite, whether Abelian or non-Abelian, have smaller groups inside of them. In this chapter we take a close look at this idea, starting with the following important definition.

**Definition 23.1.** A non-empty subset $H$ of a group $G$ with operation $*$ is a *subgroup* of $G$ if $H$ is itself a group *under the same operation $*$ as in $G$.*

It is standard to denote the fact that $H$ is a subgroup of $G$ by $H \leq G$, and more specifically by $H < G$ if $H$ is a proper sub*set* of $G$, in which case $H$ is called a *proper subgroup* of $G$.

We note first of all that a subgroup $H$ of a group $G$ (whose identity is $e$) must itself contain an identity element; let's call this identity element $f$. Then, in $H$ we have $f * f = f$, but in $G$ we have $e * f = f$ and we know $f$ has a inverse $f^{-1}$, so we get $f * f = e * f$, and multiplying on the right by $f^{-1}$, we arrive at $f = e$. Thus $H$ and $G$ share the same identity element.

Let's look at some examples of subgroups and examples which are not subgroups.

**Example 23.1.** If $G$ is any non-trivial group, then the trivial group $\{e\}$ is a proper subgroup of $G$. At the other extreme, we note that by definition $G$ is a subgroup of itself.

**Example 23.2.** Under addition, the integers $\mathbb{Z}$ are a subgroup of the rational numbers $\mathbb{Q}$, which in turn are a subgroup of the real numbers $\mathbb{R}$, which in turn are a subgroup of the complex numbers $\mathbb{C}$.

**Example 23.3.** Under multiplication, the positive real numbers $\mathbb{R}^+$ are a subgroup of the non-zero real numbers $\mathbb{R}^*$. Likewise, the cyclic group
$\langle 2 \rangle = \{\ldots, 1/8, 1/4, 1/2, 1, 2, 4, 8, \ldots\}$ (see Exercise 22.4) is a subgroup of both of these groups.

**Example 23.4.** Finite groups can possess non-trivial proper subgroups just as infinite groups do, but these situations may require a closer look. A fairly simple example is to consider the cyclic group $\langle 3 \rangle = \{3, 6, 9, 0\}$ lying inside $\mathbb{Z}_{12}$, both under addition modulo 12. It is easy to check that $\langle 3 \rangle$ is indeed a proper subgroup of $\mathbb{Z}_{12}$.

It is very important to remember that a subgroup *must* possess the exact same operation as that of the group containing it.

**Example 23.5.** The group $\mathbb{R}^+$ under multiplication is *not* a subgroup of the group $\mathbb{R}$ under addition even though the set $\mathbb{R}^+$ is a subset of the set $\mathbb{R}$. The group $\mathbb{Z}_{12}$ under addition modulo 12 is *not* a subgroup of $\mathbb{Z}$ under addition even though the set $\mathbb{Z}_{12}$ is a subset of the set $\mathbb{Z}$. And so on.

You might think that in order to check whether a subset $H$ of a group $(G, *)$ is a subgroup we need to check the four properties required for a group as stated in Definition 21.1. However, in the first place, since $H$ and $G$ must be under the same operation $*$, there is no need to re-confirm associativity. Second, the following result shows that we can check the two conditions as in Part (ii), or just check the single, slightly more complicated condition as in Part (iii), in order to confirm that a subset $H$ of $G$ is actually a subgroup of $G$.

**Theorem 23.1.** *The following conditions on a non-empty subset $H$ of a group $(G, *)$ are equivalent; i.e., each of the three conditions* (i), (ii), *and* (iii) *implies each of the others:*
  (i) $H$ *is a subgroup of the group $G$;*
  (ii) $H$ *satisfies*
      (a) (Inverses) *if $h \in H$ then $h^{-1} \in H$ and*
      (b) (Closure) *if $h, k \in H$ then $h * k \in H$;*
  (iii) $H$ *satisfies if $h, k \in H$ then $h * k^{-1} \in H$.*

Proof. This is the first time we have encountered a theorem where we wish to show that three or more conditions are equivalent. An efficient technique is to "go around the horn;" that is, show that Part (i) implies Part (ii), Part (ii) implies Part (iii), and Part (iii) implies Part (i), which completes the proof. So here we go.

Part (i) clearly implies Part (ii) since the properties in Part (ii) are properties that must hold in any group, and hence in a subgroup.

To prove that Part (ii) implies Part (iii), let $h, k \in H$. Then Part (ii)(a) implies that $k^{-1} \in H$ and hence by Part (ii)(b), we have that the product $h * k^{-1} \in H$. Thus, Part (iii) holds.

We now show that Part (iii) holding implies that Part (i) also holds.

If $h \in H$, then using Part (iii) with $k = h$, we see that

$$h * k^{-1} = h * h^{-1} = e \in H,$$

so the subgroup $H$ has an identity $e$.

Now let $h = e$ and let $g$ be an element of $H$. Then by Part (iii) with $k = g$ we have $h * k^{-1} = e * g^{-1} = g^{-1} \in H$, so that each element in $H$ has an inverse in $H$.

Finally, let $a, b \in H$. We need to show that $a * b \in H$, i.e., that the set $H$ is closed under $*$. Using Part (iii) with $h = a$ and $k = b^{-1}$ we have that

$$h * k^{-1} = a * (b^{-1})^{-1} = a * b \in H,$$

and we are done. ∎

In practice, people often choose to use the two-step process in Part (ii) of Theorem 23.1 to verify that a subset $H$ of a group $(G, *)$ is a subgroup of $G$, i.e., check that both the closure and inverse conditions hold. Let us look at a couple of examples.

**Example 23.6.** In Example 21.11 we introduced the infinite Abelian group $G$ of all polynomials in a single variable $x$ with coefficients from the real numbers $\mathbb{R}$ under polynomial (i.e., term-by-term) addition. Suppose $G_n$ is the set of all polynomials of degree less than or equal to a positive integer $n$. The set $G_n$ is clearly a subset of $G$, but we claim $G_n$ is also a subgroup of $G$. Using Theorem 23.1 Part (ii), we get closure since the sum of any two polynomials of degree less than or equal to $n$ is another such polynomial, and we get existence of inverses since if $f(x)$ is of degree $k \leq n$, then $-f(x)$ is also of degree $k$. We have then that $G_n$ is a subgroup of $G$.

**Example 23.7.** Toward the end of Chapter 12 we introduced the idea of invertible $n \times n$ matrices over the real numbers $\mathbb{R}$, and recall that a matrix is invertible if and only if its determinant is non-zero. Then in Example 21.10 we observed that the set of all such invertible matrices form an infinite non-Abelian group, commonly denoted $GL(n, \mathbb{R})$ ("general linear group"), under matrix multiplication. Let us now consider the subset $H$ of $GL(n, \mathbb{R})$ consisting of all those matrices whose determinant is 1. We claim that $H$ is a subgroup of $GL(n, \mathbb{R})$. To see this we need the facts, proved in most Linear Algebra courses, that

(1) If $M$ and $N$ are in $GL(n, \mathbb{R})$, then $\det(MN) = \det(M) \det(N)$, and
(2) $\det(M^{-1}) = 1/\det(M)$.

These two facts say that determinants of invertible matrices are "compatible" with matrix multiplication and the taking of inverses. Applying these facts to our subset $H$, if $M$ and $N$ are in $H$, then

(i) (Closure) $\det(MN) = \det(M)\det(N) = (1)(1) = 1$, and
(ii) (Inverses) $\det(M^{-1}) = 1/\det(M) = 1/1 = 1$.
We conclude then that $H$ is indeed a subgroup of $GL(n, \mathbb{R})$. In fact, $H$ is commonly denoted $SL(n, \mathbb{R})$ ("*special* linear group").

A very rich source of subgroups of a group $G$ is the collection of cyclic groups contained within $G$. (For a review of the definition of cyclic groups and various examples, see Chapter 22 and in particular Examples 22.2 through 22.6.) In fact, if we take *any* element $a \in G$ and form its cyclic group $\langle a \rangle$, then $\langle a \rangle$ is a subgroup of $G$. If $a$ happens to be a generator of $G$, then of course $\langle a \rangle = G$, but otherwise $\langle a \rangle$ is a proper subgroup of $G$.

**Example 23.8.** If $n$ is any positive integer, then

$$\langle n \rangle = \{\ldots, -3n, -2n, -n, 0, n, 2n, 3n, \ldots\}$$

is a subgroup of $\mathbb{Z}$ under addition. The subgroup $\langle n \rangle$ is often denoted $n\mathbb{Z}$. Note that when $n = 1$ we get all of $\mathbb{Z}$, but when $n > 1$, we get a proper subgroup of $\mathbb{Z}$. Thus the cyclic group $(\mathbb{Z}, +)$ has infinitely many proper cyclic subgroups.

**Example 23.9.** Let's list all the cyclic subgroups of $\mathbb{Z}_8$ under addition modulo 8:

$\langle 0 \rangle = \{0\}$
$\langle 1 \rangle = \{1, 2, 3, 4, 5, 6, 7, 0\} = \mathbb{Z}_8$
$\langle 2 \rangle = \{2, 4, 6, 0\}$
$\langle 3 \rangle = \{3, 6, 1, 4, 7, 2, 5, 0\} = \mathbb{Z}_8$
$\langle 4 \rangle = \{4, 0\}$
$\langle 5 \rangle = \{5, 2, 7, 4, 1, 6, 3, 0\} = \mathbb{Z}_8$
$\langle 6 \rangle = \{6, 4, 2, 0\}$
$\langle 7 \rangle = \{7, 6, 5, 4, 3, 2, 1, 0\} = \mathbb{Z}_8$

We note that we get the $\phi(8) = 4$ generators of $\mathbb{Z}_8$ giving us (of course) the whole group, but we get one subgroup with 4 elements, one subgroup with 2 elements, and one subgroup with 1 element.

Do you think that $\mathbb{Z}_8$ has any *non-cyclic* subgroups as well? Let's settle that question right now, making use of the Well-Ordering Principle (Theorem 5.1) and the Division Algorithm (Theorem 5.2).

**Theorem 23.2.** *Every subgroup of a cyclic group is cyclic.*

Proof. Let $G$ be a cyclic group with generator $g$ and let $H$ be a subgroup of $G$. If $H = \{e\}$, then $H$ is clearly cyclic, so we may assume there is an element $h \in H$ such that $h \neq e$.

Since the group $G$ is cyclic, we have that $h = g^n$ for some positive integer $n$. By the Well-Ordering Principle, let $m$ be the *smallest* positive integer so that $g^m$ is in $H$. We claim that $H = \langle g^m \rangle$.

Let $k$ be an arbitrary element in $H$ which is, of course, also in $G$. Thus, $k$ can be written as $k = g^s$ for some positive integer $s$. We now apply the Division Algorithm to write $s = qm + r$ where $0 \le r < m$, giving us

$$k = g^s = g^{qm+r} = (g^m)^q g^r.$$

But now since $H$ is a subgroup, $g^m \in H$ implies that $(g^m)^q$ is in $H$, so its inverse $(g^m)^{-q}$ is also in $H$. Multiplying on the left by this inverse, we arrive at $g^r = (g^m)^{-q}(k)$, that is, $g^r$ is in $H$. Since $r < m$ and by the choice of $m$, it must be the case that $r = 0$. Hence

$$k = (g^m)^q g^0 = (g^m)^q,$$

and so $k$ is a power of $g^m$. Since $k$ was an arbitrary element of $H$, we see that $g^m$ is a generator of $H$, that is, $H$ is cyclic, and we are done. ∎

You might ask if the converse of Theorem 23.2 is true, that is, if a group $G$ has the property that all of its proper subgroups are cyclic, does that imply that $G$ is cyclic? You will be asked in Exercise 23.6 to either prove this result or find a counter-example.

It turns out that if the group $G$ is finite, then to verify that a given subset $H$ of $G$ is a subgroup of $G$, we need only check closure!

**Theorem 23.3.** *If $G$ is a finite group with operation $*$, then a non-empty subset $H$ is a subgroup of $G$ if and only if $H$ is closed; i.e., if and only if whenever $h, k \in H$, then $h * k \in H$.*

Proof. If $H$ is a subgroup of $G$, then by definition it is closed under $*$. The "meat" of this theorem is the opposite implication.

So suppose that the subset $H$ of $G$ is closed under the operation $*$. We need to show that the identity $e$ of $G$ is in $H$ and that if $h$ is in $H$, then $h^{-1}$ is also in $H$. So let $h$ be an arbitrary element of $H$ and consider the set $S = \{h, h^2, h^3, \ldots\}$. By closure this set $S$ lies inside $H$, but also since $H$ is finite, it must be the case that for some positive integers $i$ and $j$ with $i < j$, $h^j = h^i$. Working in $G$, we can multiply both sides by $(h^i)^{-1} = h^{-i}$, obtaining $h^{j-i} = e$. But $h^{j-i}$ is in $S$ and hence is in $H$, so $e$ is in $H$. Now, if $j = i + 1$, then $h = e$, so we can assume that $j > i + 1$. We have then $e = hh^{j-i-1} = h^{j-i-1}h$; i.e., $h^{j-i-1}$ is the inverse of $h$ and lies in $S$, hence in $H$, and we are done. ∎

A final question we ask in this chapter about subgroups is the following: Does every non-trivial group possess non-trivial proper subgroups? If not, which ones do not have this property? The next result settles this.

**Theorem 23.4.** *Every non-trivial group $(G, *)$ possesses non-trivial proper subgroups except for the class of finite cyclic groups with a prime number of elements.*

Proof. First, let $(G, *)$ be a non-cyclic group and let $a \neq e$ be an arbitrary element of $G$. Then $\langle a \rangle$ cannot be all of $G$ since $G$ is not cyclic.

Next, suppose $G$ is an infinite cyclic group generated by $g$, i.e.,

$$G = \langle g \rangle = \{\ldots, g^{-3}, g^{-2}, g^{-1}, e, g, g^2, g^3, \ldots\}.$$

Then, for any positive integer $n > 1$, the cyclic subgroup

$$H = \langle g^n \rangle = \{\ldots, g^{-3n}, g^{-2n}, g^{-n}, e, g^n, g^{2n}, g^{3n}, \ldots\}$$

is a proper subgroup of $G$,

So now we are down to finite cyclic groups. Suppose $(G, *)$ is such a group, is generated by $g$, and has $n > 1$ elements, so that $g^n$ must be $e$. Hence

$$G = \langle g \rangle = \{g, g^2, g^3, \ldots, g^{n-2}, g^{n-1}, e\}.$$

Suppose first that $n$ is composite and that $1 < d < n$ is a divisor of $n$. Then

$$H = \langle g^d \rangle = \{g^d, g^{2d}, \ldots, g^{n/d-1}, e\}$$

is a proper subgroup of $G$ with $n/d$ elements.

However, if $n = p$, a prime number, then we know that $G$ will possess $\phi(p) = p - 1$ generators, and so every non-trivial element of $G$ generates $G$, and we obtain no non-trivial proper subgroups. This completes the proof. ∎

Theorem 23.4 shows that the finite groups $\mathbb{Z}_p$, which play a major role in modern cryptography (as we saw in Chapters 18 and 20), do stand somewhat apart from all other groups because of the relative simplicity of their internal structure.

You may have noticed, say in Example 23.9 and elsewhere, that it appears that in finite groups, the number of elements in any subgroup seems to divide the number of elements in the group itself. In the following two chapters we shall explore whether or not this potentially important property holds up.

**Exercises**

**23.1.** (a) Within the additive cyclic group $(\mathbb{Z}, +)$, write down 5 or 6 elements which lie in the intersection of the subgroups $\langle 2 \rangle = 2\mathbb{Z}$ and $\langle 3 \rangle = 3\mathbb{Z}$. Is this a subgroup of $\mathbb{Z}$? If so, label it.
(b) Repeat Part (a) with the intersection of $4\mathbb{Z}$ and $6\mathbb{Z}$.
(c) Make a conjecture: If $n$ and $m$ are integers greater than 1, then inside $(\mathbb{Z}, +)$, $n\mathbb{Z} \cap m\mathbb{Z} = \cdots$.

**23.2.** (a) Generalizing the previous exercise, prove that the intersection of two subgroups $H$ and $K$ of any group $(G, *)$ is itself a subgroup of $G$.
(b) Generalize Part (a) for the intersection of any finite number of subgroups of $G$.

**23.3.** Either prove that the union of two subgroups of a group $(G, *)$ is always a subgroup of $G$, or find a counter-example.

**23.4.** (a) If $(G, *)$ is an Abelian group, show that the set $\{x | x = x^{-1}\}$ is a subgroup of $G$.
(b) Find a counter-example to the statement in Part (a) when $G$ is non-Abelian. (Hint: Look at the smallest non-Abelian group, $S_3$, the set of all permutations of three objects, under composition.)

**23.5.** Let $(G, *)$ be a group and fix $a \in G$. Consider the set

$$N_a = \{x | xa = ax\};$$

that is, the set $N_a$ is the set of all elements in $G$ which commute with the element $a$.
(a) If $G$ is the group $S_3$, as in the previous exercise, and $a = (123)$ (in cycle notation), what is the set $N_a$? Is $N_a$ a subgroup of $S_3$?
(b) Show that, in general, $N_a$ is a subgroup of any group $G$.

**23.6.** Concerning a possible converse of Theorem 23.3: If a group $(G, *)$ has the property that all of its proper subgroups are cyclic, does that imply that $G$ is cyclic? Either prove this or give a counter-example.

**23.7.** In Example 23.9 we wrote down all the subgroups of the cyclic group $\mathbb{Z}_8$ under addition modulo 8 and noted later that the number of elements in each of them divided the number of elements in the group. Try this experiment with the non-Abelian group $S_3$ under composition of permutations by writing down all of its subgroups using cycle notation, of which there are six (including the trivial subgroup and the whole group). Do the sizes all divide the size of $S_3$?

**23.8.** The ***center*** $Z(G)$ of a group $(G, *)$ is the set of all elements in $G$ which commute with *every* element of $G$. In symbols

$$Z(G) = \{g \in G | gh = hg \text{ for all } h \in G\}.$$

(a) For any group $G$, explain why $Z(G)$ is non-empty.
(b) What is $Z(G)$ if $G$ is Abelian?
(c) Show that the center of any group $G$ is always a subgroup of $G$.
(d) What is your guess as to $Z(GL(2, \mathbb{R}))$? (Reminder: $GL(2, \mathbb{R})$) is the group of all invertible $2 \times 2$ matrices with real number entries under matrix multiplication.)

**23.9.** In an infinite cyclic group, show that every subgroup except the trivial subgroup $\{e\}$ must be infinite.

**23.10.** Define the product $AB$ of two subsets $A$ and $B$ of a group $G$ by

$$AB = \{ab | a \in A, b \in B\}.$$

(a) Show that if $G$ is finite, then $HH = H$ if and only if $H$ is a subgroup of $G$.

(Note: In showing that $H$ a subgroup implies $H = HH$, recall that to show two sets are equal, it's generally easiest to show that each is a subset of the other. For the opposite implication, make use of Theorem 23.3.)

(b) Give an example to show that in infinite groups, it may not be the case that $H = HH$ implies that $H$ is a subgroup. (Note: If your example is inside an additive infinite group, remember that you are looking at the set equality $H = H + H$.)

# CHAPTER 24

# Groups - Cosets

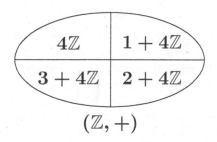

$$(\mathbb{Z}, +)$$

In the previous chapter we studied the internal structure of groups which is created by the existence of subgroups within a given group $(G, *)$. In this chapter we shall study a different but related internal structure created by what are called *cosets*. In this case we *fix* a given subgroup $H$ of a given group $G$ and form its cosets (denoted $H$, $a * H$, $b * H$, etc., see below) which have the following three properties:

(1) Every coset has the same cardinality as $H$,

(2) every coset has empty intersection with all the other cosets, and

(3) the union of all the cosets is the entire group $G$.

The cosets of $H$ in $G$ are said then to *partition* $G$ into sets all of the same cardinality. This is very powerful internal structure, and it exists in every group, finite or infinite, Abelian or non-Abelian. Here is the formal definition:

**Definition 24.1.** Let $H$ be a subgroup of a group $G$ with operation $*$ and let $a$ be an element of the group $G$ (which may or may not be an element of $H$).

We define the *left coset* $a * H$ of $a$ with respect to $H$ by

$$a * H = \{a * h | h \in H\}.$$

Similarly the *right coset* of $a$ with respect to $H$ is defined by

$$H * a = \{h * a | h \in H\}.$$

A few properties of cosets emerge right away. First, we note that $H$ is a coset of itself since $H = e * H$. Because $H$ contains the identity element $e$, we see that $a = a * e = e * a$ is an element of $a * H$ and $H * a$. The element $a$ is called a *representative* of these cosets, but it is not unique since *any* element of $a * H$ can represent that coset. This is because if $b \in a * H$, then $b = a * h$ for some $h \in H$, and so we get $b * H = a * (h * H) = a * H$. (Note: You will be asked to prove in Exercise 24.3 that if $h \in H$, then $h * H = H$.) Finally, we observe that if $a$ is not in $H$, i.e., if $a * H \neq H$, then the coset $a * H$ is *not* a subgroup since it does not contain the identity element. (If it did contain $e$, then $a * H = e * H = H$).

If $G$ is Abelian, then the left coset $a * H$ is the same as the right coset $H * a$, but if the group $G$ is non-Abelian they may turn out to be different sets (see Example 24.6 below). In general we will, for convenience, simply concentrate on left cosets; in fact, we will usually just talk about cosets with the understanding that they are left cosets. We also note that when working with an abstract group or a multiplicative example, we may use juxtaposition (e.g., $aH$), but when working with an additive example a coset will be denoted $a + H$.

Before proving that the properties (1) through (3) above hold for all groups, let's look at some examples.

**Example 24.1.** We start with the infinite cyclic group $(\mathbb{Z}, +)$. The even numbers $2\mathbb{Z}$ form an infinite cyclic subgroup of $\mathbb{Z}$. If we take any odd number, for example 1, then the coset $1 + 2\mathbb{Z} = \{\ldots, -5, -3, -1, 1, 3, 5, \ldots\}$ is the set of all odd integers and of course is a subset but *not* a subgroup of $\mathbb{Z}$. We note that both $2\mathbb{Z}$ and $1 + 2\mathbb{Z}$ are countably infinite, their intersection is empty, and their union is all of $\mathbb{Z}$.

**Example 24.2.** Let's consider $4\mathbb{Z}$, i.e., all the integer multiples of 4, which again forms a subgroup $(\mathbb{Z}, +)$. Now we will have three additional cosets, and as in Example 24.1, we choose to represent them by their "simplest" element. So those cosets are:

$1 + 4\mathbb{Z} = \{\ldots, -7, -3, 1, 5, 9, 13, \ldots\} = \{a \in \mathbb{Z} | a \equiv 1 \pmod 4\}$
$2 + 4\mathbb{Z} = \{\ldots, -6, -2, 2, 6, 10, 14, \ldots\} = \{a \in \mathbb{Z} | a \equiv 2 \pmod 4\}$
$3 + 4\mathbb{Z} = \{\ldots, -5, -1, 3, 7, 11, 15, \ldots\} = \{a \in \mathbb{Z} | a \equiv 3 \pmod 4\}.$

Again, these four sets all have the same cardinality, have no elements in common, and their union is all of $\mathbb{Z}$.

It should be clear now that inside $(\mathbb{Z}, +)$, if $n$ is any positive integer, then the subgroup $n\mathbb{Z}$ has $n$ distinct cosets (including itself). If $0 \le k < n$, then they are of the form $k + n\mathbb{Z} = \{a \in \mathbb{Z} | a \equiv k \pmod{n}\}$.

**Example 24.3.** Inside $\mathbb{R}^*$ (i.e., all non-negative real numbers) under multiplication, consider the subgroup $\mathbb{R}^+$ of all positive real numbers. It is clear that there is one additional coset, $(-1)\mathbb{R}^+$, the set of all negative real numbers, which is not itself a subgroup. Note once more that $\mathbb{R}^+$ and $(-1)\mathbb{R}^+$ have the same cardinality, do not overlap, and together form all of $\mathbb{R}^*$.

Let's look at some finite group examples now.

**Example 24.4.** Consider the subgroup $\langle 3 \rangle = \{0, 3, 6, 9\}$ of the finite cyclic group $\mathbb{Z}_{12}$ under addition modulo 12. Since 1 is not in $\langle 3 \rangle$, we can form the coset $1 + \langle 3 \rangle = \{1, 4, 7, 10\}$. Now since 2 has not yet appeared, we can form the coset $2 + \langle 3 \rangle = \{2, 5, 8, 11\}$. As expected, each coset has four elements, they do not overlap, and their union is all of $\mathbb{Z}_{12}$.

**Example 24.5.** The group $U_{20}$ is the set of all integers between 1 and 20 which are relatively prime to 20 under multiplication modulo 20. Since $\phi(20) = 8$, this group has 8 elements; specifically, $U_{20} = \{1, 3, 7, 9, 11, 13, 17, 19\}$. We note that $9^2 \equiv 1 \pmod{20}$, so $H = \{1, 9\}$ is a subgroup of $U_{20}$. What are its cosets? Well, 3 is not in $H$, so a first coset (besides $H$ itself) will be $3H = \{(3)(1), (3)(9)\} = \{3, 7\}$. Now 11 has not yet appeared, so a new coset is $11H = \{(11)(1), (11)(9)\} = \{11, 19\}$. Finally, 13 has not appeared, so we form $13H = \{(13)(1), (13)(9)\} = \{13, 17\}$. This gives us a complete set of cosets, each with two elements, none overlapping, and together making up all of $U_{20}$.

You have probably noticed the following technique we've used in these examples: To form a new coset, find an element of the group which has not yet appeared in a previous coset and let it represent the new coset. Let's do one more example, this time with a non-Abelian finite group.

**Example 24.6.** Consider the subgroup $H = \{(1), (12)\}$ in the group $S_3$ of permutations of three objects under composition (as usual, we are using cycle notation for our permutations). Forming left cosets, since $(13)$ is not in $H$, we form $(13)H = \{(13)(1), (13)(12)\} = \{(13), (123)\}$. Since $(23)$ has not yet appeared, we form another left coset $(23)H = \{(23)(1), (23)(12)\} = \{(23), (132)\}$, and we are done forming cosets.

We remark that since this is our first example of working in a non-Abelian group, it may be that left cosets and right cosets do not coincide. In fact this happens here, since the right coset $H(13) = \{(1)(13), (12)(13)\} = \{(13), (132)\}$ is not the same as the left coset $(13)H$ above. It turns out that inside a non-Abelian group $G$, some subgroups $H$ may satisfy that for all $a \in G$, $aH = Ha$, and some subgroups may not satisfy this condition. Subgroups which *do* satisfy this condition are called *normal* subgroups, and they play a very important role in group theory, though we will not pursue this idea further in this text (except in Exercise 24.8).

Let us now summarize what we have observed in each of these examples and verify that it all holds true for *any* group $(G, *)$. Again, this shows the power of abstraction!

**Theorem 24.1.** *Let $(G, *)$ be a group and let $H$ be a subgroup of $G$. Then*
(i) *The subgroup $H$ is a coset of itself.*
(ii) *If $b \in a * H$, then the coset $b * H = a * H$; that is, any element of a coset can be used to represent that coset.*
(iii) *Unless $a \in H$, the coset $a * H$ is not a subgroup of $G$.*
(iv) *Any two cosets of $H$ have the same cardinality; in particular, if $G$ is finite, they all (including $H$) have the same number of elements.*
(v) *If $a, b \in G$, and $a * H \neq b * H$, then $(a * H) \cap (b * H) = \emptyset$;*
(vi) *The union of all of $H$'s cosets (including $H$ itself) is the entire group $G$.*

Proof. Properties (i), (ii), and (iii) were discussed and proved in general in the paragraph immediately following Definition 24.1 (but see also Exercise 24.3). We now prove the latter three properties, which were first stated in the opening paragraph of this chapter. Throughout this proof, we will, for the sake of simplicity, denote $h * k$ by $hk$ and $a * H$ by $aH$.

For Property (iv), we show that every coset $aH$ has the same cardinality as the subgroup $H$, from which it follows that all cosets of $H$ have the same cardinality (by the transitive property of cardinality equality). Let $f : H \to aH$ be defined by $f(h) = ah$. In Exercise 24.4 you are asked to show that $f$ is a bijection, so by definition the sets $H$ and $aH$ have the same cardinality (see Chapter 9 if you need some reminders). Finally, for finite sets, "same cardinality" means "same number of elements."

For Property (v), we do a contrapositive proof (as introduced in Chapter 3). Suppose there exists an $x \in G$ such that $x \in aH \cap bH$. Then for some $h_1$ and $h_2$ in $H$, we have $ah_1 = x = bh_2$, and multiplying both ends on the right by $h_1^{-1}$ (which is in $H$ since $H$ is a subgroup!), we get $a = b(h_2 h_1^{-1})$, i.e., we get $a \in bH$ (since $H$ is closed!). Then by Property (ii), $aH = bH$, and we have established Property (v).

Finally, for Property (vi), we know that any $a \in G$ is in the coset $aH$, so the union of all the cosets (including $H$ itself) must be all of $G$. ∎

Let's look at one more finite example.

**Example 24.7.** The group $S_5$ of all permutations of 5 objects under composition in a non-Abelian group with $5! = 120$ elements (the $S_n$ groups get very large very quickly as $n$ goes up). Let $H = \langle (12345) \rangle$ be the cyclic subgroup of $S_5$ generated by $(12345)$; then by computing powers of the generator, we get

$$
\begin{aligned}
H &= \{(12345), (12345)^2, (12345)^3, (12345)^4, (12345)^5\} \\
&= \{(12345), (13524), (14253), (15432), (1)\}.
\end{aligned}
$$

(Check this calculation.) We know now that $H$ must have $120/5 = 24$ distinct cosets (including $H$ itself) to fill out all of $S_5$. We shall refrain from writing them all down (!), but in Exercise 24.7 we ask you, just for practice in using cycle notation, to compute just one of them.

Theorem 24.1 is important across all of group theory, but it has special importance in understanding the structure of *finite* groups, with applications well outside of abstract algebra. We shall finish our relatively brief study of groups by examining these ideas in the following chapter.

**Exercises**

**24.1.** Inside the group $\mathbb{Z}_{15}$ under addition modulo 15, let $\langle 5 \rangle$ be the cyclic subgroup generated by 5. Name and then write out all the cosets of $\langle 5 \rangle$ in $\mathbb{Z}_{15}$, including $\langle 5 \rangle$ itself, as was done in Example 24.4.

**24.2.** Inside the group $U_{21}$ of integers in the range $\{1, \ldots, 21\}$ which are relatively prime to 21 under multiplication modulo 21, let $H$ be the cyclic subgroup generated by 4. Compute $H$ and all its other cosets.

**24.3.** Suppose $H$ is a subgroup of a group $(G, *)$ and $h \in H$. Prove that $h * H = H$. (Note: To prove $h * H \subseteq H$ you only need closure of $H$, but to prove $H \subseteq h * H$, you should actually make use of all four properties which groups (and hence subgroup) possess!)

**24.4.** Concerning the proof of Property (iv) in Theorem 24.1, suppose that $H$ is a subgroup of the group $G$ and that $a \in G$. Define the function $f : H \to aH$ by $f(x) = ax$. Prove that $f$ is a bijection.

**24.5.** The group $\mathbb{R}$ of real numbers is a subgroup of the group $\mathbb{C}$ of complex numbers under addition.
(a) List three distinct cosets (in the form $x + \mathbb{R}$) of $\mathbb{R}$ in $\mathbb{C}$. (Recall that to form a new coset, take an element of the group which has not yet appeared in a coset and let it represent the new coset.)
(b) Picturing the complex numbers (as usual) as the plane with horizontal real axis and imaginary vertical axis, describe geometrically the cosets of $\mathbb{R}$ in $\mathbb{C}$.

**24.6.** Let $H$ be the cyclic subgroup generated by $(1, 1)$ in the Cartesian product group $\mathbb{Z} \times \mathbb{Z}$ under component-wise addition.
(a) Write out five or so elements of $H$. Thinking geometrically, the set $\mathbb{Z} \times \mathbb{Z}$ is all integer points in the plane (i.e., lots of dots). Describe $H$ geometrically.
(b) Write out five or so elements of the coset $(1, 0) + H$. Describe this coset geometrically. More generally, describe any coset $(a, b) + H$ where $a \neq b$.

**24.7.** Concerning Example 24.7, compute all the permutations in the left coset $(12)\langle (12345) \rangle$ within the group $S_5$.

**24.8.** (a) Within the non-Abelian finite group $S_3$ of permutations of three objects under composition, name and then write out all the left cosets of the permutation $\langle (123) \rangle$ in $S_3$, including $\langle (123) \rangle$ itself, as was done in Example 24.6.

(b) Repeat Part (a) with right cosets.

(c) Explain why whenever $(G, *)$ is a finite group with an even number $n$ of elements and $H$ is a subgroup with $n/2$ elements, then for every $a \in G$, the left coset $a * H$ equals the right coset $H * a$.

**24.9.** In Example 23.7 we introduced the subgroup $SL(2, \mathbb{R})$ of the group $GL(2, \mathbb{R})$ of all invertible $2 \times 2$ matrices over $\mathbb{R}$ under matrix multiplication. Specifically, $SL(2, \mathbb{R})$ consists of all those matrices whose determinant has the value 1. Suppose the matrix $A = \begin{pmatrix} 1 & -4 \\ -2 & 3 \end{pmatrix}$. What do all the matrices in the left coset $A(SL(2, \mathbb{R}))$ have in common?

**24.10.** We know that any coset which is not the given subgroup cannot itself be a subgroup since it does not contain the identity element $e$. However, what about other properties of subgroups?

(a) Find an example of a coset $a * H$ (with $a \notin H$) of a subgroup $H$ inside a group $(G, *)$ which has the property that if $x \in a * H$, then $x^{-1} \in a * H$ also (i.e., $a * H$ contains all its inverses).

(b) Prove that if the full set of $H$'s cosets is $\{H, a * H\}$, then $a * H$ has the property described in Part (a).

(c) Prove that if $a * H$ has the property described in Part (a), then $a * H$ cannot be closed under $*$.

---

# CHAPTER 25

# Groups - Lagrange's Theorem

---

$$|G| < \infty, H \leq G \Rightarrow |H|\,\big|\,|G|$$

The uniform structure inside a group $(G, *)$ provided by the set $\{a * H\}$ of cosets of any of its subgroups $H$ bears particular fruit when $G$ is finite, either Abelian or non-Abelian. We now arrive at a fundamentally important result in finite group theory, made possible by this uniformity and known as *Lagrange's Theorem* in honor of its discoverer Joseph L. Lagrange (1736-1813). First, however, we need some terminology.

**Definition 25.1.** The **order** $|G|$ of a finite group $(G, *)$ is the number of elements in $G$. If $a \in G$, the **order** $|a|$ of $a$ is the order of the cyclic group $\langle a \rangle$ generated by $a$, and so $|a|$ is the smallest positive integer such that $a^{|a|} = e$.

**Example 25.1.** $|\mathbb{Z}_n| = n$. $|U_n| = \phi(n)$. $|S_n| = n!$. $|\mathbb{Z}_n \times \mathbb{Z}_m| = nm$. For $3 \in \mathbb{Z}_{12}$, $|3| = 4$. For $3 \in U_8$, $|3| = 2$. For $(123) \in S_n(n \geq 3)$, $|(123)| = 3$. For $(1,1) \in \mathbb{Z}_4 \times \mathbb{Z}_6$, $|(1,1)| = 12$. Make sure you "agree" with all of these assertions.

We can now state and prove our main result, which follows easily from our work on cosets in the previous chapter.

**Theorem 25.1. (Lagrange's Theorem)** *Let $(G, *)$ be a finite group and let $H$ be a subgroup of $G$. Then the order $|H|$ of $H$ divides the order $|G|$ of $G$. More specifically, if the subgroup $H$ has $m$ distinct cosets (including itself) in $G$, then*

$$|G| = m|H|.$$

Proof. By Parts (iv), (v) and (vi) of Theorem 24.1, the $m$ cosets of $H$ partition the group $G$ into pair-wise disjoint subsets, all of which have the same number of elements $|H|$ as $H$. Then, by Part (vi) of Theorem 24.1,

$$G = a_1 H \cup \cdots \cup a_m H$$

for some elements $a_1, a_2, \ldots, a_m \in G$, it follows that

$$|G| = \underbrace{|H| + |H| + \cdots + |H|}_{m \text{ terms}},$$

that is,

$$|G| = m|H|,$$

which completes the proof. ∎

**Example 25.2.** If $G$ is a group with $|G| = 60$ and if $H$ is a proper subgroup of $G$, then by Lagrange's Theorem the order $|H|$ of $H$ must be in the set $\{1, 2, 3, 4, 5, 6, 10, 12, 15, 20, 30\}$. We note that Lagrange's Theorem does *not* say that there will be subgroups of each of these orders in $G$; it says only that these are the sole possibilities for $|H|$. We shall explore this idea later in this chapter.

**Example 25.3.** Inside the additive cyclic group $\mathbb{Z}_{12}$, $|\langle 2 \rangle| = 6$, $|\langle 3 \rangle| = 4$, $|\langle 4 \rangle| = 3$, $|\langle 6 \rangle| = 2$, and $|\langle 0 \rangle| = 1$. Hence in *this* case every proper divisor of 12 is in fact the order of at least one proper subgroup, and we know from Lagrange's Theorem that no other orders are possible.

We now state some results which follow from Lagrange's Theorem. This first one is a simple observation about finite groups.

**Corollary 25.2.** *If $G$ is a group with a prime number $p$ of elements, then $G$ must be cyclic.*

Proof. Suppose $g \neq e$ is an element of $G$, so $|g| > 1$. By Lagrange's Theorem, $|g|$ divides $|G|$, so we must have $|g| = p$, i.e., $G$ is generated by $g$ and hence is cyclic. ∎

We remark that, as in Theorem 23.4, this implies that $G$ has no non-trivial proper subgroups.

We now demonstrate the power of group theory in general and Lagrange's Theorem in particular by giving new, surprisingly short proofs of the two fundamentally important theorems from number theory which we studied in Chapters 17 and 19. In those chapters our proofs employed only number theory concepts; here we employ some number theory concepts, but the primary tool (might we say "the punch line"?) is Lagrange's Theorem.

**Corollary 25.3. (Fermat's Theorem)** *If $p$ is a prime and $p$ does not divide an integer $a$, then $a^{p-1} \equiv 1 \pmod{p}$.*

Proof. Since the integer $a$ is not divisible by the prime $p$, the least non-negative residue $r$ of $a$ modulo $p$ is non-zero and $r$ lies in the multiplicative group $\mathbb{Z}_p^*$, whose order is $p - 1$. By Lagrange's Theorem, $|r| = |\langle r \rangle|$ divides $p - 1$, so

$$r^{p-1} = (r^{|r|})^k \equiv 1^k = 1 \pmod{p} \text{ for some } k \in \mathbb{Z}^+.$$

Finally, by Exercise 15.9, $a \equiv r \pmod{p}$ implies that $a^{p-1} \equiv r^{p-1} \pmod{p}$, so we arrive at

$$a^{p-1} \equiv 1 \pmod{p},$$

and we are done. ∎

**Corollary 25.4. (Euler's Theorem)** *If $a$ and $n$ are relatively prime with $n \geq 2$, then $a^{\phi(n)} \equiv 1 \pmod{n}$.*

Proof. As was true in Chapter 19, the proof runs pretty much parallel to the proof just given of Fermat's Theorem, but now Lagrange's Theorem is applied to the group $U_n$, whose order is $\phi(n)$. See Exercise 25.10.

We now turn to the issue which was raised above in Example 25.2, namely, to what extent is the *converse* of Lagrange's Theorem true? That converse, if true, would answer "yes" to the following question : Given a divisor $k$ of the order $|G|$ of a finite group $G$, does $G$ possess a subgroup $H$ of order $k$? We saw in Example 25.3 that for the additive cyclic group $\mathbb{Z}_{12}$, the answer is indeed yes. Hence one might ask if the answer is yes for *all* finite cyclic groups.

**Theorem 25.5.** *If $(G, *)$ is a finite cyclic group of order $|G|$ and if a positive integer $k$ divides $|G|$, then $G$ has a subgroup $H$ of order $k$.*

Proof. Suppose that $G$, being cyclic, is generated by $g \in G$. Since $k$ divides $|G|$, we claim that the set

$$H = \{g^{|G|/k}, g^{2|G|/k}, \ldots, g^{(k-1)|G|/k}, e\}$$

is a subgroup of $G$ which is clearly of order $k$. By Theorem 23.3 we need only check that $H$ is closed under $*$. But for $1 \leq i, j < k$,

$$g^{i|G|/k} * g^{j|G|/k} = g^{(i+j)|G|/k}.$$

If $i + j \leq k$ we have closure. Otherwise write $i + j = r + k$ where now $1 \leq r < k$, and we get

$$g^{(i+j)|G|/k} = g^{(r+k)|G|/k} = g^{r|G|/k} * e = g^{r|G|/k},$$

and again we have closure. Hence $H$ is a subgroup of $G$ of order k, and we are done. ∎

We remark that for any finite cyclic group, each of its subgroups is actually unique (i.e., for each $k$ dividing $|G|$, there is one and *only one* subgroup of that order). We ask you to prove this fact in Exercise 25.5. However, among *non-cyclic* groups this uniqueness is definitely not the case. For a simple example, inside the group $U_8 = \{1, 3, 5, 7\}$ under multiplication modulo 8, the three subgroups $\langle 3 \rangle$, $\langle 5 \rangle$ and $\langle 7 \rangle$ are all of order 2. (Check this.)

Okay, so since finite cyclic groups satisfy our hoped-for converse of Lagrange's Theorem, what about the much broader class of finite Abelian groups? Well, though we cannot prove it here, it turns out that *all finite Abelian groups have the exact same structure as some finite Cartesian product of cyclic groups.* (This is known as the Fundamental Theorem of Finite Abelian Groups.) Again using $U_8$ as a simple example, it (and all Klein 4-groups) have the same structure as $\mathbb{Z}_2 \times \mathbb{Z}_2$. So let us look at the Cartesian product of cyclic groups in the following lemma. (Recall that a lemma is a "helping theorem.")

**Lemma 25.6.** *Suppose the positive integer $k$ divides the order $nm$ of the group $G = \mathbb{Z}_n \times \mathbb{Z}_m$ under component-wise modular addition. Then $G$ contains a subgroup $H$ of order $k$.*

Proof. Since $k$ divides $nm$, we write $k = k_n k_m$ where $k_n$ divides $n$ and $k_m$ divides $m$. Then by Theorem 25.5 we know $\mathbb{Z}_n$ has a subgroup $H_n$ of order $k_n$ and $\mathbb{Z}_m$ has a subgroup $H_m$ of order $k_m$, so $H = H_n \times H_m$ is a subgroup of $G$ of order $k_n k_m = k$, as desired. $\blacksquare$

This lemma can of course be extended to the Cartesian product of any finite number of cyclic groups. Using this and the (for us unproven) Fundamental Theorem of Finite Abelian Groups, we arrive at the following theorem:

**Theorem 25.7.** *If $(G, *)$ is a finite Abelian group of order $|G|$ and if a positive integer $k$ divides $|G|$, then $G$ has a subgroup $H$ of order $k$.*

This tells us then that if we wish to discover a counter-example to our possible converse of Lagrange's Theorem, we need to look in the class of finite *non-Abelian* groups. Our usual focus in that class has been the groups $S_n$ of all permutations of $n$ objects under composition. It turns out that there are counter-examples lurking there, but to discover them we need to go into a bit more detail (but not too much we hope) about these permutation groups.

A permutation in $S_n$ is called **even** if it can be written as a product of an even number of (not necessarily disjoint) 2-cycles. It turns out that the set of all even permutations in $S_n$ forms a subgroup, denoted $A_n$, of order $n!/2$ (i.e., half of the permutations in $S_n$ are even and half are odd, where "odd" has the expected definition). We note that $A_n$ is closed since the composition of two even permutations will again be even, so by Theorem 23.3 $A_n$ is indeed a subgroup. Note that in fact the odd permutations are the unique other coset of $A_n$ in $S_n$. Let us look at an example.

**Example 25.4.** In the group $S_4$, of order 24, check that the following permutations are even:
   (i) The identity (1), since $(1) = (12)(12)$,
   (ii) (123), since $(123) = (13)(12)$, and
   (iii) (12)(34).

How many even permutations are there in $S_4$? Well, there are eight 3-cycles, each listed next to its inverse,

$$\{(123),(132),(124),(142),(134),(143),(234),(243)\};$$

there are 3 products of two disjoint 2-cycles, each of which is its own inverse,

$$\{(12)(34),(13)(24),(14)(23)\};$$

and the identity (1), for a total of 12 permutations, as expected.

It is worth noting that the terminology "even" and "odd" for permutations can be confusing since they are referring to the number of 2-cycles which can represent them, *not* to the *order* of the permutation. For example, a 3-cycle such as (123) is even, as we saw above, but has *odd order*, specifically $|(123)| = 3$. On the other hand, the even permutation (12)(34) has even order, specifically 2. So try to keep straight the order of a permutation as opposed to it being "even" or "odd."

We are now in a position to find the counter-example to the converse of Lagrange's Theorem we seek; that is, a finite group $G$ of order $|G|$ and a divisor $k$ of $|G|$ such that $G$ has no subgroup of order $k$. We note that if we had gone somewhat deeper into group theory and learned about what are called "factor groups," we could give a very short proof of the following result. As it is, our proof will be a bit messier but not hard. The crux of the matter will be that $A_4$ contains too many 3-cycles (8 of them) to allow for the existence of a certain subgroup.

**Theorem 25.8.** *The group $A_4$, of order 12, has no subgroup of order 6.*

Proof. Just suppose $H$, of order 6, were a subgroup of $A_4$. Since $H$ must contain the identity (1), we need 5 more elements. Since $H$ must contain inverses of its elements, $H$ can contain either two 3-cycles or four 3-cycles. We examine each of these cases:

Case 1: If $H$ contains two 3-cycles, say, without loss of generality, (123) and (132), to fill $H$ out we need to include all three pairs of 2-cycles. But now

$$(12)(34)(123) = (1)(243) = (243)$$

so $H$ is not closed.

Case 2: If $H$ contains four 3-cycles, starting, say, with (123) and (132), we need one pair of 2-cycles, say (12)(34). Now since $(12)(34)(123) = (243)$ as above, to maintain closure we must pick as our other pair of 3-cycles (234) and (243). But now, for example

$$(12)(34)(234) = (124)(3) = (124)$$

and again $H$ fails to be closed. Hence $H$ cannot be a subgroup, and we are done. ∎

Thus we conclude that the converse of Lagrange's Theorem is not true in general (but is true for finite Abelian groups). Theorem 25.8 is a special case of the following, whose proof requires more advanced techniques, and hence we state without proof.

**Theorem 25.9.** *For $n \geq 4$, the group $A_n$, of order $n!/2$, has no subgroup of order $n!/4$.*

**Example 25.5.** The subgroup $A_5$ of $S_5$ contains 60 even permutations: twenty-four 5-cycles (e.g., (12345)), twenty 3-cycles (e.g., (123)), fifteen pairs of disjoint 2-cycles (e.g., (12)(34)), and the identity (1). A similar but significantly more involved proof like the one we gave of Theorem 25.8 (or a much shorter one using "factor groups") shows that $A_5$ has no subgroup of order 30.

We come then to the end of our exploration of groups, a type of algebraic object which has a single operation. There is much more to be learned about them, and that can be done in any course in Abstract Algebra (or on your own!). Next we shall explore *rings*, which are algebraic objects with *two* binary operations which are related via a "distributive law." As you may suspect, a number of the groups we have studied (e.g., $\mathbb{Z}$, $\mathbb{Q}$, $\mathbb{R}$, $\mathbb{C}$, $\mathbb{Z}_n$, $n \times n$ matrices over $\mathbb{R}$, etc.) also form rings.

## Exercises

**25.1.** Suppose the group $G$ has 90 elements. According to Lagrange's Theorem, what are the possible orders of all the subgroups of $G$? (Hint: There are 12 such orders.)

**25.2.** According to Lagrange's Theorem, what are the possible orders of all the subgroups of the multiplicative group $U_{60}$? (Hint: There are 5 such orders.)

**25.3.** (a) According to Lagrange's Theorem, what are the possible orders of all the subgroups of the additive cyclic group $\mathbb{Z}_{28}$?
(b) Write out the elements of one subgroup of $\mathbb{Z}_{28}$ of each of the orders in Part (a).

**25.4.** (a) According to Lagrange's Theorem, what are the possible orders of all the subgroups of the multiplicative group $U_{20}$?
(b) Write out the elements of one subgroup of $U_{20}$ of each of the orders in Part (a).

**25.5.** Suppose $(G, *)$ is a cyclic group generated by $g$. As in the proof of Theorem 25.5, if $k$ divides $|G|$, then

$$H = \{g^{|G|/k}, g^{2|G|/k}, \ldots, g^{(k-1)|G|/k}, e\}$$

is a subgroup of $G$ of order $k$. Prove that $H$ is unique; i.e., $H$ is the sole subgroup of order $k$ in $G$. (Hint: Suppose $H_1$ is another such subgroup; then by Theorem 23.2 $H_1$ is cyclic, say generated by $h$. Since $|G|$ is cyclic, $h = g^r$ for some $1 \leq r < |G|$. Show that $r$ must be a multiple of $|G|/k$, so $h \in H$. Conclude that $H_1 = H$.)

**25.6.** (a) Following up on Exercise 25.2 above, prove that $U_{20}$ is not cyclic by checking the orders of each of its elements.
(b) Since $U_{20}$ is not cyclic, it may possess multiple subgroups of a given order. In fact $U_{20}$ has three different subgroups of order 2. Find them.
(c) $U_{20}$ has three different subgroups of order 4. Find them. (Hint: One of the three subgroups is not cyclic. What is its structure called?)

**25.7.** Consider the group $S_3$ of all permutations of 3 objects under composition.
(a) What are the possible orders of non-trivial proper subgroups of $S_3$?
(b) Write out the elements of each of these subgroups. (Hint: There are 3 subgroups of one of the orders and only 1 of the other.)

**25.8.** (a) Inside $S_4$, explain why the left coset $(12)A_4$ of the subgroup $A_4$ of all even permutations must consist entirely of odd permutations.
(b) Compute two sample elements of the coset $(12)A_4$ in their simplest form: (i) $(12)(123)$ and (ii) $(12)((12)(34))$.
(c) Write out, in their simplest forms, all 12 permutations in the coset $(12)A_4$, i.e., all odd permutations in $S_4$.

**25.9.** (a) In Example 25.5 we implied that inside $S_5$ the 5-cycle $(12345)$ is an even permutation; that is, it can be written in the form of an even number of (not necessarily disjoint) 2-cycles. Show that this is true of the 5-cycle $(12345)$.
(b) We have noticed that even permutations can have even or odd order. Show that in fact $(12345)$ has odd order by writing out all the elements of $\langle (12345) \rangle$.

**25.10.** Using the "group theory proof" of Fermat's Theorem (Corollary 25.3) as a guide, prove Euler's Theorem (Corollary 25.4).

# CHAPTER 26

# Rings

$$a, b, c \in R \Rightarrow a(b + c) = ab + ac$$

As we continue our introduction to some key ideas in abstract algebra, we now move from focusing on groups, which possess a single binary operation, to classes of sets whose algebraic structure involves *two* binary operations. Nonetheless, we continue to make use of the ideas we encountered in our study of groups. The first class we now study is that of objects called *rings*.

The two operations of a ring $R$ are thought of as addition and multiplication. Under addition, the non-empty set $R$ must form an Abelian group, but this is not necessarily true of multiplication in $R$. We do require that under multiplication $R$ is closed and that multiplication is associative; we do *not* require that there be an identity for multiplication, nor that multiplication be commutative. Finally though, we *do* require that multiplication and addition be "compatible," in a sense which will look very familiar to you, namely the requirement of a distributive law.

Just as we did with a group $(G, *)$, we will often take the liberty of simply writing $ab$ in place of $a \cdot b$ to indicate the product of the elements $a$ and $b$.

We now give a formal definition of a ring.

**Definition 26.1.** A non-empty set $R$ is a ***ring*** if there are two operations, addition and multiplication, for which the following properties hold:

R1: for all $a, b \in R, a + b \in R$ (addition is closed);

R2: for all $a, b, c \in R, a + (b + c) = (a + b) + c$ (addition is associative);

R3: there is an element $0 \in R$ so that, for all $a \in R$, $a + 0 = 0 + a = a$ (identity for addition);

R4: for each $a \in R$ there is an element $-a \in R$ so that $a + (-a) = -a + a = 0$ (inverses for addition);

R5: for all $a, b \in R$, $a + b = b + a$ (addition is commutative);

R6: for all $a, b \in R$, $ab \in R$ (multiplication is closed);

R7: for all $a, b, c \in R, a(bc) = (ab)c$ (multiplication is associative);

R8: for all $a, b, c \in R, a(b+c) = ab+ac$, and $(a+b)c = ac+bc$ (multiplication distributes over addition).

You should not be intimidated by all of these conditions for a set to be a ring. First of all, Properties R1, R2, R3, R4, and R5 simply say that the set $R$ under addition forms an Abelian group. Next, Properties R6 and R7 tell us that $R$ is closed under multiplication and that operation is associative, *but* under multiplication there may be no identity element, some or all of the elements of $R$ may not have multiplicative inverses, and the operation may not be commutative. Finally, Property R8 tells us how addition and multiplication are connected to each other.

If the multiplicative operation in the ring is commutative, i.e., if $ab = ba$ for all $a, b \in R$, then $R$ is said to be a ***commutative*** ring. Further, if the ring $R$ contains an element 1 so that, for all $a \in R$, we have $a(1) = (1)a = a$ (i.e., is a multiplicative identity), then the element 1 is called a ***unity*** element for the ring $R$ and $R$ is said to be a ***ring with unity***. (You were asked to prove in Exercise 12.6 that if a set with a binary operation has an identity element, that element is unique, so we shall assume this as we move forward.) We note here that if $R$ has *no* unity element, then by definition no elements of $R$ can possess multiplication inverses. On the other hand, if $R$ does have a unity element, then any element $a \in R$ may or may not possess a multiplicative inverse.

It's time for some examples of rings. As you will quickly notice, most of these examples involve sets and operations with which we are already quite familiar.

**Example 26.1.** The quintessential example of a ring, and arguably the motivation for the required properties of an abstract ring, is the set $\mathbb{Z}$ of integers with the usual operations of addition and multiplication. We've known for quite a while that under addition $\mathbb{Z}$ is an Abelian group. We also know that $\mathbb{Z}$ is closed under multiplication and that multiplication is associative. Lastly, as you know from your algebra and calculus days, the distributivity property holds for all real numbers and hence for the integers. Thus Properties R1-R8 all hold. Concerning other (not required) properties, multiplication of integers is commutative, the number 1 is of course the multiplicative identity, but only two elements (which two?) of $\mathbb{Z}$ possess multiplicative inverses. Hence $\mathbb{Z}$ is a commutative ring with unity.

**Example 26.2.** We have seen that the set $\mathbb{Q}$ of all rational numbers, the set $\mathbb{R}$ of all real numbers, and the set $\mathbb{C}$ of all complex numbers satisfy all the properties to qualify as rings under addition and multiplication. Each set has 0 as the additive identity and 1 as the multiplicative identity, and in each case multiplication is commutative, so again these are examples of commutative rings with unity. However, a definite difference between these three rings and $\mathbb{Z}$ above is that in $\mathbb{Z}$ only two elements have multiplicative inverses, but in $\mathbb{Q}$, $\mathbb{R}$ and $\mathbb{C}$ *every element except 0* has a multiplicative inverse. There is a special and very important category for such rings; they are called *fields*, and we shall study them in Chapter 29.

**Example 26.3.** As we studied in Chapter 12, the set of all $n \times n$ matrices with real entries under the usual operations of matrix addition and multiplication forms a ring. The additive identity is the zero matrix and the unity element is the identity matrix (i.e., all ones on the main diagonal and all zeros elsewhere). The elements with multiplicative inverses are the invertible matrices (i.e, matrices whose determinants are non-zero). Note finally that since in general for matrices $AB \neq BA$, this is our first example of a *non-commutative* ring.

**Example 26.4.** Let $R$ be a ring and let $R[x]$ denote the set of all polynomials in a single variable $x$ whose coefficients are in the ring $R$. Define the sum and product of two polynomials in $R[x]$ via the usual way of adding and multiplying polynomials, with the coefficients being computed using the ring operations in $R$. With these definitions, the set $R[x]$ forms a ring. In Exercise 26.8, we ask you to verify that it is a ring, to decide if this is a ring with unity or not and if so to identity the invertible elements, and finally to decide if this is a commutative ring or not.

**Example 26.5.** We have not seen here as yet an example of a *finite* ring, but we are in fact quite familiar with a whole class of them, the sets $\mathbb{Z}_n$ ($n \geq 2$) under addition and multiplication modulo $n$. All the required ring properties are clearly satisfied, including distributivity, because of the compatibility of both integer addition and multiplication with reduction modulo $n$ (by Lemma 15.1). These rings are clearly commutative and contain a unity element. Concerning multiplicative inverses, we know that $a \in \mathbb{Z}_n$ is invertible if and only if $a$ and $n$ are relatively prime (we have called this set of invertible elements $U_n$). It follows that $\mathbb{Z}_n$ is a field (i.e., every element except 0 is invertible), as $\mathbb{Q}$, $\mathbb{R}$ and $\mathbb{C}$ are, if and only if $n = p$, a prime number.

**Example 26.6.** How about a finite non-commutative ring? Here we can combine Examples 26.3 and 26.5 to get the set of all $k \times k$ matrices over $\mathbb{Z}_n$. Since there are $n$ choices of what to put into each of $k^2$ entries, the order of these rings is $n^{k^2}$. We ask you in Exercise 26.13 to examine somewhat closely the case of $2 \times 2$ matrices over the (very small) ring $Z_2$.

**Example 26.7.** Finally, are there rings *without* unity? We ask you to verify in Exercise 26.3 that the sets $n\mathbb{Z} = \{\ldots, -3n, -2n, n, 0, n, 2n, 3n, \ldots\}$ for $n \geq 2$ are such rings.

Returning to an abstract ring $R$, we now state two facts which will probably strike you as being of the "well, of course" variety, but that's because you have "always" known them in the cases of our standard rings of numbers ($\mathbb{Z}$, $\mathbb{R}$, etc.). That a property holds in some examples of a ring does *not* imply that it will hold in *every* ring. This is, again, an advantage of abstraction; you can do one proof which applies to many, many cases.

**Lemma 26.1.** *In the ring $R$*
  (a) $0 \cdot a = a \cdot 0 = 0$, *and*
  (b) $(-1) \cdot a = a \cdot (-1) = -a$.

Proof of (a). We prove that $0 \cdot a = 0$, the proof of $a \cdot 0 = 0$ being essentially the same. Using the definition of 0 and the distributive property, we get

$$0 \cdot a = (0 + 0) \cdot a = 0 \cdot a + 0 \cdot a.$$

Now adding the additive inverse $-(0 \cdot a)$ of $0 \cdot a$ to both ends, making use of the definitions of 0 and of additive inverses, *and* using associativity of addition, we get

$$0 = -(0 \cdot a) + (0 \cdot a) = (-(0 \cdot a) + 0 \cdot a) + 0 \cdot a = 0 + 0 \cdot a = 0 \cdot a,$$

as was to be proved. ∎

We leave the proof of Lemma 26.1 Part (b) to you in Exercise 26.4. We note that the proof given here for Part (a) makes use of six of the eight required properties of rings (including closure of both operations, e.g., that $0 \cdot a$ is in $R$); only R5 (commutativity of addition) and R7 (associativity of multiplication) are unused.

In our study of groups, it was required that every element possess an inverse with respect to the operation. In cases where we started with a set and an operation in which not all elements had inverses, we basically threw those non-invertible elements out. For example, with $\mathbb{R}$ under multiplication, we threw out 0 and used the set $\mathbb{R}^*$, which did then form a group. For another example, in the finite set $\mathbb{Z}_n$ under multiplication modulo $n$, we tossed the elements which are not relatively prime to $n$, arriving at the set $U_n$, which is then a group. For a last example, among all $n \times n$ matrices over $\mathbb{R}$ under the operation of matrix multiplication, we needed to toss all matrices whose determinant is 0 in order to form a group.

In a ring, on the other hand, there is no requirement that under multiplication every element have a multiplication inverse, so there is no need to throw out elements in order to satisfy that the set is a ring (as long as the eight necessary properties are satisfied, of course). So now the elements of the set $R$ sort themselves into invertible and non-invertible with respect to multiplication. Further, among the non-invertible ones, there is a class which we now define.

**Definition 26.2.** An element $r \in R$, $r \neq 0$, is a ***zero divisor*** if $R$ contains an element $u \neq 0$ such that either $r \cdot u = 0$ or $u \cdot r = 0$.

We note that if $r$ is a zero divisor, then it must indeed be non-invertible in $R$ under multiplication. For if $r$ were both invertible and a zero divisor, and if $u$ is as in the definition above, then

$$u = 1 \cdot u = (r^{-1} \cdot r) \cdot u = r^{-1} \cdot (r \cdot u) = r^{-1} \cdot 0 = 0,$$

where the last equality is due to Lemma 26.1 Part (a). We remark here that any time you read a string of equalities (or inequalities, etc.) like this, it's a good practice to pinpoint the reason for *each* equality. Do so right now on this string above.

**Example 26.8.** In $\mathbb{Z}_{20}$ under multiplication modulo 20, the eight elements $\{1, 3, 7, 9, 11, 13, 17, 19\}$ are invertible, the other twelve (including 0) then being non-invertible. Which of these eleven non-zero elements are zero divisors? Well, working modulo 20, we get $2 \cdot 10 = 0$, $4 \cdot 5 = 0$, $6 \cdot 10 = 0$, $8 \cdot 5 = 0$, $12 \cdot 5 = 0$, $14 \cdot 10 = 0$, $15 \cdot 4 = 0$, $16 \cdot 5 = 0$ and $18 \cdot 10 = 0$ (check each of these!), which shows that all eleven are zero divisors.

**Example 26.9.** In the real numbers $\mathbb{R}$ under multiplication, all non-zero elements are invertible, so there are no zero divisors. Likewise in $\mathbb{Z}$ under multiplication, since the product of any two non-zero integers is non-zero, there are no zero divisors.

Was it a coincidence that in $\mathbb{Z}_{20}$ (Example 26.8 above), every non-zero element was either invertible or a zero divisor? This next result shows that this occurs in every *finite* ring with unity.

**Theorem 26.2.** *If $R$ is a finite ring with unity, then every non-zero element in $R$ is either invertible or a zero divisor.*

Proof. (Note: This proof is a bit tricky but pretty nice. Read it carefully.) Assume that a ring $R$ with unity has $n > 1$ elements and let $r$ be a non-zero element of $R$. Consider the set $\{1 = r^0, r, r^2, \ldots, r^n\}$ of powers of $r$, which has $n + 1$ elements. Since the ring $R$ contains only $n$ distinct elements, by the Box Principle (see Chapter 3) two of the powers must be equal. By the Well-ordering Principle (Chapter 5), let $i$ be the smallest exponent in the set for which $r^i = r^{i+j}$ for some $j > 0$. Then we must have $r^{i+j} - r^i = 0$, so that $r^i(r^j - 1) = 0$.

We now consider two cases:
(1) If $i = 0$, then $r^j - 1 = 0$, so $r(r^{j-1}) = 1$, which says that $r$ is invertible in $R$.
(2) On the other hand if $i > 0$, by the minimality of $i$, we must have $r^{i-1}(r^j - 1) \neq 0$. However,

$$r(r^{i-1}(r^j - 1)) = r^i(r^j - 1) = 0,$$

and thus in this case $r$ is a zero divisor in the ring $R$, and we are done. ∎

Asking now about the situation in infinite rings, we saw in Example 26.8 that in both $\mathbb{Z}$ and $\mathbb{R}$, all non-zero elements are either invertible or neither invertible nor zero divisors. Is it possible for an infinite ring to have elements of all three types (i.e., (1) invertible, (2) zero divisor, and (3) neither invertible nor a zero divisor)? The answer is yes, as the following example shows.

**Example 26.10.** Consider the ring $\mathbb{Z}_4[x]$ of all polynomials in a single variable $x$ with coefficients from the finite ring $\mathbb{Z}_4$ (see Example 26.4). Then for every positive integer $k$ (and doing all arithmetic modulo 4)
(1) $(2x^k + 1)(2x^k + 1) = 0x^{2k} + 0x^k + 1 = 1$, so $2^k + 1$ is invertible,
(2) $(2x^k)(2x) = 0x^{k+1} = 0$, so $2x^k$ is a zero divisor,
(3) If $f(x)$ is any non-zero polynomial in $\mathbb{Z}_4[x]$, then $x^k f(x)$ is of degree at least 1, i.e., is not equal to either 0 or 1. Hence $x^k$ is neither invertible nor a zero divisor.

We see then that the infinite ring $\mathbb{Z}_4[x]$ has infinitely many of each of our three types of elements.

We finish this chapter by introducing an important concept for rings, and in particular for fields, which will be discussed in more detail in Chapter 29.

**Definition 26.3.** The ***characteristic*** of a ring with unity is the least positive integer $n$ such that
$$n \cdot 1 = \underbrace{1 + 1 + \cdots + 1}_{n \text{ copies of } 1} = 0.$$
If no such $n$ exists, the characteristic is said to be 0.

An alternative way to describe *characteristic* is that it is the order of the additive cyclic group $\langle 1 \rangle$ generated by the unity element 1 if that group is finite, and it is 0 if that group is infinite. Clearly then, the rings $\mathbb{Z}$, $\mathbb{Q}$, $\mathbb{R}$ and $\mathbb{C}$ have characteristic 0, whereas the rings $\mathbb{Z}_n$ have characteristic $n$. This concept seems quite straightforward, but, as stated above, it takes on more significance in the cases where our ring is actually a field, in which case the characteristic turns out to be either 0 or a prime number $p$.

All algebraic objects possess "sub-objects;" for example, groups have subgroups, etc. As we continue our brief study of rings, the next obvious step is to learn about "subrings." It turns out that more useful sub-objects in rings are called "ideals," which are subrings which have the additional property that they can "absorb" multiplication by elements from the whole ring. For the details, read on!

**Exercises**

**26.1.** (a) Consider the set $2\mathbb{Z}$ of all even integers under the usual operations of addition and multiplication. Is $2\mathbb{Z}$ a ring? If so, verify briefly that all eight

of the properties R1-R8 of Definition 26.1 are satisfied. If it is *not* a ring, say specifically all of the properties which fail.
(b) Answer the same questions as in Part (a) for the set $2\mathbb{Z}+1$ of all odd integers.

**26.2.** (a) Consider the set $S$ of all $3 \times 3$ upper triangular matrices with real entries, i.e., matrices of the form

$$\begin{pmatrix} a & b & c \\ 0 & d & e \\ 0 & 0 & f \end{pmatrix}$$

under the usual operations of matrix addition and multiplication.. Is $S$ a ring? If so, verify briefly that all eight of the properties R1-R8 of Definition 26.1 are satisfied. If it is not a ring, say specifically all of the properties which fail.
(b) Answer the same questions as in Part (a) for the set $T$ of all $3 \times 3$ upper triangular matrices with real entries which have ones on the main diagonal, i.e. matrices of the form

$$\begin{pmatrix} 1 & a & b \\ 0 & 1 & c \\ 0 & 0 & 1 \end{pmatrix}.$$

**26.3.** Prove that for every $n \geq 2$, the set $n\mathbb{Z} = \{\ldots, -3n, -2n, n, 0, n, 2n, 3n, \ldots\}$ is a commutative ring without unity.

**26.4.** Prove Part (b) of Lemma 26.1, i.e., prove that $(-1)a$ is the additive inverse $-a$ of $a$. (Hint: Use Part (a) of the lemma, the fact that $0 = 1 + (-1)$, the distributive law, and the fact that additive inverses are unique (see Exercise 12.8).)

**26.5.** You learned long ago the formula $(x + y)^2 = x^2 + 2xy + y^2$ for squaring a binomial, where $x$ and $y$ are numbers of some sort. However:
(a) Show that this formula does *not* hold for the following two matrices over $\mathbb{Z}$

$$\begin{pmatrix} 1 & 0 \\ 0 & 0 \end{pmatrix} \qquad \begin{pmatrix} 0 & 1 \\ 0 & 0 \end{pmatrix}$$

under the usual operations of matrix addition and multiplication.
(b) Using the distributive law, and remembering that rings need not be commutative, compute the correct formula for the square of a binomial in any ring.
(c) Apply your new formula to the matrices in Part (a), confirming that it holds for this example (as it must).

**26.6.** (a) Prove that if $p$ is a prime number, then for $1 \leq k \leq n-1$, the binomial coefficient $\binom{p}{k}$ is divisible by $p$ (see Examples 1.2 and 1.3).
(b) Use Part (a) to prove that in the ring $\mathbb{Z}_p$ with $p$ a prime, for any elements $a, b \in \mathbb{Z}_p$, $(a + b)^p = a^p + b^p$.

**26.7.** As was done in Example 26.8, list the elements in the ring $\mathbb{Z}_{18}$ of integers modulo 18 which are zero divisors, showing why each one has this property.

**26.8.** Show that the set $R[x]$ in Example 26.4 (where $R$ is any ring) is itself a ring. Is it a ring with unity or not? If so, what are its invertible elements? Is it a commutative ring or not? What is its characteristic? (Hint: It depends!)

**26.9.** If $A$ is a ring with unity element 1, and $A$ contains at least two elements, show that $1 \neq 0$. (Hint: Use Lemma 26.1.)

**26.10.** Find a polynomial $f$ in $\mathbb{Z}_3[x]$ which induces the zero function on $\mathbb{Z}_3$; i.e., $f$ maps every element of the ring $\mathbb{Z}_3$ of integers modulo 3 to zero.

**26.11.** Let $R$ be a ring such that the additive group $(R, +)$ is a cyclic group. Prove that $R$ is a commutative ring.

**26.12.** Given two rings $A$ and $B$, it is easy to verify that their Cartesian product $A \times B$ is itself a ring. If both $A$ and $B$ have unity, then so will $A \times B$. What is the characteristic of $A \times B$ if:
(a) the characteristic of $A$ is 0 and the characteristic of $B$ is $n \in \mathbb{Z}^+$?
(b) the characteristic of $A$ is $m$ and the characteristic of $B$ is $n$, with $m, n \in \mathbb{Z}^+$?

**26.13.** As suggested in Example 26.6, consider the ring $A$ of all $2 \times 2$ matrices over the ring $\mathbb{Z}_2$.
(a) How many matrices does $A$ have in it?
(b) By Theorem 26.2, every element of $A$ is either invertible or a zero divisor. Find a pair of the zero divisors.
(c) Find a (non-identity) invertible matrix and its inverse.

**26.14.** Let $F = \{0, 1, a, b\}$. With the following operation tables, one can verify that $F$ is a ring.

| + | 0 | 1 | a | b |
|---|---|---|---|---|
| 0 | 0 | 1 | a | b |
| 1 | 1 | 0 | b | a |
| a | a | b | 0 | 1 |
| b | b | a | 1 | 0 |

| · | 0 | 1 | a | b |
|---|---|---|---|---|
| 0 | 0 | 0 | 0 | 0 |
| 1 | 0 | 1 | a | b |
| a | 0 | a | b | 1 |
| b | 0 | b | 1 | a |

(a) We shall skip a full verification that the distributive law holds in $F$, but, for example, verify that $a \cdot (b + 1) = a \cdot b + a \cdot 1$.
(b) Looking at the addition table, we have seen the structure of $F$ before, with each element being its own additive inverse. What is this structure called?
(c) Looking at the multiplication table, verify that $F$ is a commutative ring with unity, and in fact that the structure of the non-zero elements of $F$ is that of a multiplicative cyclic group of order 3, generated by either $a$ or $b$. It follows then that $F$ is a field (since both $F$ and $F^*$ are Abelian groups).
(d) What is the characteristic of $F$?
(Note: In fact, $F$ turns out to be the unique finite field of four elements, and it is normally denoted by $\mathbb{F}_4$. See Chapter 29 to learn more.)

CHAPTER **27**

---

# Subrings and Ideals

---

$$(2 + 10\mathbb{Z}) \cdot (3 + 10\mathbb{Z}) = 6 + 10\mathbb{Z}$$

All algebraic objects have "sub-objects." In Chapter 23 we were introduced to the concept of subgroups within a group, and in the two chapters following that we learned about the cosets which a subgroup generates and about the fundamental fact, known as Lagrange's Theorem, that in a finite group the order of any subgroup divides the order of the group. Now that we have developed the idea of a ring, we should not be surprised that rings possess *subrings*. In rings, however, there is a small twist in that rings possess very similar but, it turns out, sometimes more useful sub-objects called *ideals*. We shall shortly learn the difference between these two structures, and later in the chapter we shall discuss how their difference come into play in the study of rings. So let's define these sub-objects.

**Definition 27.1.** Let $R$ be a ring. A non-empty subset $H$ of $R$ is a *subring* if $H$ itself forms a ring with the same operations of addition and multiplication that are used in $R$.

Further, a subring $I$ of a ring $R$ is an *ideal* of $R$ if the product of any element in $R$ and any element in $I$ is always an element of $I$. Thus the subring $I$ is an ideal if for each $i \in I$ and for each $a \in R$, we have that $ia \in I$ and $ai \in I$.

We note that, in particular, for a subset $H$ of a ring $R$ to be a subring, the sum and product of two elements of $H$ must be in $H$. The condition to be satisfied if $H$ is to be an *ideal* is even stronger; namely the subset $H$ must still be closed under multiplication if one of the two elements is in $R$ but is outside of $H$.

Because a subring $H$ or an ideal $I$ of a ring $R$ is automatically an additive subgroup of $R$, we know that Lagrange's Theorem will apply to finite rings as well as finite groups; that is, if $R$ is finite, the order of $H$ or $I$ divides the order of $R$.

Starting then with some of the examples of rings we developed in Chapter 26, let's find some subrings and ideals inside those rings.

**Example 27.1.** For any ring $R$, at the one extreme $\{0\}$ and at the other extreme $R$ itself are both ideals in $R$ (check the former one). Beyond that, the ring $R$ may or may not possess non-trivial proper subrings or ideals. For example, in group theory we saw that $\mathbb{Z}_p$, the integers under addition modulo $p$ where $p$ is prime, has no "intermediate" subgroups, and so as a ring, $\mathbb{Z}_p$ has no intermediate subrings or ideals.

**Example 27.2.** In Example 26.2 we saw that the ring of integers $\mathbb{Z}$ is contained in the ring of rational numbers $\mathbb{Q}$, which is contained in the ring of real numbers $\mathbb{R}$, which is contained in the ring of the complex numbers $\mathbb{C}$. Hence each of these (except $\mathbb{C}$) is a proper subring of the others "above" it. We note however, that because of what you are asked to show in Exercise 27.2, none of these subrings are ideals.

**Example 27.3.** Inside the ring of integers $\mathbb{Z}$, we know from group theory that for each $n \geq 2$, the set $n\mathbb{Z}$ of all multiples of $n$ is a proper additive subgroup. Moreover, for any $k \in \mathbb{Z}$ and any $na \in n\mathbb{Z}$, $k(na) = n(ka)$ is an element of $n\mathbb{Z}$, and so $n\mathbb{Z}$ is not only a subring of $\mathbb{Z}$, but in fact is an ideal of $\mathbb{Z}$.

**Example 27.4.** Following up on the previous example of $\mathbb{Z}$ and using the fact that modular arithmetic is compatible with both integer addition and integer multiplication (see Lemma 15.1 Parts (i) and (iii)), we can see that every additive subgroup of the ring $\mathbb{Z}_n$ of integers modulo $n$ will be an ideal in $\mathbb{Z}_n$. For example, consider the additive subgroup $\langle 3 \rangle = \{3, 6, 9, 0\}$ inside $\mathbb{Z}_{12}$. Suppose we multiply this subgroup by, say, 7; we get

$$\{7(3), 7(6), 7(9), 7(0)\} \equiv \{9, 6, 3, 0\} \pmod{12},$$

i.e., $\langle 3 \rangle$ "absorbs" multiplication by 7, or any element of $\mathbb{Z}_{12}$, and so $\langle 3 \rangle$ is an ideal of $\mathbb{Z}_{12}$.

**Example 27.5.** Consider the ring $\mathbb{R}[x]$ of polynomials in a single variable $x$ with real coefficients.
(a) Let $H_k$ be the set of all polynomials in $\mathbb{R}[x]$ of degree greater than or equal to $k$ together with the zero polynomial.
(b) Let $H_{\text{even}}$ be the set of all polynomials in $\mathbb{R}[x]$ all of whose non-zero terms have even-numbered exponents (e.g., $x^6 + 4x^2 - 7$).
In Exercise 27.6 you are asked to verify that both of these subsets are subrings, but only one of them is an ideal.

**Example 27.6.** In Exercise 26.2 Part (a) you were asked to verify that the set of upper triangular $3 \times 3$ matrices (i.e., matrices with all zeros below the main diagonal) with real entries is a ring. Let us call this set $UT$. (If you didn't do that exercise then, do it now.) Hence $UT$ is a subring of the ring of all $3 \times 3$ matrices over $\mathbb{R}$, but it is not an ideal. We can verify this by a single example: Say $A = \begin{pmatrix} 0 & 1 & 0 \\ 0 & 0 & 0 \\ 0 & 0 & 0 \end{pmatrix}$, which is in $UT$, and $B = \begin{pmatrix} 0 & 0 & 0 \\ 0 & 0 & 0 \\ 1 & 0 & 0 \end{pmatrix}$, then

$BA = \begin{pmatrix} 0 & 0 & 0 \\ 0 & 0 & 0 \\ 0 & 1 & 0 \end{pmatrix}$, which is not in $UT$. Alternatively, one can simply apply Exercise 27.2.

Just as was the case with subgroups, verifying that a subset of a ring is a subring or an ideal does not require us to check all of the eight properties a ring must possess (Definition 26.1). In particular, there is no need to recheck associativity of either operation, nor to recheck distributivity. In analogy to Theorem 23.1 Part (iii) on subgroups, a shortest route to verification of subring or ideal status is the following:

**Theorem 27.1.** (i) *A non-empty set $H$ of a ring $R$ is a subring of $R$ if and only if $a - b$ and $ab$ are in $H$ for all $a, b \in H$.*
(ii) *A non-empty subset $I$ of a ring $R$ is an ideal in $R$ if and only if $a - b \in I$ for all $a, b \in I$ and $xa$ and $ax$ are in $I$ for every $x \in I$ and every $a \in R$.*

Proof. All of this follows directly the definition of a ring (Definition 26.1) and of a subring and an ideal (Definition 27.1) except for the sufficiency of checking that $a - b$ is in $H$ for all $a, b \in H$ (or in $I$) to verify in one step that ring properties R2, R3 and R4 are satisfied for $H$ (again, see Definition 26.1). We ask you to supply the details on this point in Exercise 27.4. ∎

You may be wondering why we bother defining ideals in the theory of rings when it would seem that subrings are an obvious analogue to subgroups in the the theory of groups. The answer here is not simple, but we hope to give you a brief and informal sense of the utility of ideals. The key idea here is structures in abstract algebra called *factor objects*. In groups we can form "factor groups," in rings we can form "factor rings," and so on. Since under addition a ring $R$ is an Abelian group, given any subgroup $H$ of $(R, +)$, we can form the factor group, denoted $R/H$ which has the following two properties:
(1) The elements of $R/H$ are all of (additive) cosets of $H$ in $R$, and
(2) the operation is "coset addition," meaning that if $a + H$ and $b + H$ are two cosets in $R/H$, then $(a + H) + (b + H) = (a + b) + H$, i.e., the sum of two cosets is the coset which contains the sum of their representatives.

While being careful about this definition of coset addition, one must verify that this operation is *well-defined*, meaning that if we choose different representatives besides $a$ and $b$ for our two cosets, we would still arrive in the coset containing $a + b$. We shall not go into that detail here. Time for an example.

**Example 27.7.** Though we didn't mention it at the time, the figure at the beginning of Chapter 24 is a picture of the four element factor group $\mathbb{Z}/4\mathbb{Z}$. We copy that picture here, with small changes to emphasize our new way of looking at it:

$$
\begin{array}{c|c}
0 + 4\mathbb{Z} & 1 + 4\mathbb{Z} \\
\hline
3 + 4\mathbb{Z} & 2 + 4\mathbb{Z}
\end{array}
$$

$$(\mathbb{Z}/4\mathbb{Z}, +)$$

The coset addition of this group is very simple; in fact, it should be clear that we are looking at a group exactly like the additive cyclic group $\mathbb{Z}_4$. More generally, we can see that the $n$ distinct cosets

$$\{0 + n\mathbb{Z}, 1 + n\mathbb{Z}, 2 + n\mathbb{Z}, \ldots, (n-1) + n\mathbb{Z}\}$$

of $n\mathbb{Z}$ in $\mathbb{Z}$ form a finite additive group $\mathbb{Z}/n\mathbb{Z}$ which is exactly like the additive cyclic group $\mathbb{Z}_n$.

Returning now to an abstract ring $R$ containing an additive subgroup $H$ for which we can form the factor group $R/H$, what if we would like to have $R/H$ be a factor *ring*; that is, if we would like $R/H$ to be equipped with both a coset addition *and* a coset multiplication? Well, certainly then we would expect that $H$ should be a subring; i.e., it should be closed under $R$'s multiplication as well as being an additive subgroup. What we want then is for the *product* of the cosets $a + H$ and $b + H$ to arrive at the coset $a \cdot b + H$. Will this work if $H$ is a subring? Well, using the distributive law, and noting that, for example, by $a \cdot H$ we mean $\{a \cdot h | h \in H\}$, we have

$$(a + H)(b + H) = a \cdot b + a \cdot H + H \cdot b + H \cdot H.$$

Here comes the punch line! Look at the two middle terms on the right hand side. Are they within $H$, which would seem to be necessary for our coset multiplication to work properly? The answer is: If $H$ is a subring, possibly *no*, but if $H$ is an *ideal*, then *yes*! That is exactly the additional condition a subring must satisfy to be an ideal: it must "absorb" multiplication by arbitrary elements of $R$ on both the left and the right. ·

Let's look at two examples, one where $H$ is a subring but not an ideal, and one where $H$ *is* an ideal.

**Example 27.8.** We saw in Example 27.2 that the integers $\mathbb{Z}$ are a subring but not an ideal of the real numbers $\mathbb{R}$. In the additive factor group $\mathbb{R}/\mathbb{Z}$, let's try to define a coset multiplication. Consider the cosets

$$0.2 + \mathbb{Z} = \{\ldots, -1.8, -0.8, 0.2, 1.2, 2.2, \ldots\}$$

and
$$0.3 + \mathbb{Z} = \{\ldots, -1.7, -0.7, 0.3, 1.3, 2.3, \ldots\}.$$

We want their product to land in the coset $0.06 + \mathbb{Z}$, but by the distributive law, we get

$$(0.2 + \mathbb{Z}) \cdot (0.3 + \mathbb{Z}) = 0.06 + 0.2 \cdot \mathbb{Z} + 0.3 \cdot \mathbb{Z} + \mathbb{Z} \cdot \mathbb{Z}.$$

Is this the "correct" coset? Well, a sample element in the right-hand side, using $1 \in \mathbb{Z}$, is $0.06 + 0.2 + 0.3 + 1 = 1.56$, which is clearly *not* in the coset $0.06 + \mathbb{Z}$. What went wrong? Precisely that $\mathbb{Z}$ is not an ideal in $\mathbb{R}$.

**Example 27.9.** We observed in Example 27.3 that, for example, $10\mathbb{Z}$ is an ideal in the ring $\mathbb{Z}$. Now let's multiply the cosets $2 + 10\mathbb{Z}$ and $3 + 10\mathbb{Z}$ and see if we land in the coset $6 + \mathbb{Z}$:

$$(2 + 10\mathbb{Z}) \cdot (3 + 10\mathbb{Z}) = 6 + 2 \cdot 10\mathbb{Z} + 3 \cdot 10\mathbb{Z} + 10\mathbb{Z} \cdot 10\mathbb{Z},$$

and we see that elements in the latter three terms on the right-hand side clearly lie in $10\mathbb{Z}$, so the coset multiplication does indeed work correctly, precisely because $10\mathbb{Z}$ is an ideal.

We hope that you now have some sense of why it is necessary in the theory of rings to go beyond subrings and define ideals as well. We shall not pursue topics involving factor objects further in this text, but if and when you dive somewhat deeper into abstract algebra, factor objects, along with sub-objects of course, will play an important role.

We next focus on a particularly important category of rings known as *integral domains*. These are rings which do not contain zero divisors, and again the integers $\mathbb{Z}$ are a fundamental example.

**Exercises**

**27.1.** Write out *all* the ideals in the ring $\mathbb{Z}_{20}$ of integers modulo 20.

**27.2.** Show that if $R$ is a ring with unity element 1 and if an ideal $I$ of $R$ contains 1, then $I = R$. (Or, stating the contrapositive, any proper ideal of $R$ does not contain the unity element.)

**27.3.** Show that the set of all $2 \times 2$ matrices with even entries is an ideal in the ring of $2 \times 2$ matrices with integer entries.

**27.4.** Concerning the proof of Theorem 27.1, prove that if $a - b$ is in $H$ (or in $I$) for all $a, b \in H$, then the following ring properties hold for $H$:
R3: $0 \in H$,
R4: $b \in H$ implies $-b \in H$, and hence
R2: $a, b \in H$ implies that $a + b \in H$ (closure under addition).

**27.5.** (a) If $I$ and $J$ are ideals in a ring $R$, show that $I \cap J$ is an ideal of $R$.
(b) Is it true that $I \cup J$ must also be an ideal of $R$? Either prove it or give a counter-example.

**27.6.** As introduced in Example 27.5, verify that the subsets $H_k$ and $H_{even}$ of the polynomial ring $\mathbb{R}[x]$ are both subrings, but that only one of them is an ideal.

**27.7.** (a) Give an example of a proper subring $H$ of the real numbers $\mathbb{R}$ which satisfies $H \cap \mathbb{Q} = \emptyset$ ($\mathbb{Q}$ is the ring of rational numbers).
(b) Show that your subring $H$ in Part (a) is not an ideal of $\mathbb{R}$.
(c) Prove that the ring of real numbers $\mathbb{R}$ contains no non-trivial proper ideals.

**27.8.** Assume that $I$ and $J$ are ideals in a ring $R$. Show that the set

$$I + J = \{i + j | i \in I, j \in J\}$$

is an ideal in the ring $R$.

**27.9.** Assume that $I$ and $J$ are ideals in a ring $R$. Consider the set

$$I \cdot J = \{i \cdot j | i \in I, j \in J\}.$$

(a) In the ring $\mathbb{Z}_{12}$ of integers modulo 12, let $I = \langle 3 \rangle$ and $J = \langle 4 \rangle$. Write down the elements of the set $I \cdot J$. Is this set an ideal in $\mathbb{Z}_{12}$?
(b) In the ring $\mathbb{Z}$, let $I = 3\mathbb{Z}$ and $J = 4\mathbb{Z}$. Describe the set $I \cdot J$. Is this set an ideal in $\mathbb{Z}$?
(c) Do you think that in general $I \cdot J$ is always an ideal of an arbitrary ring $R$? Without doing a proof or finding a counter-example, what might be an impediment to a general proof? (What property do $\mathbb{Z}_{12}$ and $\mathbb{Z}$ share that some other rings do not?)

**27.10.** Let $R$ be a commutative ring and let $r \in R$. Show that the set

$$P_r = \{xr | x \in R\}$$

is an ideal in $R$. The ideal $P_r$ is called the **_principal ideal_** generated by the element $r$.

**27.11.** Consider the ideal $I = \langle 5 \rangle$ in $\mathbb{Z}_{20}$ (see Exercise 27.1).
(a) Write down all the additive cosets of $I$ in $\mathbb{Z}_{20}$ (e.g., $0 + I = \{5, 10, 15, 0\}$, etc.). These cosets are the five *elements* of the factor ring $\mathbb{Z}_{20}/I$.
(b) Verify, for example, that $(2 + I)(3 + I) = 1 + I$, that is, if you take any representative from $2 + I$ and multiply it by any representative of $3 + I$, you will end up with a representative of $1 + I$. (You are checking that coset multiplication is "well-defined" in this case.)

# CHAPTER 28

## Integral Domains

$$a \cdot b = 0 \Rightarrow a = 0 \text{ or } b = 0$$

Given a ring $R$ and a polynomial $f$ with coefficients in $R$, how many solutions does the equation $f(x) = 0$ have in $R$? The answer depends on both $f$ and $R$; that is, even the same polynomial $f$ can have a different number of solutions depending on a change in the ring $R$. Let's look at a hopefully familiar example followed by a less familiar one.

**Example 28.1.** Consider the quadratic polynomial equation $x^2 - 3 = 0$, and note that coefficients are in the integers $\mathbb{Z}$, but also in the real numbers $\mathbb{R}$ and in the complex numbers $\mathbb{C}$. We can see that in both $\mathbb{R}$ and $\mathbb{C}$, there are two solutions $\pm\sqrt{3}$, but in $\mathbb{Z}$ there are no solutions. On the other hand, the equation $x^2 + 3 = 0$ has no solutions in both $\mathbb{Z}$ and $\mathbb{R}$, but it has two solutions $\pm i\sqrt{3}$ in $\mathbb{C}$. Note, however, that by the Quadratic Formula (Example 1.1) *in these three rings any quadratic polynomial equation can have at most two solutions.* We shall be able to generalize this idea nicely in Chapter 35.

**Example 28.2.** Consider the quadratic polynomial equation $x^2 + 6x + 8 = 0$ and note that the coefficients can be viewed as coming from $\mathbb{Z}$ or from $\mathbb{Z}_{12}$, the integers modulo 12. By factoring, the equation has two solutions in $\mathbb{Z}$: $-2$ and $-4$. However, in $\mathbb{Z}_{12}$, there actually are *four* solutions: 2, 4, 8 and 10 (check!!). Hence it is possible in certain rings for a quadratic polynomial equation to have more than two solutions.

So what's going on in these two examples? Well, in Example 28.1 we are working with sets of numbers ($\mathbb{Z}$ and $\mathbb{R}$) with which you have long been familiar ($\mathbb{C}$ not so much), and you learned that in those sets if the product of some numbers is 0, then at least one of them is 0. Using the language we now have (Definition 26.2), and as we observed in Example 26.9, neither $\mathbb{Z}$ nor $\mathbb{R}$ contain any *zero divisors*. However, in Example 28.2 we encountered the finite ring $\mathbb{Z}_{12}$, and we learned in Chapter 26 that the ring $\mathbb{Z}_n$ contains zero divisors if and only if $n$ is not prime. Indeed, in $\mathbb{Z}_{12}$ it is easy to check (do so) that 2, 3, 4, 6, 8, 9 and 10 are all zero divisors since they all fail to be relatively prime to 12. In the example, then, we have $x^2 + 6x + 8 = (x+2)(x+4) = 0$, so $-2 = 10$ and $-4 = 8$ are clearly solutions in $\mathbb{Z}_{12}$, but so are 2 and 4 since $(2+2)(2+4) = (4)(6) = 0$ and $(4+2)(4+4) = (6)(8) = 0$ in $\mathbb{Z}_{12}$. The bottom line is that the existence of zero divisors will lead to additional solutions.

We see then that it is important in analyzing the existence of solutions to a polynomial equation to know if zero divisors are present in our ring $R$. Rings that do *not* have zero divisors are an important class to which we now give a name.

**Definition 28.1.** An ***integral domain*** is a commutative ring with unity which does not contain any zero divisors.

Be warned that some authors omit the conditions of commutativity, unity or both when discussing integral domains. Using our definition, let's identify some integral domains.

**Example 28.3.** As we saw in our discussion of zero divisors in Chapter 26, our standard infinite rings of numbers $\mathbb{Z}$, $\mathbb{Q}$ (the rational numbers), $\mathbb{R}$ and $\mathbb{C}$ are all integral domains. Among finite rings, again, it follows directly from Theorem 26.2 that $\mathbb{Z}_n$ is an integral domain if and only if $n$ is prime. Finally, we ask you in Exercise 28.6 to confirm that if $R$ is a ring and $R[x]$ is the ring of polynomials in the single variable $x$ with coefficients in $R$, then $R[x]$ is an integral domain if and only if $R$ is itself an integral domain.

Given an arbitrary ring $R$, we can ask to what extent *cancellation* is possible. By this we mean: If $a, b, c \in R$ and if $ab = ac$, when can we be sure that we can cancel the element $a$, concluding that $b = c$? Well, to start with, it's obvious that $a$ must be non-zero. It's also clear that if $a$ has a multiplicative inverse $a^{-1}$ in $R$, then multiplying both sides of the equality on the left by $a^{-1}$ and using associativity, etc., we can compute that

$$b = (1)b = (a^{-1}a)b = a^{-1}(ab) = a^{-1}(ac) = (a^{-1}a)c = (1)c = c.$$

As we have remarked before, when confronted with a string of equalities like this, it is an excellent exercise to state to yourself exactly what justifies each one. Do that now!

But is it *necessary* that the element $a$ we wish to cancel possess a multiplicative inverse in $R$? The answer is clearly "no," as a simple example in the integers $\mathbb{Z}$ can illustrate: 2 does not have a multiplicative inverse in $\mathbb{Z}$, but nonetheless for any integers $b$ and $c$, we know that $2b = 2c$ implies that $b = c$.

The point, as you might expect, is that 2 may not be invertible, *but* it is also not a zero divisor. This then leads us to the following "if and only if" result which shows that the ability to always cancel a non-zero element is a central feature of integral domains.

**Theorem 28.1.** *Let $R$ be a commutative ring with unity. The following two statements about $R$ are equivalent:*
(i) *$R$ is an integral domain.*
(ii) *For all $a, b, c \in R$ with $a \neq 0$, if $ab = ac$, then $b = c$.*

Proof.
(i) implies (ii): If $ab = ac$, then by the distributive law $ab - ac = a(b - c) = 0$. Since $R$ is an integral domain, and therefore since $a$ is not a zero divisor, we must have that $b - c = 0$. Hence $b = c$.
(ii) implies (i): Let $a \in R$ be a non-zero element of $R$, and suppose that for some $b \in R$, $ab = 0$. By Lemma 26.1 Part (a), we have then that $ab = a(0)$, so by our assumption that $a$ can be cancelled, we arrive at $b = 0$. Thus $a$ is by definition not a zero divisor, and we conclude that $R$ is indeed an integral domain. ∎

Returning now to solutions of polynomial equations over rings, let's look at an example showing that things can go wrong if cancellation is incorrectly applied.

**Example 28.4.** Consider the polynomial equation $2x^2 - 8x + 6 = 0$. When viewed as an equation in $\mathbb{Z}$, we can easily solve it by first factoring out and then cancelling the common factor of 2, obtaining $x^2 - 4x + 3 = 0$, which then factors into $(x - 1)(x - 3) = 0$, revealing the two solutions of 1 and 3.

But now let's view our original equation $2x^2 - 8x + 6 = 0$ as being over the finite ring $\mathbb{Z}_{10}$ of integers modulo 10. You should check, simply by substitution, that 1, 3, 6 and 8 are solutions. (We knew additional solutions, namely 6 and 8 in this case, might appear because $\mathbb{Z}_{10}$ contains zero divisors.) But now, as we did in the integer case, factor out and cancel the 2, getting $x^2 - 4x + 3 = 0$. Again by substituting in values, you should discover that this equation has *only* the two solutions 1 and 3; that is, by cancelling the factor of 2, we lost two of the four correct solutions! The reason, of course, is that 2 is a zero divisor in $\mathbb{Z}_{10}$, and so cannot be cancelled.

We end this brief chapter by reminding you that every ring is, to start with, an Abelian group under addition, whereas the multiplicative structure may well not be that of a group. Our final step then is to focus on the extremely important class of rings in which the non-zero elements also form an Abelian group under multiplication. Of course, as required in all rings, the two operations, and hence in this case these two groups, must be related via the distributive law. Such rings are called *fields*, which we shall study in the next chapter.

## Exercises

**28.1.** Suppose $f(x) = 6x + 4$. Write down all the solutions of the equation $f(x) = 0$ in the ring
(a) $\mathbb{Z}$ of integers,
(b) $\mathbb{R}$ of real numbers,
(c) $\mathbb{Z}_8$ of integers modulo 8.

**28.2.** Suppose $g(x) = x^2 + 4$. Write down all the solutions of the equation $g(x) = 0$ in the ring
(a) $\mathbb{R}$ of real numbers,
(b) $\mathbb{C}$ of complex numbers,
(c) $\mathbb{Z}_7$ of integers modulo 7,
(d) $\mathbb{Z}_8$ of integers modulo 8.

**28.3.** Give an example of a subring $S$ of an integral domain $R$ which fails to itself be an integral domain.

**28.4.** Prove that if $R$ is an integral domain, then any subring of $R$ which contains the unity element of $R$ is an integral domain.

**28.5.** Give two examples of integral domains $R$ which have the property that none of their proper subrings are integral domains. Justify your answers.

**28.6.** If $R$ is a ring and $R[x]$ is the ring of polynomials in the single variable $x$ with coefficients in $R$, prove that $R[x]$ is an integral domain if and only if $R$ is itself an integral domain.

**28.7.** We know that for any prime $p$, the finite ring $\mathbb{Z}_p$ of integers modulo $p$ is an integral domain. However, the finite ring of $2 \times 2$ matrices with entries in $\mathbb{Z}_p$ fails two of the three conditions to be an integral domain, even when $p = 2$. Verify this by giving examples of those two conditions failing.

**28.8.** Try to solve the equation $3x^2 + 9x + 6 = 0$ over the finite ring $\mathbb{Z}_{12}$ of integers modulo 12 in two ways:
(a) (Correct) By substituting the values 0 through 11 into the equation.
(b) (Incorrect - why?) By factoring out and cancelling the 3, then solving the remaining equation by substitution.

**28.9.** Solve the equation $3x^2 + 9x + 6 = 0$ over the finite ring $\mathbb{Z}_{11}$ of integers modulo 11 using factoring and cancellation. Why is this a correct method while working in this ring?

**28.10.** The first author was presented in Algebra I class many years ago with the following "proof" that $2 = 1$. Find the error.

Suppose that $a$ and $b$ are non-zero integers such that $a = b$. Multiply both sides by $a$, getting $a^2 = ab$. Subtract $b^2$ from both sides, obtaining $a^2 - b^2 = ab - b^2$. Factor each side, getting $(a - b)(a + b) = (a - b)(b)$. Cancel $a - b$ on both sides, obtaining $a + b = b$. Since $a = b$, this last equation says that $2b = b$. Cancel $b$ on both sides, arriving at $2 = 1$. QED

# CHAPTER 29

# Fields

$$\mathbb{Q} \subset \mathbb{Q}[\sqrt{p}] \subset \mathbb{R} \subset \mathbb{C}$$

When we were introduced to rings in Chapter 26, we learned that there are eight properties a set $R$ must have in order to be a ring. These boil down to the following: $R$ has two binary operations, addition and multiplication, which are related via the distributive law. Under addition $R$ is an Abelian group, but under multiplication all that's required is that the operation be closed and associative. It is not required that multiplication be commutative, nor that there be a unity element, and hence nor that multiplicative inverses be present (since the existence of inverses depends on the existence of a unity element). However, we then encountered numerous examples of rings which possess additional properties beyond the required ones.

(i) In many examples the multiplication is in fact commutative,

(ii) many examples do in fact contain a unity element, and

(iii) some examples have the property that even if not every non-zero element possesses a multiplicative inverse, at least there are no zero divisors (see Definition 26.2).

Rings which satisfy *all three* of these additional properties are, as you know, called integral domains.

**Example 29.1.** (a) A ring which possesses only the required properties (but not (i), (ii) and (iii) above) is the set of $2 \times 2$ matrices over the ring $2\mathbb{Z}$ of even integers. (Check this.)

(b) A ring which possesses only the additional property of commutativity is the set $\{0, 2, 4, 6\}$ under addition and multiplication modulo 8. (Check.)

(c) A ring which possesses only the additional property of a unity element is the set of $2 \times 2$ matrices over the ring $\mathbb{Z}$ of integers.

(d) A ring which possesses all three of the additional properties listed above is, of course, the integers $\mathbb{Z}$.

If you liked this example, be sure to work through Exercise 29.2 as well.

The final step then in achieving maximal algebraic structure in a ring $R$ is to require not only that $R$ be an Abelian group under addition, but that its non-zero elements $R^*$ be an Abelian group under multiplication. Hence we have the following important definition.

**Definition 29.1.** A *field* is a commutative ring with a unity element with the additional property that each non-zero element has a multiplicative inverse.

**Example 29.2.** We are already quite familiar with numerous fields.

(a) Among our standard infinite rings of numbers, the integers $\mathbb{Z}$ are of course not a field (but are the "runner-up" type, being an integral domain). However, the rational numbers $\mathbb{Q}$, the real numbers $\mathbb{R}$, and the complex numbers $\mathbb{C}$ are all fields since in each case, as we have seen, every non-zero element possesses a multiplicative inverse.

(b) Among our finite rings $\mathbb{Z}_n$, we know that if $n$ is composite, then $\mathbb{Z}_n$ contains zero divisors and hence is neither an integral domain nor a field. On the other hand, if $n = p$ is prime, then we know that every non-zero element possesses a multiplicative inverse, so $\mathbb{Z}_p$ is indeed a field. Since there are infinitely many primes (see Example 3.4), this means that there are infinitely many distinct finite fields.

**Example 29.3.** How about a field which we have not seen before? First consider the set $\mathbb{Z}[\sqrt{2}]$ of all real numbers of the form $a + b\sqrt{2}$, where $a$ and $b$ are integers. You are asked to check in Exercise 29.3 that with the usual operations of addition and multiplication, the set $\mathbb{Z}[\sqrt{2}]$ is an integral domain but is not a field. However, if we now consider the larger set $\mathbb{Q}[\sqrt{2}]$ consisting of all real numbers of the form $x + y\sqrt{2}$ where $x$ and $y$ are *rational* numbers, we claim that this set is in fact a field. The only tricky point to check is that every non-zero element has a multiplicative inverse. To this end, assume that $x + y\sqrt{2}$ is non-zero (so that at least one of $x$ or $y$ is not 0). By Exercise 29.4 we have then that $x^2 - 2y^2 \neq 0$. You should confirm now (pencil to paper!) that if $w = x/(x^2 - 2y^2)$ and $z = -y/(x^2 - 2y^2)$, then $(x + y\sqrt{2})(w + z\sqrt{2}) = 1$, i.e., the element $w + z\sqrt{2}$ is indeed the multiplicative inverse of the element $x + y\sqrt{2}$.

Hence we have found a field which lies inside (i.e., is a subset of) the field of real numbers $\mathbb{R}$ and itself contains the field of rational numbers $\mathbb{Q}$ (when $y = 0$). Finally, we note that more generally, by Exercise 29.4, the set $\mathbb{Q}[\sqrt{p}]$ is a field for all primes $p$, so we have discovered infinitely many new infinite fields, all being subsets of $\mathbb{R}$.

The concept of "sub-object" applies to fields just as it does to sets, groups and rings. Here is the definition you would expect.

**Definition 29.2.** A non-empty subset $S$ of a field $F$ is a **subfield** if $S$ is itself a field under the same operations as in the field $F$.

**Example 29.4.** We have seen lots of examples of subfields already in this chapter. The real numbers $\mathbb{R}$ are a subfield of the complex numbers $\mathbb{C}$. We just showed in Example 29.3 that $\mathbb{R}$ contains infinitely many subfields $\mathbb{Q}[\sqrt{p}]$ for $p$ a prime, and each of those contains the rational numbers $\mathbb{Q}$. We can summarize this as follows:

$$\mathbb{Q} \subset \mathbb{Q}[\sqrt{p}] \subset \mathbb{R} \subset \mathbb{C}.$$

Let us focus now on *finite* sets which form fields. We have already observed that the set $\mathbb{Z}_n$ of integers under addition and multiplication modulo $n$ form a field if and only if $n = p$ is prime. Are there other finite fields besides this infinite collection? The answer turns out to be yes, but the examples are not obvious. Let's examine this question a bit.

One observation is that among finite sets, a search for fields or for integral domains turns out to be the same search, for the following reason.

**Theorem 29.1.** *A finite set $D$ is a field if and only if it is an integral domain.*

Proof. This follows directly from Theorem 26.2, in which we proved (with some effort!) that every non-zero element of a finite ring is either invertible or a zero-divisor. Hence if $D$ is a field, it contains no zero-divisors and hence is an integral domain. Likewise if $D$ is an integral domain, then all of its non-zero elements must be invertible and so is a field. ■

We note that the existence of the ring of integers $\mathbb{Z}$, an integral domain which is not a field, shows that Theorem 29.1 definitely does not apply to infinite sets.

Continuing our search for finite fields besides the collection $\mathbb{Z}_p$ of integers modulo a prime $p$, we in fact discovered one back in Exercise 26.14. Here is that field, consisting of 4 elements.

**Example 29.5.** Suppose $F$ is the set $\{0, 1, a, b\}$ with the following operation tables:

| + | 0 | 1 | a | b |
|---|---|---|---|---|
| 0 | 0 | 1 | a | b |
| 1 | 1 | 0 | b | a |
| a | a | b | 0 | 1 |
| b | b | a | 1 | 0 |

| · | 0 | 1 | a | b |
|---|---|---|---|---|
| 0 | 0 | 0 | 0 | 0 |
| 1 | 0 | 1 | a | b |
| a | 0 | a | b | 1 |
| b | 0 | b | 1 | a |

In that exercise you verified that $F$ is in fact a ring, that the additive structure is that each element is its own additive inverse (i.e., is a Klein-four group), and that each of the three non-zero elements possesses a multiplicative inverse. Hence $F$ is our first example of a finite field with a composite number of elements! We observe that this field $F$ is definitely *not* the same as the ring $\mathbb{Z}_4$ of integers modulo 4 (since in $\mathbb{Z}_4$ the additive structure is cyclic). As stated in Exercise 26.14, this field is normally denoted $\mathbb{F}_4$.

In Chapter 26 we defined the concept of the *characteristic* of a ring. (See Definition 26.3). A glance at the addition table of the field $\mathbb{F}_4$ above reveals that its characteristic is 2. This is one example of the following general result.

**Theorem 29.2.** *If $F$ is a field, then the characteristic of $F$ is either 0 or a prime number $p$.*

Proof. Just suppose that the characteristic of $F$ is a composite integer $n$, say $n = ab$ with $a, b > 1$. Because the characteristic (if not 0) is the least positive integer such that $n(1) = \underbrace{1 + 1 + \cdots + 1}_{n \text{ copies of } 1} = 0$, we know that $a(1) = \underbrace{1 + 1 + \cdots + 1}_{a \text{ copies of } 1} \neq 0$
and $b(1) = \underbrace{1 + 1 + \cdots + 1}_{b \text{ copies of } 1} \neq 0$. But by multiple applications of the distributive
law, $(a(1))(b(1)) = n(1) = 0$. Thus the elements $a(1)$ and $b(1)$ in $F$ are zero-divisors, contradicting the assumption that $F$ is a field, and we are done. ∎

We note that for every field $F$ of characteristic 0, using the notation of the proof just above, for every $n \in \mathbb{Z}^+$ we would have $n(1) \neq 0$, which forces $f$ to be infinite. Hence every finite field $F$, no matter how many elements it has, must be of characteristic $p$ for some prime $p$. It would seem then that $F$ may be "built out of" $\mathbb{Z}_p$ in some way. For example, our four element field $\mathbb{F}_4$ in Example 29.5 has the same additive structure as the Cartesian product $\mathbb{Z}_2 \times \mathbb{Z}_2$, i.e., it is, in some way, built out of $\mathbb{Z}_2$. This does in fact turn out to always be the case, as stated in the following theorem, the proof of which is somewhat beyond the scope of this text. If you would like to know the details, see, for example, [6].

**Theorem 29.3.** (i) *If $p$ is a prime and $m$ is a positive integer, there is a finite field containing exactly $p^m$ distinct elements. Its additive structure is the same as that of the Cartesian product $\underbrace{\mathbb{Z}_p \times \mathbb{Z}_p \times \cdots \times \mathbb{Z}_p}_{m \text{ copies of } \mathbb{Z}_p}$, and its multiplicative*
*structure is that of a cyclic group of order $p^m - 1$.*
(ii) *Every finite field is of the type in* Part (i).

Finite fields are normally denoted $\mathbb{F}_{p^m}$, or also $\mathbb{F}_q$ where $q = p^m$. We note that when $m = 1$, $\mathbb{F}_p$ and $\mathbb{Z}_p$ are denoting the same field, but when $m > 1$, the field $\mathbb{F}_{p^m}$ is *not* the same as the ring $\mathbb{Z}_{p^m}$, because in particular their characteristics (and hence additive structures) are not the same.

**Example 29.6.** (a) Since $125 = 5^3$, there exists a finite field $\mathbb{F}_{125}$ with 125 elements. The additive structure is the same as that of $\mathbb{Z}_5 \times \mathbb{Z}_5 \times \mathbb{Z}_5$, and the multiplicative structure is that of a cyclic group of order 124.

Theorem 29.3 tells us then that up to 50, say, there are finite fields of orders 2, 3, $4 = 2^2$, 5, 7, $8 = 2^3$, $9 = 3^2$, 11, 13, $16 = 2^4$, 17, 19, 23, $25 = 5^2$, $27 = 3^3$, 29, 31, $32 = 2^5$, 37, 41, 43, 47 and $49 = 7^2$, and no other orders besides these.

Finite fields are heavily used in modern digital communications. For one of many examples, in Chapter 18 we used the cyclic multiplicative structure of $\mathbb{F}_p$ ($= \mathbb{Z}_p$) to guarantee the existence of primitive roots (i.e., generators), which were then used to implement the Diffie-Hellman Key Exchange system. For a wider discussion of these types of applications, see again, for example, [6].

Having been introduced to three central types of algebraic objects: groups, rings and fields, we shall now encounter a fourth type known as *vector spaces*. A vector space $V$ is an Abelian group under addition which does *not* possess a binary multiplicative operation, but does allow multiplication of its elements "from the outside" by the elements of some field $F$. The elements of the field are called "scalars," and their multiplicative operation is called "scalar multiplication." Vector spaces, as discussed in the next chapter, will be the last type of algebraic object to be introduced in this text.

### Exercises

**29.1.** Prove that in a field $F$, if $a \neq 0$ and $b \in F$, then the equation $ax = b$ has a unique solution in the field $F$.

**29.2.** Following up on Example 29.1, in each case give an example of a ring which
(a) is commutative with unity but contains zero divisors,
(b) is commutative with no zero divisors but has no unity element,
(c) has no zero divisors and has a unity element, but is not commutative. (Note: Don't lose sleep trying to come up with a ring like this one. If you're curious, google Quaternions.)

**29.3.** Concerning Example 29.3, confirm that with the usual operations of addition and multiplication, the set $\mathbb{Z}[\sqrt{2}]$ is an integral domain, but is not a field.

**29.4.** Also in reference to Example 29.3, prove that if $x$ and $y$ are rational numbers, not both 0, and if $p$ is a prime number, then $x^2 \neq py^2$. (Hint: Just suppose $x^2 = py^2$, which forces both $x$ and $y$ to be non-zero. Write $x = a/b$ and $y = c/d$ where $a, b, c, d \in \mathbb{Z}$. Clear the fractions, creating an equation in integers. Now count the possible number of factors of $p$ on each side, reaching a contradiction. Note: If this argument looks somewhat familiar, it is a slightly more complicated version of a standard contradiction proof that if $p$ is prime, $\sqrt{p}$ is irrational.)

**29.5.** Let $F$ be a field. Show that the polynomial ring $F[x]$ consisting of all polynomials whose coefficients are in the field $F$ is an integral domain, but is not a field.

**29.6.** (a) Among the infinite fields we have identified in this chapter ($\mathbb{Q}$, $\mathbb{Q}\sqrt{p}$ ($p$ prime), $\mathbb{R}$ and $\mathbb{C}$), in which of these does the equation $x^2 - 5 = 0$ have a solution or solutions? What are those solutions?
(b) Answer the same question for the equation $x^2 + 5 = 0$.

**29.7.** For any prime $p \neq 2$, show that in the field $\mathbb{Z}_p$ of integers modulo $p$, the sum of the elements in $\mathbb{Z}_p$ is 0.

**29.8.** Show that if $F$ is a field of characteristic 2, then $-a = a$ for any $a \in F$.

**29.9.** Factor the polynomial $x^4 + 1$ over the field $\mathbb{Z}_2$ of integers modulo 2.

**29.10.** (a) In which of our infinite fields do there exist solutions to the equation $x^2 = -1$? What are those solutions?
(b) In which of the finite fields $\mathbb{Z}_5$, $\mathbb{Z}_7$, $\mathbb{Z}_{11}$ and $\mathbb{Z}_{13}$ are there elements which are solutions of the equation $x^2 = -1$? (Recall that in the finite field $\mathbb{Z}_n$, -1 is simply the element $n - 1$.)
(c) Make a conjecture about what will be true of the odd prime $p$ in order for the equation $x^2 = -1$ to have solutions in $\mathbb{Z}_p$. (Hint: Look at $p$ modulo 4.) If you have the courage, test out your conjecture on $\mathbb{Z}_{17}$ and $\mathbb{Z}_{19}$

**29.11.** (a) Use the Binomial Theorem (Example 1.3) to show that if a field $F$ has characteristic $p$ a prime, then for any $a, b \in F$, $(a + b)^p = a^p + b^p$. (Wow, none of those bothersome middle terms!)
(b) More generally, show that if a field $F$ has characteristic $p$ a prime, then for any $a, b \in F$ and any $n \in \mathbb{Z}^+$, $(a + b)^{p^n} = a^{p^n} + b^{p^n}$.

**29.12.** Consider the set $F$ consisting of the following four $2 \times 2$ matrices with entries from $\mathbb{Z}_2$, the ring of integers modulo 2:

$$\begin{pmatrix} 0 & 0 \\ 0 & 0 \end{pmatrix}, \begin{pmatrix} 1 & 0 \\ 0 & 1 \end{pmatrix}, \begin{pmatrix} 1 & 1 \\ 1 & 0 \end{pmatrix}, \begin{pmatrix} 0 & 1 \\ 1 & 1 \end{pmatrix}.$$

(a) Show that $F$ is a field. (Note: There is a fair amount to show here. You must first show that $F$ is a subring of the ring of all $2 \times 2$ matrices over $\mathbb{Z}_2$ by showing that it is closed under both addition and multiplication. Having done so, $F$ clearly contains the unity element, but you must establish commutativity of multiplication. Finally, you must show that each non-zero matrix has a multiplicative inverse.)
(b) What is the characteristic of $F$? Have we seen this field, in a different form, already?

**29.13.** (a) Prove that the field $\mathbb{F}_p$ of integers modulo the prime $p$ contains no proper subfields.
(b) Conjecture as to what proper subfields will the finite field $\mathbb{F}_{16}$ have. (Hint: Working in the integers, how does the polynomial $x^4 - 1$ factor? Which of those factors are of the form $x^k - 1$? Now apply this to the case $x = 2$.)
(c) Generalizing, what proper subfields will the finite field $\mathbb{F}_{p^n}$ have? (Hint: For what $k$ does $x^n - 1$ have a factor of the form $x^k - 1$?)

**29.14.** (a) Show that the intersection $S \cap T$ of two subfields $S$ and $T$ of a field $F$ is a subfield of $F$.
(b) What familiar field is $\mathbb{Q}[\sqrt{2}] \cap \mathbb{Q}[\sqrt{3}]$? Justify your answer. (Hint: Suppose $x$ is an element of $\mathbb{Q}[\sqrt{2}] \cap \mathbb{Q}[\sqrt{3}]$, then for some $a, b, c, d \in \mathbb{Q}$, $x = a + b\sqrt{2} = c + d\sqrt{3}$. Show that unless $b$ and $d$ are both zero, we get a contradiction.)

# CHAPTER 30

## Vector Spaces

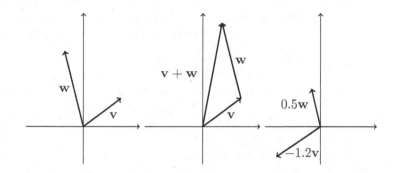

If you have taken a course in linear algebra (or physics, etc.), you are no doubt familiar with the concept of vectors. Geometrically, we can think of a vector as an object which has both a direction and a magnitude. The most basic setting for encountering vectors is in the real $xy$-plane (i.e., $\mathbb{R} \times \mathbb{R} = \mathbb{R}^2$), or in real $xyz$-space (i.e. $\mathbb{R} \times \mathbb{R} \times \mathbb{R} = \mathbb{R}^3$). On the left above we show two vectors $\mathbf{v}$ and $\mathbf{w}$ lying in $\mathbb{R}^2$, always drawn as starting from the origin. (We shall denote vectors using **bold** font).

So what natural operations can we perform on our two vectors above? Two such operations come to mind:

1. We can add them. Geometrically, this means placing one of the vectors (say $\mathbf{w}$) to start at the end of the other vector $\mathbf{v}$. The resulting vector, starting from the origin, is $\mathbf{v} + \mathbf{w}$, as illustrated in the middle above.

2. We can change their magnitude by multiplying by some real number. This is illustrated on the right above.

With these two operations, the set of all vectors in $\mathbb{R}^2$ is an example of what we call a *vector space*. The changing of magnitude via multiplication is called *scalar multiplication*, and the multipliers (i.e., elements of $\mathbb{R}$ in this case) are called *scalars*.

In the example above we have viewed vectors geometrically, but we can also view them algebraically, and this is the point of view we adopt, not surprisingly, in abstract algebra. So what are our vectors $\mathbf{v}$ and $\mathbf{w}$ algebraically, meaning what are the coordinates at the terminal points of the vectors? Well, we can't tell from our given graphs since there are no scales on the axes, but let's say that $\mathbf{v} = (2, 1.5)$ and $\mathbf{w} = (-1, 4)$. Hence, as expected, $\mathbf{v} + \mathbf{w} = (1, 5.5)$, $-1.2\mathbf{v} = (-2.4, -1.8)$ and $0.5\mathbf{w} = (-0.5, 2)$.

Having viewed this basic example, it is time to make a definition of an abstract algebraic object just as we've done with groups, rings and fields. The main ingredients of a "vector space" seem to be existence of the two operations we saw above (addition of the vectors and "scalar multiplication" of the vectors by elements of some field). We will need, of course, to be more specific about what properties those operations have and how they interact (e.g., distributive law or laws). So here we go with the formal definition.

**Definition 30.1.** Given a field $F$, a **vector space** $V$ **over** $F$ is an additive Abelian group for which we also have a **scalar multiplication** of elements of $V$ by elements of $F$ which must satisfy the following properties.

V1: for each $\mathbf{v} \in V$ and $\lambda \in F$, $\lambda\mathbf{v} \in V$
($V$ is closed under scalar multiplication);
V2: for each $\mathbf{v} \in V$ and $\lambda, \mu \in F$, $(\lambda\mu)\mathbf{v} = \lambda(\mu\mathbf{v})$
(associativity of scalar multiplication);
V3: for each $\mathbf{v} \in V$, $1\mathbf{v} = \mathbf{v}$
(unity element acts on $V$ as it does in $F$);
V4: for each $\mathbf{v} \in V$ and $\lambda, \mu \in F$, $(\lambda + \mu)\mathbf{v} = \lambda\mathbf{v} + \mu\mathbf{v}$
(first distributive law);
V5: for each $\mathbf{v}, \mathbf{w} \in V$ and $\lambda \in F$, $\lambda(\mathbf{v} + \mathbf{w}) = \lambda\mathbf{v} + \lambda\mathbf{w}$
(second distributive law).

The elements of $V$ are called **vectors**, while the elements of $F$ are called **scalars**. Again, throughout this chapter we denote the names of vectors (but not their coordinates if any) in bold type, and we often denote scalars using Greek letters such as $\lambda$ ("lambda") and $\mu$ ("mu").

Beyond the familiar example of $V = \mathbb{R}^2$ over $\mathbb{R}$ under coordinate-wise addition of real numbers and real number multiplication of the scalar times each coordinate (i.e., $\lambda(x_1, x_2) = (\lambda x_1, \lambda x_2)$) which we already discussed, there are numerous other natural examples of vector spaces.

**Example 30.1.** The vector space $V = \mathbb{R}^2$ over $\mathbb{R}$ can easily be generalized to $V = \mathbb{R}^n$ over $\mathbb{R}$ for any positive integer $n$. In this more general setting, we again use coordinate-wise addition, and scalar multiplication, as expected, is simply

$$\lambda(x_1, \ldots, x_n) = (\lambda x_1, \ldots, \lambda x_n).$$

Physicists find $\mathbb{R}^3$ and $\mathbb{R}^4$ (with the fourth dimension often representing time) particularly useful.

**Example 30.2.** The complex numbers $\mathbb{C}$ are a vector space over the real numbers $\mathbb{R}$. Because $\mathbb{R}$ is a subfield of $\mathbb{C}$, the five properties V1-V5 are automatically satisfied since all the operations are occurring within $\mathbb{C}$. In particular, property V1 is satisfied since the product of a real number and a complex number is a complex number. Similarly, the real numbers $\mathbb{R}$ are a vector space over the rational numbers $\mathbb{Q}$.

Example 30.2 contains two cases of a general result which we now state as a theorem.

**Theorem 30.1.** *If $S$ is a subfield of a field $F$, then $F$ is a vector space over $S$.*

Proof. Because multiplication in the field $F$ is closed and associative, and because distributivity holds for its two operations, properties V1 through V5 are all satisfied for scalar multiplication by the elements of $S$. ∎

You are asked in Exercise 30.7 to show that the following statement is false: If $S$ is a subfield of a field $F$, then $S$ is a vector space over $F$.

Another source of examples of vector spaces is the following: Suppose that $G$ is an Abelian group which is in some way "built out of" a field $F$. It could be a Cartesian product of copies of $F$, or it could be sets of matrices "over" F (meaning all matrix entries are from $F$), or it could be a set of polynomials whose coefficients come from $F$, and so on. In fact, we should observe right here that several of the examples we've seen so far fall into this pattern: $\mathbb{R}^2$ and $\mathbb{R}^n$ are Cartesian products of copies of $\mathbb{R}$, and the *additive* structure of the complex numbers $\mathbb{C}$ is simply that of $\mathbb{R} \times \mathbb{R}$. So let's explore some more examples of this idea.

**Example 30.3.** By Theorem 30.1, the finite field $\mathbb{F}_{p^n}$ is a vector space over its subfield $\mathbb{F}_p$. However, we can also view this as an example of the idea in the previous paragraph since the additive structure of $\mathbb{F}_{p^n}$ is that of the Cartesian product of $n$ copies of $\mathbb{F}_p$.

**Example 30.4.** When studying matrices and matrix rings in Chapters 12 and 26, we concentrated on $n \times n$ (i.e., square) matrices since only those allow a closed matrix multiplicative operation. (In fact, recall that if $n \neq m$, one cannot multiply two $n \times m$ matrices at all!) However, to possibly qualify to be a vector space, we need only be concerned with the additive structure of our set, and we can indeed *add* two $n \times m$ matrices whether or not $n = m$. Hence if we take the set $A = (a_{ij})$ of all $n \times m$ matrices over any field $F$ (i.e., for all $i, j$, $a_{ij} \in F$), then $A$ will indeed be a vector space over $F$ with scalar multiplication defined, as expected, by $\lambda A = \lambda(a_{ij}) = (\lambda a_{ij})$.

**Example 30.5.** The set $F[x]$ of all polynomials whose coefficients are in a field $F$, with addition defined by adding polynomials in the usual way, is a vector space over $F$. Scalar multiplication by an element $\lambda$ in $F$ is defined by simply multiplying each coefficient of a polynomial by the scalar $\lambda$. Likewise, the set $F_n[x]$ of all polynomials with coefficients in $F$ whose degree is at most $n$, using the same operations of polynomial addition and scalar multiplication as above, will form a vector space over $F$. We note that $F_n[x]$ has the exact same vector space structure as the set of all $1 \times (n+1)$ matrices over $F$ (since a polynomial of degree $n$ has $n + 1$ coefficients).

Now that we know the definition of a vector space and have plenty of examples under our belts, we will next examine some key properties of vector spaces and (as you might expect) learn about *subspaces* of vector spaces in the following two chapters.

### Exercises

**30.1.** In the abstract vector space $V$ (which is an additive Abelian group), we denote its zero element by bold **0**. Show that for any vectors $\mathbf{v}, \mathbf{w}$ in $V$ over a field $F$ (whose zero element we denote by regular 0) and any scalar $\lambda$ in $F$, the following hold:
  (a) $0\mathbf{v} = \mathbf{0}$;
  (b) $\lambda\mathbf{0} = \mathbf{0}$.
(Hint: Use vector space properties V4 and V5 from Definition 30.1.)

**30.2.** In the abstract vector space $V$ over a field $F$, show that the following hold for any $\mathbf{v}$ and $\mathbf{w} \in V$ and $\lambda \in F$:
  (a) $-1\mathbf{v} = -\mathbf{v}$;
  (b) $\lambda(\mathbf{v} - \mathbf{w}) = \lambda\mathbf{v} - \lambda\mathbf{w}$.
(Hint: Use Exercise 30.1 and vector space properties V2, V3 and V5.)

**30.3.** Which of the following form vector spaces over the given field? Justify your answers.
(a) The set of all real numbers of the form $a + b\sqrt{2} + c\sqrt[3]{3}$ where $a, b, c \in \mathbb{Q}$, the field of rational numbers, over $\mathbb{Q}$.
(b) The set of all polynomials of degree greater than 5 over a field $F$.
(c) The set of all $2 \times 2$ matrices with entries from the finite field $\mathbb{F}_{27}$, over the finite field $\mathbb{F}_3$.

**30.4.** Which of the following form vector spaces over the given field? Justify your answers.
(a) The set of all polynomials with zero constant terms over a field $F$.

(b) The subset $\{0, x+2, 2x+4, 3x+1, 4x+3\}$ of the ring of polynomials $\mathbb{Z}_5[x]$, over the field $\mathbb{Z}_5$.

(c) The set of all functions $f : \mathbb{R} \to \mathbb{R}$ which have the property that for all $x \in \mathbb{R}$, $f(x+1) = f(x)$, over the field of real numbers.

**30.5.** Suppose $\lambda$ and $\mu$ are in a field $F$ and $\mathbf{v}$ is a non-zero element of $V$, a vector space over $F$. Prove that if $\lambda \mathbf{v} = \mu \mathbf{v}$, then $\lambda = \mu$.

**30.6.** Suppose $\lambda$ is a non-zero element in a field $F$ and $\mathbf{v}$ and $\mathbf{w}$ are elements of $V$, a vector space over $F$. Prove that if $\lambda \mathbf{v} = \lambda \mathbf{w}$, then $\mathbf{v} = \mathbf{w}$. (Note: Surprisingly, this is easier to prove than is Exercise 30.5.)

**30.7.** Show by giving two counter-examples that the following statement is false: If $S$ is a subfield of a field $F$, then $S$ is a vector space over $F$.

**30.8.** Prior to Example 30.3, we noted that the additive structure of the complex numbers $\mathbb{C}$ can be viewed as the Cartesian product $\mathbb{R} \times \mathbb{R}$. Explain why even though the real numbers do form a vector space over the rational numbers $\mathbb{Q}$ (by Theorem 30.1), we cannot view the additive structure of $\mathbb{R}$ as a Cartesian product of copies of $\mathbb{Q}$. (Hint: See Chapter 9.)

**30.9.** Let $V$ and $W$ be vector spaces over a field $F$. A function $f : V \to W$ is a called *linear mapping* if $f(\mathbf{v} + \mathbf{u}) = f(\mathbf{v}) + f(\mathbf{u})$ for all $\mathbf{v}, \mathbf{u} \in V$ and $\lambda f(\mathbf{v}) = f(\lambda \mathbf{v})$ for all $\mathbf{v} \in V$ and all $\lambda \in F$.

Let $L(V, W)$ denote the set of all linear mappings from the vector space $V$ to the vector space $W$. Defining addition of mappings by $(f+g)(\mathbf{v}) = f(\mathbf{v}) + g(\mathbf{v})$, and scalar multiplication as above, show that the set $L(V, W)$ is a vector space over the field $F$. (Note: A key item to check is closure of both addition and scalar multiplication; i.e., is the sum of two linear mappings a linear mapping, and is a scalar multiple of a linear mapping a linear mapping?)

**30.10.** Returning to our first example in this chapter of a vector space while also looking ahead, consider the two vectors $\mathbf{v} = (3, 1)$ and $\mathbf{w} = (2, -4)$ in the vector space $\mathbb{R}^2$ over $\mathbb{R}$.

(a) Sketch $\mathbf{v}$ and $\mathbf{w}$.

(b) Suppose that $(x, y)$ is an arbitrary vector in $\mathbb{R}^2$. Show that $(x, y)$ can be written as a "linear combination" of $\mathbf{v}$ and $\mathbf{w}$; that is, show that there exist scalars $\lambda$ and $\mu$ such that $(x, y) = \lambda \mathbf{v} + \mu \mathbf{w}$. Do this by solving the two simultaneous equations $x = 3\lambda + 2\mu$ and $y = \lambda - 4\mu$ for $\lambda$ and $\mu$ in terms of $x$ and $y$.

(c) Using Part (b), write (for example) the vector $(-5, 2)$ as a linear combination of $\mathbf{v}$ and $\mathbf{w}$.

(Note: Because any vector in $\mathbb{R}^2$ can be expressed as a linear combination of $\mathbf{v}$ and $\mathbf{w}$, these two vectors form what is called a "basis" for $\mathbb{R}^2$. We shall study this concept in the next chapter.)

CHAPTER **31**

## Vector Space Properties

$$(\lambda, \mu) = \lambda(1, 0) + \mu(0, 1)$$

Having learned the definition of a vector space and seen numerous examples, let's now take a close look at some of their most important properties. We start with a simple example.

**Example 31.1.** In Chapter 30 we started with the two vectors $\mathbf{v}$ and $\mathbf{w}$ in $\mathbb{R}^2$ which are pictured there. Two key questions we can ask about them and which we shall explore and be able to generalize in this chapter are:
(1) Do $\mathbf{v}$ and $\mathbf{w}$ "depend" on each other in some way, or are they "independent?"
(2) Given any vector $\mathbf{u}$ in $\mathbb{R}^2$, can we write $\mathbf{u}$ in terms of $\mathbf{v}$ and $\mathbf{w}$?
Dealing first with Question 1, by "depend" we mean: Can one of them, say $\mathbf{w}$, be written in the form $\lambda\mathbf{v}$ for some scalar $\lambda$ in $\mathbb{R}$? The answer here is obviously no, since, geometrically, $\lambda\mathbf{v}$ lies on the line through the origin on which $\mathbf{v}$ lies, and $\mathbf{w}$ is not on that line. Hence we say $\mathbf{u}$ and $\mathbf{w}$ are "independent."
Now with Question 2, by the words "in terms of" we mean: Can an arbitrary vector $\mathbf{u} = (s, t)$ in $\mathbb{R}^2$ be written in the form $\mathbf{u} = \lambda\mathbf{v} + \mu\mathbf{w}$ for some scalars $\lambda$ and $\mu$ in $\mathbb{R}$? This is called writing $\mathbf{u}$ as a *linear combination* of $\mathbf{v}$ and $\mathbf{w}$. Working algebraically now, since $\mathbf{v} = (2, 1.5)$ and $\mathbf{w} = (-1, 4)$, we seek $\lambda$ and $\mu$ by solving two simultaneous linear equations (one involving the first coordinates of $\mathbf{v}$ and $\mathbf{w}$; the second involving their second coordinates) as follows:

$$s = 2\lambda + (-1)\mu \text{ and } t = 1.5\lambda + 4\mu.$$

Work through this solution (Do it! Good practice). You should arrive at

$$\lambda = \frac{2}{7}s + \frac{1}{14}t \text{ and } \mu = -\frac{3}{7}s + \frac{1}{7}t.$$

211

For these values of $\lambda$ and $\mu$, we have

$$\mathbf{u} = \lambda\mathbf{v} + \mu\mathbf{w},$$

and since $\mathbf{u}$ was arbitrary, we have established the answer to Question 2. We say then that the independent vectors $\mathbf{v}$ and $\mathbf{w}$ *span* all of $\mathbb{R}^2$.

**Example 31.2.** Staying in the vector space $\mathbb{R}^2$ over $\mathbb{R}$ but generalizing, let's assume that $\mathbf{u} = (s, t)$, $\mathbf{v} = (a, b)$ and $\mathbf{w} = (c, d)$ are all arbitrary vectors in $\mathbb{R}^2$ and ask under what conditions is $\mathbf{u}$ a linear combination of $\mathbf{v}$ and $\mathbf{w}$? This means we seek scalars $\lambda$ and $\mu$ such that

$$s = \lambda a + \mu b \text{ and } t = \lambda c + \mu b.$$

Solving these simultaneous equations (very carefully), and labeling the quantity $ad - bc$ as $D$, we arrive at

$$\lambda = \frac{d}{D}s - \frac{b}{D}t \text{ and } \mu = -\frac{c}{D}s + \frac{a}{D}t.$$

We observe then that there are solutions for $\lambda$ and $\mu$ *if and only if $D = ad - bc$ is non-zero!* Does the quantity $ad - bc$ look familiar? It should, since, as we learned in Chapter 12, it is the determinant of the matrix $\begin{pmatrix} a & b \\ c & d \end{pmatrix}$; that is, the matrix whose rows are the coordinates of the vectors $\mathbf{v}$ and $\mathbf{w}$. Hence we get solutions for $\lambda$ and $\mu$ if and only if that matrix is non-singular (i.e., if its determinant is non-zero). In this case the rows of the matrix (and so the vectors they represent) are said to be *linearly independent*. All this carries over to general vector spaces, as we shall now see.

**Definition 31.1.** Let $\mathbf{v}_1, \ldots, \mathbf{v}_n$ be a set of $n$ vectors in a vector space $V$ defined over a field $F$. We say that this set of vectors is *linearly dependent* if there are scalars $\lambda_1, \ldots, \lambda_n \in F$, not all zero, so that

$$\lambda_1\mathbf{v}_1 + \cdots + \lambda_n\mathbf{v}_n = \mathbf{0}.$$

If the set is not linearly dependent, it is *linearly independent*. Equivalently, the set of vectors is linearly independent over the field $F$ if, whenever

$$\lambda_1\mathbf{v}_1 + \cdots + \lambda_n\mathbf{v}_n = \mathbf{0},$$

it must be the case that $\lambda_1 = \cdots = \lambda_n = 0$.

The definition of linear dependence makes perfect sense as follows: Suppose

$$\lambda_1\mathbf{v}_1 + \cdots + \lambda_k\mathbf{v}_k + \cdots + \lambda_n\mathbf{v}_n = \mathbf{0}$$

where $\lambda_k \neq 0$. Then we can solve for $\mathbf{v}_k$, obtaining

$$\mathbf{v}_k = (-\lambda_1/\lambda_k)\mathbf{v}_1 + \cdots + (-\lambda_{k-1}/\lambda_k)\mathbf{v}_{k-1} + (-\lambda_{k+1}/\lambda_k)\mathbf{v}_{k+1} + \cdots + (-\lambda_n/\lambda_k)\mathbf{v}_n,$$

i.e., $\mathbf{v}_k$ can be written as a linear combination of the other vectors, and so is "dependent" on them.

Let's look at an example in $\mathbb{R}^3$ (i.e. "real $xyz$-space").

**Example 31.3.** We claim that the vectors $(1,1,1)$, $(1,2,1)$ and $(3,5,3)$ are linearly dependent over the field $\mathbb{R}$. In this case it's easy enough simply to observe that $(1,1,1) + 2(1,2,1) = (3,5,3)$, showing the dependence of $(3,5,3)$ on the other two explicitly. Another approach would be to compute the determinant of the matrix $\begin{pmatrix} 1 & 1 & 1 \\ 1 & 2 & 1 \\ 3 & 5 & 3 \end{pmatrix}$. We did not discuss computing the determinant of a $3 \times 3$ matrix in Chapter 12, but it is in fact

$$3 \det \begin{pmatrix} 1 & 1 \\ 1 & 2 \end{pmatrix} - 5 \det \begin{pmatrix} 1 & 1 \\ 1 & 1 \end{pmatrix} + 3 \det \begin{pmatrix} 1 & 1 \\ 2 & 1 \end{pmatrix} = 3(1) - 5(0) + 3(-1) = 0.$$

A third approach would be to set up three simultaneous equations in $\lambda_1$, $\lambda_2$ and $\lambda_3$ and discover (using, for example, Gauss-Jordan elimination) non-zero solutions, the simplest of which, we already observed, is $\lambda_1 = 1$, $\lambda_2 = 2$ and $\lambda_3 = -1$. By the way, the geometric implication of these three vectors being linearly dependent is that they all lie in the same plane through the origin in $\mathbb{R}^3$.

The following definition covers three key concepts in the structure of vector spaces.

**Definition 31.2.** A set $\{\mathbf{v}_1, \ldots, \mathbf{v}_n\}$ of vectors in a vector space $V$ is said to **span** the vector space $V$ if any vector $\mathbf{u} \in V$ can be written in the form

$$\mathbf{u} = \lambda_1 \mathbf{v}_1 + \cdots + \lambda_n \mathbf{v}_n$$

for some set of scalars $\{\lambda_1, \ldots, \lambda_n\}, \lambda_i \in F$; i.e., if every vector in $V$ can be written as a linear combination of the vectors $\mathbf{v}_1, \ldots, \mathbf{v}_n$.

In addition, a set of vectors forms a **basis** for the vector space $V$ if it is both linearly independent and the set spans $V$. If $V$ has a basis consisting of $n$ vectors, then the **dimension** of $V$ is $n$, and $V$ is called *finite dimensional*. Otherwise $V$ is called *infinite dimensional*.

We will not prove the following (not surprising) result here. For a proof, see, for example, [5].

**Theorem 31.1.** *Any two bases of a finite dimensional vector space over a field $F$ contain the same number of vectors.*

**Example 31.4.** Looking back at Example 31.1, we see that the two vectors $\mathbf{v}$ and $\mathbf{w}$ form a basis for the 2-dimensional vector space $\mathbb{R}^2$ over $\mathbb{R}$. On the other hand, the three vectors in Example 31.3 do *not* form a basis for the 3-dimensional vector space $\mathbb{R}^3$ over $\mathbb{R}$, but any two of them will form a basis for a 2-dimensional "subspace" of $\mathbb{R}^3$ (again, geometrically, a plane through the origin in 3-space). We discuss the concept of subspaces in the next chapter.

**Example 31.5.** Generalizing the previous example, we know that for any positive integer $n$, $\mathbb{R}^n$ is an $n$-dimensional vector space over $\mathbb{R}$. For each $n$, the simplest basis is the set of $n$-dimensional vectors

$$\{(1, 0, 0, \ldots, 0), (0, 1, 0, \ldots, 0), \ldots, (0, 0, 0, \ldots 0, 1)\}.$$

This set is called the ***standard basis*** for $\mathbb{R}^n$. We note also that if $F$ is *any* field, then the Cartesian product $F \times F \times \cdots \times F = F^n$ is an $n$-dimensional vector space over $F$, and if the unity element of $F$ is denoted 1, then the same set as above forms the standard basis for $F^n$ over $F$. You are asked to verify this in Exercise 31.6 below.

**Example 31.6.** The complex numbers $\mathbb{C}$ are a 2-dimensional vector space over the real numbers $\mathbb{R}$, the simplest basis being the pair $\{1, i\}$. However, as we examined in Exercise 30.8, $\mathbb{R}$ is definitely an infinite-dimensional vector space over the rational numbers $\mathbb{Q}$. See also Exercise 31.12 below.

**Example 31.7.** The set $F[x]$ of all polynomials over a field $F$ is an infinite dimensional vector space over $F$. If the unity element of $F$ is denoted 1, the simplest basis is the infinite set of polynomials $\{1, x, x^2, \ldots, x^n, \ldots\}$. Likewise, the set $F_n[x]$ of all polynomials in $F[x]$ whose degree is at most $n$ over a field $F$ has dimension $n + 1$, with the simplest basis being the set $\{1, x, x^2, \ldots, x^n\}$. Verify yourself that this is indeed a basis.

**Example 31.8.** If $F$ is any field, the set of all $n \times m$ matrices with entries from $F$ is an $nm$-dimensional vector space over $F$. Again, if the unity element of $F$ is denoted by 1, the simplest basis is the set of $nm$ distinct matrices in which one entry is 1 and the remaining $nm - 1$ entries are 0. Once again, check this assertion.

As referred to in Example 31.3 above, vector spaces, like all other algebraic objects, have "sub-objects", called, of course, *subspaces*. We shall study this concept in the following chapter.

**Exercises**

**31.1.** Are the vectors $(2, 3)$ and $(4, 9)$ linearly independent or linearly dependent over the field $\mathbb{R}$ of real numbers? Justify your answer.

**31.2.** Prove that two vectors in a vector space over a field $F$ are linearly dependent over $F$ if and only if one of them is a scalar multiple of the other.

**31.3.** Are the matrices $A = \begin{pmatrix} 2 & 1 \\ 3 & 4 \end{pmatrix}$ and $B = \begin{pmatrix} 1 & -1 \\ 2 & -1 \end{pmatrix}$ linearly independent or linearly dependent over the field of real numbers? (Hint: See the previous exercise.)

**31.4.** If $\{\mathbf{u}, \mathbf{v}, \mathbf{w}\}$ is a set of linearly independent vectors in a vector space over a field $F$, show that the set $\{\mathbf{u}, \mathbf{u} + \mathbf{v}, \mathbf{u} + \mathbf{v} + \mathbf{w}\}$ is also linearly independent.

**31.5.** Consider the three vectors $(2, 3, 4), (3, 4, 5)$, and $(4, 5, 6)$ over the field of real numbers. Are these vectors linearly independent or not? Justify your answer. (Note: Several approaches are suggested in Example 31.3.)

**31.6.** Following up on Example 31.5, let $F$ be a field whose unity element is denoted 1 and let $n \geq 2$ be a positive integer. Show that the "standard basis" for $F^n$ consisting of the $n$ distinct vectors containing a single 1 and the rest all zeros really does form bases for $F^n$ over $F$, i.e.; show that this is a set of linearly independent vectors which span $F^n$.

**31.7.** Show that any set of $n + 1$ vectors in a vector space of dimension $n$ must be a set of linearly dependent vectors.

**31.8.** If $p$ is a prime number, we know that $V = F_p \times F_p$ is a 2-dimensional vector space over the finite field $F_p$. Prove that the two vectors $(a, b)$ and $(c, d)$ in $V$ are linearly independent (and hence form a basis of $V$) if and only if $ad - bc$ is not congruent to 0 modulo $p$.

**31.9.** Let $p \geq 7$ be a prime number. Over which finite fields $\mathbb{F}_p$ are the vectors $(6, 5)$ and $(5, 6)$ in $F_p \times F_p$ linearly independent?

**31.10.** Let $p \geq 7$ be a prime number. Show that the vectors $(5, 5, 4), (1, 2, 3)$ and $(4, 0, 3)$ are linearly independent over every field $\mathbb{F}_p$ except for one. Which field is that? (Hint: See Example 31.3.)

**31.11.** For $a, b, c \in \mathbb{R}$, let $V$ be the set of all matrices of the form $\begin{pmatrix} a & b \\ b & c \end{pmatrix}$. Show that $V$ is a vector space over $\mathbb{R}$. What is the dimension of $V$ over $\mathbb{R}$? Find a basis for $V$ over $\mathbb{R}$.

**31.12.** We pointed out in Example 31.6 that the real numbers $\mathbb{R}$ are an infinite dimensional vector space over the rational numbers $\mathbb{Q}$.
(a) Show that two distinct real numbers $x$ and $y \neq 0$ are linearly independent over $\mathbb{Q}$ if and only if their ratio $x/y$ is irrational.
(b) $\mathbb{R}$, being a vector space over $\mathbb{Q}$, must have a basis, but it is not at all clear what that basis might be. See what you can find out about this question by googling something like "Basis of real numbers as vector space over rational numbers."

# CHAPTER 32

## Subspaces of Vector Spaces

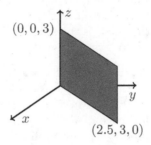

We know from our previous studies that every type of algebraic object we have encountered can contain *sub-objects*, which are subsets of the given set which themselves are the same type of algebraic object *under the same operation or operations*. That last part is important, and a simple example to keep in mind is that although $\mathbb{Z}_n = \{0, 1, \ldots, n-1\}$ is a *subset* of the integers $\mathbb{Z}$, it is *not* a *subring* of $\mathbb{Z}$ since addition and multiplication modulo $n$ are different operations than integer addition and multiplication. With that reminder in place, we now have the following definition.

**Definition 32.1.** Let $V$ be a vector space over a field $F$. A non-empty subset $U$ of $V$ is a *subspace* of $V$ if $U$ itself forms a vector space over the field $F$ using the same operations as in the vector space $V$.

Thus, if the subset $U$ is itself a group under addition and is closed under scalar multiplication, then $U$ forms a subspace of $V$. As usual, let's look at some examples.

**Example 32.1.** Suppose $V$ is a vector space over a field $F$ and suppose $\mathbf{v}$ is a non-zero vector in $V$. Then it is easy to check (do it!) that the set $U = \{\lambda \mathbf{v} | \lambda \in F\}$ is a one-dimensional subspace of $V$. If, for example, $V = \mathbb{R}^3$ and $\mathbf{v} = (a, b, c)$ then geometrically $U$ is the line through the origin and through the point $(a, b, c)$ in $xyz$-space.

**Example 32.2.** Consider the two vectors $\mathbf{v} = (0, 0, 3)$ and $\mathbf{w} = (2.5, 3, 0)$ in $\mathbb{R}^3$, as pictured at the opening of this chapter. By Exercise 31.2 (or various other methods discussed in Chapter 31), $\mathbf{v}$ and $\mathbf{w}$ are linearly independent and hence span a two-dimensional subspace $U$ of $\mathbb{R}^3$. Geometrically, $U$ (partially shown in the figure at the start of the chapter) is the plane through the origin which contains both $\mathbf{v}$ and $\mathbf{w}$.

**Example 32.3.** We have seen that the complex numbers $\mathbb{C}$ form a two-dimensional vector space over the real numbers $\mathbb{R}$. The vector $\mathbf{v} = (1, i)$ then generates a one-dimensional subspace $U = \{(\lambda, \lambda i) | \lambda \in \mathbb{R}\}$. Viewing $\mathbb{C}$ as the complex plane, $U$ is the diagonal line through the origin and the point $(1, i)$.

**Example 32.4.** Consider the vector space $V = F_n[x]$ of all polynomials of degree less than or equal to $n$ with coefficients from a field $F$. Consider the subset $U$ of all polynomials in $V$ in which the exponent of every non-zero term is an even number. It is easy to check (do so!) that $U$ is an Abelian group under polynomial addition and that $U$ is closed under scalar multiplication. Hence $U$ is a subspace of $V$ whose dimension you are asked to compute in Exercise 32.2. Finally, note that if $W$ is the subset with "odd" replacing "even" for the exponents of the non-zero terms, then $W$ is *not* a subspace of $V$. Why?

**Example 32.5.** Suppose $V$ is the set of all $n \times n$ matrices over a field $F$. Let $U$ denote the subset of $V$ consisting of all matrices in which every element off the main diagonal is equal to 0 (such matrices are called *diagonal matrices*). It's easy to see that $U$ is a subspace of $V$, being both an Abelian group under matrix addition and closed under scalar multiplication. Its dimension over $F$ is $n$. Similarly, the subset $W$ of $V$ consisting of all matrices which have all zeros *below* the main diagonal (such matrices are called *upper triangular matrices*) forms a subspace of $V$. You are asked to compute its dimension in Exercise 32.2.

**Example 32.6.** We have observed several times that the real numbers $\mathbb{R}$ form an infinite dimensional vector space over the rational numbers $\mathbb{Q}$. Let $B$ be the infinite set $\{\sqrt{2}, \sqrt{3}, \sqrt{5}, \ldots, \sqrt{p}, \ldots\}$. (That is, $B = \{\sqrt{p} | p \text{ prime}\}$.) You can check, using an argument similar to the one you used to show that $\sqrt{2}$ is irrational (Exercise 5.9), that the elements of $B$ are linearly independent over $\mathbb{Q}$. Now if $U$ is the subset of $\mathbb{R}$ spanned by the set $B$, $U$ is a subspace of $\mathbb{R}$. The subspace $U$ is clearly infinite dimensional over $\mathbb{Q}$, but it must nonetheless be a *proper* subspace of $\mathbb{R}$. Using an argument exactly like the "zig-zag" argument we used in Example 9.3 that the positive rational numbers $\mathbb{Q}^+$ are countable, we can see that the Cartesian product of a countable number of countable sets is itself countable. However, the set of all real numbers $\mathbb{R}$ is, as we know, uncountable.

As we have now seen illustrated in numerous examples, it seems clear that since a subspace $U$ is a subset of the vector space $V$, the dimension of $U$ must be at most that of the dimension of $V$. For the case of finite dimensional vector spaces, let us state this formally as a theorem.

**Theorem 32.1.** *If $V$ is a vector space of dimension $n$ over a field $F$, then the dimension $m$ of a subspace $U$ of $V$ is at most $n$, i.e., $m \leq n$.*

Proof. Let $\{\mathbf{v}_1, \mathbf{v}_2, \ldots, \mathbf{v}_m\}$ be a basis for the subspace $U$ over $F$. Suppose that $m > n$. Since each vector $\mathbf{v}_i$ is also in $V$, we have a set of at least $n + 1$ vectors which are linearly independent. But by Exercise 31.7, our set of $m$ vectors must be linearly *dependent* since the dimension of $V$ is $n$. This contradiction forces us to conclude that $m \leq n$. ∎

Since we have now come to the end of our introduction to types of sets with algebraic structure, it may be helpful to review some of the basics of what we have learned. The chart below is by no means complete, of course, but it can perhaps help you recall most of the different types of structures and some basic examples of each type. Take a little time to go over this chart and make sure that you are comfortable with all the types and examples given there.

| Algebraic Type | Basic Example(s) |
|---|---|
| finite Abelian group | $(\mathbb{Z}_n, +)$, $(U_n, \cdot)$ |
| infinite Abelian group | $(\mathbb{Z}, +)$, $(\mathbb{R}^*, \cdot)$ |
| finite non-Abelian group | $S_n$ (permutations of $n$ objects) |
| infinite non-Abelian group | invertible $n \times n$ matrices over $\mathbb{R}$ |
| finite commutative ring | $\mathbb{Z}_n$ |
| infinite commutative ring | $\mathbb{Z}$, $\mathbb{R}[x]$ (polynomials over $\mathbb{R}$) |
| non-commutative ring | $n \times n$ matrices over $\mathbb{R}$ |
| finite field | $\mathbb{F}_p = \mathbb{Z}_p$, $\mathbb{F}_{p^n}$ |
| infinite field | $\mathbb{Q}$, $\mathbb{R}$, $\mathbb{C}$ |
| finite dimensional vector space | $\mathbb{R}^n$ over $\mathbb{R}$ |
| infinite dimensional vector space | $\mathbb{R}$ over $\mathbb{Q}$ |

We now finish this text by taking a close look at polynomials. We know that the set $R[x]$ of all polynomials over a ring $R$ (meaning with coefficients from $R$), or the set $R_n[x]$ of just those polynomials of degree less than or equal to a fixed positive integer $n$, are themselves rings. We also know that if the polynomials are over a field $F$, then they form a vector space over $F$. But polynomial rings especially have interesting *multiplicative* structure depending on what ring $R$ or field $F$ they are over, and this structure is what we now wish to examine.

**Exercises**

**32.1.** Consider the vector space $V = \mathbb{R}^3$ over $\mathbb{R}$, which in "calculus language" is $xyz$-space.
(a) What is the calculus language term for the subspace $U$ of $V$ which is spanned by the vector $(0, 1, 0)$?
(b) Same question for $U$ spanned by the vector $(0, 0, 0)$.
(c) Same question for $U$ spanned by the vectors $(1, 0, 0)$ and $(0, 0, 1)$.

**32.2.** (a) Following up in Example 32.4, what is the dimension of the subspace $U$ of polynomials whose non-zero terms all have even exponents. (Note: The answers are slightly different depending on whether $n$ is even or odd.)

(b) Following up on Example 32.5, what is the dimension of the subspace of the upper triangular $n \times n$ matrices over the field $F$?

**32.3.** Let $S = \{(x, y, z) \mid x, y, z \in \mathbb{R}, x = 2y + 3z\}$. Show that the set $S$ is a subspace of the vector space $\mathbb{R}^3$ over $\mathbb{R}$. Find a set of vectors which forms a basis for the subspace $S$.

**32.4.** Assume that $V$ is a vector space of dimension 5 over a field $F$. Further assume that $U$ and $W$ are subspaces of $V$, both of dimension 3. Show that $U \cap W \neq \{\mathbf{0}\}$; i.e., show that $U \cap W$ must contain a non-zero vector. (Hint: Any six vectors in $V$ must be linearly dependent.)

**32.5.** Let $V$ be the vector space $\mathbb{R}^3$ over the field of real numbers. Consider the set $S$ of vectors $(x, y, z)$ in $V$ with the property that $x^2 + y^2 = z^2$. Is $S$ a subspace of $V$? If so, what is its dimension? (Note: It's fun to try to picture what the set $S$ looks like in $xyz$-space. Think of two "rounded" cones. If you have access to some good mathematical software, it can help.)

**32.6.** If $H$ and $K$ are subspaces of a vector space $V$ over a field $F$, show that the set $H + K = \{\mathbf{h} + \mathbf{k} \mid \mathbf{h} \in H \text{ and } \mathbf{k} \in K\}$ is a subspace of $V$.

**32.7.** (a) Show that any vector space $V$ of dimension greater than or equal to 1 always has at least two distinct subspaces.

(b) For a prime $p$, how many subspaces does $\mathbb{F}_p \times \mathbb{F}_p$ have over $\mathbb{F}_p$?

**32.8.** (a) Prove that, as a vector space over $\mathbb{R}$, $\mathbb{R}^2$ has infinitely many subspaces. Geometrically, describe them.

(b) More generally, prove that if $V$ is a vector space of dimension at least 2 over an infinite field $F$, then $V$ has infinitely many subspaces.

**32.9.** (a) Prove that the intersection of two subspaces of a vector space $V$ over a field $F$ forms a subspace of $V$ over $F$.

(b) Show by counter-example that the union of two subspaces of a vector space $V$ over a field $F$ need not form a subspace of $V$.

**32.10.** Concerning our summary of types of algebraic structures at the end of the chapter, note that the set of real numbers $\mathbb{R}$ has the "distinction" of being a group, a ring, a field, and a vector space over a proper subset of itself. Give two more examples of sets we have studied which share this distinction.

# CHAPTER 33

# Polynomials

$$g(x) = q(x)f(x) + r(x), \deg(r) < \deg(f)$$

You became acquainted with polynomials quite early in your math education, as they played a central role in your courses in algebra and calculus. In algebra, some effort goes into understanding how to multiply two polynomials, and much more effort goes into learning to factor (i.e, undo multiplying) polynomials. This latter effort is aimed at finding ways to *solve polynomial equations*, a topic in which we will be very interested here. Indeed, in this text, our very first example (Example 1.1) concerns the famous Quadratic Formula, which tell us how to solve any polynomial equation of the form $ax^2 + bx + c = 0$. Moving on to calculus, our first and simplest differentiable functions (i.e., functions which possess a derivative function) are polynomials, and "everyone's favorite" derivative rule, the Power Rule (Example 2.1 in our text), tells us what the derivative of the polynomial $x^n$ is (for $n \in \mathbb{Z}^+$). Suffice it to say then that polynomials play a very important role in much of mathematics.

In this and the following two chapters, we will concentrate primarily on two key ideas about polynomials:

(1) As long as the coefficients of our polynomials come from an integral domain (i.e., a ring which possesses no zero divisors) or more especially a field, *polynomials behave very much like the integers* $\mathbb{Z}$. For example, there are polynomial versions of the Euclidean Algorithm and the Division Algorithm, and, like the integers, polynomials have *unique factorization*.

(2) Concerning polynomial equations, we will look closely at how the solutions to a given polynomial equation will differ when we view the coefficients as being from different rings. For example, the polynomial equation

$(x^2 - 2)(x^2 + 2) = 0$ has no solutions in the integers $\mathbb{Z}$, no solutions in the rational numbers $\mathbb{Q}$, two solutions in the real numbers $\mathbb{R}$, and four solutions in the complex numbers $\mathbb{C}$. We already discussed this idea in Chapters 1, 11 and 28, but we'll pursue it further at the end of this chapter and again in Chapter 35. In particular, we shall prove below (Theorem 33.5) that if $f$ is a polynomial of degree $n$ over a field $F$, then the equation $f(x) = 0$ can have at most $n$ solutions in $F$. (Such solutions are called *roots* of $f$.) So in our example above, the equation $(x^2 - 2)(x^2 + 2) = 0$ seeks the roots of the polynomial $f(x) = (x^2 - 2)(x^2 + 2)$ (of degree 4), and Theorem 33.5 assures us that there can be no more than 4 roots no matter what field $F$ the polynomial $f$ is over.

Here is a formal definition of a polynomial and some of the common terms involved:

**Definition 33.1.** Let $R$ be a commutative ring. A ***polynomial*** over $R$ (i.e., with coefficients from $R$) is an expression of the form

$$f(x) = a_n x^n + a_{n-1} x^{n-1} + \cdots + a_1 x + a_0$$

where the coefficients $a_n, \ldots, a_0$ are in $R$, $x$ is an *indeterminate* or *variable* which can take on values from the ring $R$, and $n$ is a non-negative integer. A polynomial is called ***monic*** if the coefficient of the highest power of $x$ is 1, i.e., in our notation above, if $a_n = 1$. Finally, a polynomial $f(x)$ has ***degree*** $n$ if the coefficient $a_n$ of the highest power of $x$ is a non-zero element in the ring $R$. We define the degree of the zero polynomial to be $-1$.

We note that a polynomial $f$ over a ring $R$ can be viewed as a function from $R$ to $R$; that is, $f : R \to R$ takes $a \in R$ to $f(a) \in R$. In a similar vein as point (2) above, the "same" polynomial can act quite differently depending on its underlying ring. For example, over the finite field $\mathbb{F}_{13}$ image of the polynomial $f(x) = x^{10}$ has 7 elements ($\{0, 1, 3, 4, 9, 10, 12\}$), but over $\mathbb{Z}_{11}$, by Fermat's Theorem (Theorem 17.1), the image has only 2 elements ($\{0, 1\}$).

As we have done throughout the text, the set of all polynomials over the commutative ring $R$ is denoted $R[x]$. We first observed in Example 26.4 that $R[x]$ is itself a ring under the "usual" operations of polynomial addition and multiplication. We note that polynomial addition is very simple since it is just a matter of adding (using addition in $R$) the coefficients of "like" terms, i.e., terms with the same power on the variable. Polynomial multiplication is more complicated, just as integer multiplication is more complicated than integer addition. In both cases multiplication essentially involves repeated use of the distributive law followed by gathering of like terms. For example, in the integers we have

$$(31)(12) = (3(10) + 1)((10) + 2) = (3(10) + 1)((10)) + (3(10) + 1)(2)$$

$$= 3(100) + 1(10) + 6(10) + 2 = 3(100) + 7(10) + 2 = 372,$$

whereas in polynomials, we have

$$(3x + 1)(x + 2) = (3x + 1)(x) + (3x + 1)(2) = 3x^2 + x + 6x + 2 = 3x^2 + 7x + 2.$$

This computation is, of course, the FOIL (i.e., "first, outer, inner, last") computation you enjoyed so much in junior high!

Because the multiplicative structure on $R[x]$ is significantly more complicated than the additive structure, it is also significantly more interesting, as we shall see as we move ahead.

If we assume that the underlying ring $R$ is an integral domain or a field, the lack of zero divisors guarantees that if the degree of $f$ is $n$ and the degree of $g$ is $m$, then the degree of the product $fg$ must be $n + m$. This is, of course, not true if $R$ contains zero divisors. For example, if $R = \mathbb{Z}_{15}$, then

$$(5x + 2)(3x + 1) = 0x^2 + 11x + 2 = 11x + 2.$$

In fact, you were asked to prove in Exercise 28.6 that $R[x]$ is itself an integral domain if and only if $R$ is an integral domain. However, it is *not* true that if $F$ is a field, then $F[x]$ is also a field. This is obvious since any polynomial of degree 1 or greater does not possess a multiplicative inverse in $F[x]$. Again, as we noted previously, $R[x]$ (or $F[x]$) closely resembles the integers $\mathbb{Z}$ in its algebraic structure. Put briefly, *polynomial rings are built very much like the integers*. Let's start to pursue this idea now.

In what follows we shall now assume that the elements of our polynomial ring $F[x]$ have their coefficients in a *field* $F$. We say that the polynomial $f \in F[x]$ **divides** a polynomial $g \in F[x]$ if there exists a polynomial $k \in F[x]$ such that $g = fk$, exactly as in the integers $\mathbb{Z}$. For what comes a bit later, we note that if $f$ divides $g$ and if $a$ is a non-zero element of the field $F$, then $af$ also divides $g$ since $g = (af)(a^{-1}k)$. For a simple example over the real numbers $\mathbb{R}$, clearly $x - 2$ divides $x^2 - 4$, but so does $2(x - 2)$ since $x^2 - 4 = (2(x - 2))(\frac{1}{2}(x + 2))$. An analogy to integers here is that 3 divides 12, but so does $-3$.

We introduced the Division Algorithm for the *integers* $\mathbb{Z}$ in Chapter 3 and then stated and proved it in Theorem 5.2. The proof required that we measure the "size" of a given positive integer, which obviously is its absolute value, and then applied the Well-Ordering Principle to assure that the remainder $r$ was strictly less than the divisor $a$. It's a slightly tricky proof which is worth looking back at (again, Theorem 5.2). In this polynomial setting we need a measure of the "size" of a non-zero polynomial. Well, a perfect measure is its *degree* since that number lies in the set $\{0, 1, 2, \ldots\}$ of non-negative integers. Hence the proof can employ the Well-Ordering Principle applied to degrees rather than to values!

**Theorem 33.1. (Division Algorithm for Polynomials)** *Let $F$ be a field. Let $f, g$ be two polynomials in $F[x]$ where $f \neq 0$. Then there are unique polynomials $q$ and $r$ with the degree of $r$ less than the degree of $f$ so that $g = fq + r$.*

**Proof.** This proof mirrors the proof of the Division Algorithm for integers (Theorem 5.2) but is a bit more complicated because we are measuring the degrees of our objects instead of the objects themselves. If $f$ divides $g$, i.e., if $g = fk$ for some polynomial $k \in F[x]$, then we clearly we have that $r$ is the zero polynomial (whose degree is $-1$) and we're done. Assuming then that $f$ does not divide $g$, we define $D$ to be the following set of polynomials in $F[x]$:

$$D = \{g - fk \mid k \in F[x]\}.$$

By our assumption, the set $D$ consists of polynomials whose degrees are nonnegative integers (i.e., $D$ does not contain the zero polynomial), and it is obviously non-empty, containing, for example, $g$ if we select $k$ to be the zero polynomial.

Hence, by the Well-Ordering Principle applied to the degrees of its elements, $D$ contains a polynomial, say $r$, such that $\deg(r) \leq \deg(h)$ for all $h = g - fk \in D$. Since $r$ is in $D$, $r$ must be of the form $r = g - fq$ for some polynomial $q$, i.e., $g = fq + r$. To finish the proof we must now show that $\deg(r) < \deg(f)$. Suppose not, that is, suppose $\deg(r) \geq \deg(f)$. Let $k = \deg(r) - \deg(f)$ and suppose the leading coefficient of $r$ is $a \in F$ and the leading coefficient of $f$ is $b \in F$. Consider the polynomial $r_2 = r - ab^{-1}x^k f$. Because the leading term of $r$ has cancelled out in $r_2$, $\deg(r_2) < \deg(r)$. Moreover,

$$r_2 = r - ab^{-1}x^k f = g - fq - ab^{-1}x^k f = g - f(q + ab^{-1}x^k),$$

which is an element of the set $D$. The existence of $r_2$ in $D$ contradicts our choice of $r$ as an element of smallest degree in $D$, so we conclude that $\deg(r) < \deg(f)$, and are done. (Whew again!) ∎

Not surprisingly, the polynomial $q$ is called the *quotient* and the polynomial $r$ is called the *remainder*. Let's look at an example.

**Example 33.1.** Using (the somewhat painful) long division of polynomials and working over the field of real numbers, if we divide $g(x) = x^4 + 3x + 2$ by $f(x) = 2x^2 - 1$, we obtain $x^4 + 3x^2 + 2 = (2x^2 - 1)(\frac{1}{2}x^2 + \frac{1}{4}) + 3x + \frac{9}{4}$, so that $q(x) = \frac{1}{2}x^2 + \frac{1}{4}$ and $r(x) = 3x + \frac{9}{4}$ (check it!). Note that the degree of the remainder $r$ is less than the degree of the divisor $f$.

Just as any two integers have a greatest common divisor (which we require to be positive, please note), any two polynomials $f$ and $g$ in $F[x]$ ($F$ a field) will have greatest common divisors $d$. We say "divisors" since, as we pointed out already, if $d$ divides both $f$ and $g$ and if $a$ in a non-zero element of $F$, then $ad$ also divides $f$ and $g$. Hence, just as we require that the gcd of two integers be positive to guarantee its uniqueness, we shall require that the gcd of two polynomials be *monic* (see Definition 33.1) to insure its uniqueness. So we have the following definition:

**Definition 33.2.** The **greatest common divisor** of two polynomials $f$ and $g$ in $F[x]$ ($F$ a field), denoted $\gcd(f,g)$, is the monic polynomial of greatest degree among those polynomials that divide both $f$ and $g$.

So how can we compute a polynomial gcd? Well. exactly as with integers, one method is to try to use factoring. Here's a simple example over the reals $\mathbb{R}$.

**Example 33.2.** Let $g(x) = x^3 + 3x^2 - 2x - 6$ and $f(x) = x^2 - 9$. Factoring $g$ by grouping (remember that one?), we get

$$g(x) = x^2(x+3) - 2(x+3) = (x+3)(x^2-2) = (x+3)(x-\sqrt{2})(x+\sqrt{2}).$$

Clearly $f(x) = (x-3)(x+3)$, and so $\gcd(f,g) = x+3$.

That was easy enough, but, just as is true with integers, factoring polynomials can be quite difficult, or worse. We learned some time ago that there is a relatively quick, surefire method for computing the gcd of two integers, and, not surprisingly, it works for polynomials too!

**Theorem 33.2.** (**Euclidean Algorithm for Polynomials**) *Let $f$ and $g$ be polynomials over a field $F$. If $f$ divides $g$ and the leading coefficient of $f$ is $a \in F$, then the $\gcd(f,g) = a^{-1}f$. Otherwise, using Theorem 33.1 repeatedly, there are non-zero remainder polynomials $r_1, \ldots, r_n$ of strictly decreasing degrees so that*

$$
\begin{aligned}
g &= fq_1 + r_1 \\
f &= r_1 q_2 + r_2 \\
r_1 &= r_2 q_3 + r_3 \\
&\vdots \\
r_{n-2} &= r_{n-1} q_n + r_n \\
r_{n-1} &= r_n q_{n+1}.
\end{aligned}
$$

*Then if the leading coefficient of $r_n$ is $b \in F$, $\gcd(f,g) = b^{-1}r_n$.*

This algorithm is exactly parallel to the integer case (Theorem 13.3). It is guaranteed to terminate because our Division Algorithm insures that the degree of each subsequent remainder $r_k$ is less than the degree of the previous one, and then the Well-Ordering Principle kicks in. As in Exercise 13.7, finally, it is not hard to show that $r_n$ is a common divisor of $f$ and $g$, and that it is the one of greatest degree.

Let's run this algorithm on the polynomials in Example 33.2.

**Example 33.3.** We have $g(x) = x^3 + 3x^2 - 2x - 6$ and $f(x) = x^2 - 9$. Omitting the details of polynomial long division again, we get

$$
\begin{aligned}
x^3 + 3x^2 - 2x - 6 &= (x^2 - 9)(x+3) + 7x + 21 \\
x^2 - 9 &= (7x + 21)(\tfrac{1}{7}x - \tfrac{3}{7}) + 0
\end{aligned}
$$

225

The gcd is the monic version of the last non-zero remainder, so $\gcd(f, g) = \frac{1}{7}(7x + 21) = x + 3$, as we previously saw by factoring. Even this computation is not fun, *but* it is guaranteed to find the desired gcd in a finite (and usually small) number of steps. Some good mathematical software can be quite helpful!

Let's look at one more example of using the Euclidean Algorithm to compute a gcd, this time over a finite field.

**Example 33.4.** We seek the gcd of $g(x) = x^3 + x^2 + x + 1$ and $f(x) = x^2 + 2x$ over the finite field $\mathbb{F}_5$, once more omitting the details of the long division. However, you should check that the equations, calculated via modulo 5 arithmetic, are correct.

$$
\begin{aligned}
x^3 + x^2 + x + 1 &= (x^2 + 2x)(x + 4) + 3x + 1 \\
x^2 + 2x &= (3x + 1)(2x) + 0
\end{aligned}
$$

The gcd is the monic version of the last non-zero remainder, so

$$
\gcd(f, g) = 2(3x + 1) = x + 2.
$$

We shall now close out this chapter with a lemma and two theorems which follow from our "Polynomial Division Algorithm" (Theorem 33.1).

**Lemma 33.3.** *If $f$ is a polynomial over a field $F$ and $a$ is in $F$, then $f(a)$ is the remainder when $f(x)$ is divided by $x - a$.*

**Proof.** By Theorem 33.1, $f(x) = (x - a)q(x) + r(x)$ where the degree of $r(x)$ is less than the degree of $x - a$ (which is 1). By evaluating $f$ at $x = a$, the result follows. ∎

The next two results provide important information concerning the *roots* of polynomials over a field. Recall that an element $a$ of the field $F$ is called a **root** of a polynomial $f \in F[x]$ if $f(a) = 0$. The first result follows from Lemma 33.3. We note that this result is frequently used in calculus for the field $\mathbb{R}$ of real numbers, but it also holds over any field $F$.

**Theorem 33.4.** *Let $f$ be a polynomial whose coefficients lie in a field $F$, and let $a$ be an element in $F$. Then $f(a) = 0$ if and only if $x - a$ divides $f(x)$.*

**Proof.** See Exercise 33.8. Remember that an "if and only if" theorem requires two arguments. ∎

Finally then, we state and prove a very important result concerning the possible number of roots of a polynomial of degree $n$ over a field $F$.

**Theorem 33.5.** *Let $f$ be a non-zero polynomial of degree $n \geq 1$ whose coefficients lie in a field $F$. Then $f$ has at most $n$ roots in $F$.*

**Proof.** Our proof is by induction on the degree $n$ of the polynomial $f$. (Remember that whenever you want to prove that some property holds for every positive integer, you should consider trying induction.)

For the base case, if the polynomial $f$ has degree 1, then it is a linear polynomial, say $f(x) = ax + b$ with $a \neq 0$, so that $x = -\frac{b}{a}$ is the unique root of $f$ in the field $F$.

Assume now that $f$ is a polynomial of degree $n > 1$. If $f$ has no roots in $F$, then we are done. Otherwise, if $a \in F$ is a root of $f$, so that $f(a) = 0$, then by Theorem 33.4 we have $f(x) = (x - a)g(x)$ for some polynomial $g$. If $f$ has only the root $a$ in $F$, then we are again done since $1 < n$. Otherwise, if there exists a root $c$ of $f$ with $c \neq a$, then $0 = f(c) = (c - a)g(c)$, so $c$ must be a root of the polynomial $g$. (Recall that $F$, being a field, contains no zero divisors.) Thus the only roots of $f$ are $a$ and any roots $c$ of $g$. But $g$ is of degree $n - 1$, so by the induction hypothesis $g$ has at most $n - 1$ roots, and hence $f$ has at most $n$ roots, and we are done. ∎

Let's look at one more example.

**Example 33.5.** Consider the polynomial $f(x) = (x - 3)^2(x^2 + 1)$. Since $f$ is of degree 4, it can have at most 4 roots in any field. Over the real numbers $\mathbb{R}$ there is a little twist in that 3 appears twice. This is what's called a *multiple root*, and it must be counted by its multiplicity, i.e., twice in this case. On the other hand, the factor $x^2 + 1$ has no roots in $\mathbb{R}$, so, over $\mathbb{R}$, $f$ has two roots: $\{3, 3\}$. But what if we view $f$ as being over the complex numbers $\mathbb{C}$ instead? Well, now the factor $x^2 + 1$ does indeed have two distinct roots, namely $i$ and $-i$, so, over $\mathbb{C}$, $f$ has 4 roots: $\{3, 3, i, -i\}$. You may have noticed, by the way, that in the example we gave at the beginning of this chapter and in this example, two polynomials of degree 4 had all 4 roots in the complex numbers $\mathbb{C}$. This is not a coincidence; we'll discuss this point in Chapter 35.

We shall now, in the next chapter, turn to another example of the parallels between polynomials over a field and the integers: *unique factorization* into "primes."

**Exercises**

**33.1.** Let $f(x) = (x^3 - 3x)(x^2 + 4)$ be a polynomial over the field $F$. Find all the roots of $f$ in $F$ (i.e., all the solutions in $F$ of $f(x) = 0$) if
  (a) $F = \mathbb{Q}$ (the rational numbers).
  (b) $F = \mathbb{R}$ (the real numbers).
  (c) $F = \mathbb{C}$ (the complex numbers).

**33.2.** Let $g(x) = (x^2 - 2)(x^2 + x + 1)$ be a polynomial over the field $F$. Find all the roots of $g$ in $F$ (i.e., all the solutions in $F$ of $g(x) = 0$) if
  (a) $F = \mathbb{Q}$ (the rational numbers).
  (b) $F = \mathbb{R}$ (the real numbers).
  (c) $F = \mathbb{C}$ (the complex numbers).

**33.3.** (a) For the polynomial $f(x) = x^2 + x + 1$, find all the roots of $f$ in the finite field $\mathbb{Z}_7$. (Note: Just evaluate $f$ at each element of $\mathbb{Z}_7$.)
(b) Do the same as Part (a) with the polynomial $h(x) = x^2 + 3x + 1$.

**33.4.** (a) For the polynomial $g(x) = x^2 + 3x + 2$, find all the roots of $g$ in the finite field $\mathbb{Z}_5$. (Note: Again, just evaluate $g$ at each element of $\mathbb{Z}_5$, either before or after factoring.)
(b) For the same polynomial $g$ in Part (a), find all of its roots in the finite ring $\mathbb{Z}_6$. Why does this not contradict Theorem 33.5?

**33.5.** Find the quotient $q$ and the remainder $r$ guaranteed by the Division Algorithm (Theorem 33.1) when $g(x) = 2x^3 - x^2 + 5x + 3$ is divided by $f(x) = x + 2$ over the real numbers $\mathbb{R}$. Remember that $r$ must be of smaller degree than $f$.

**33.6.** Find the quotient $q$ and the remainder $r$ guaranteed by the Division Algorithm (Theorem 33.1) when $g(x) = x^3 + 6x^2 + 5x + 4$ is divided by $f(x) = x^2 + 2x$ over the finite field $\mathbb{F}_7$. Remember that $r$ must be of smaller degree than $f$.

**33.7.** Use factorization to determine $\gcd(f, g)$ over the real numbers $\mathbb{R}$ for the polynomials $f(x) = 4x^2 - 1$ and $g(x) = 2x^3 + 7x^2 + 3x$. Remember that a polynomial gcd must be a monic polynomial.

**33.8.** Prove Theorem 33.4.

**33.9.** Use the Euclidean Algorithm for polynomials (Theorem 33.2) to show that the two polynomials $f(x) = x^2 + x - 2$ and $g(x) = x^3 + 3x^2 + 2x + 1$ over $\mathbb{R}$ are *relatively prime*, i.e., that their gcd is 1. (Remember that if the computation results in a final non-zero remainder which is a constant, the gcd is defined to be the "monic constant," i.e., 1.)

**33.10.** Find all the roots of $f(x) = x^8 - 1$ over the complex numbers $\mathbb{C}$. (Hint: Start by factoring as much as you can; that should yield 4 roots. But, as you may suspect, there are 4 more. If $a = \frac{1}{\sqrt{2}}(1+i)$, what is $a^2$? Note that geometrically $a$ is the point in the first quadrant of complex plane lying on the unit circle (as do 1 and $i$) and at a 45 degree angle to the real axis. By the way, for a slightly different approach to this question, see Example 11.4.)

# CHAPTER 34

# Polynomials - Unique Factorization

$$x^4 - 1 = (x - 1)(x + 1)(x^2 + 1)$$

Continuing with our theme of finding similarities between the structure of the ring $\mathbb{Z}$ of integers and that of the ring $F[x]$ of polynomials over a field $F$, we turn now to unique factorization. We proved in Theorem 14.4 that every positive integer $n \geq 2$ has a *unique* representation as a product of prime number factors. It may not surprise you then that every polynomial $f$ in $F[x]$ can be written as a unique product of "prime" polynomial factors. However, we shall call these polynomials *irreducible* rather than "prime." Here is the definition we need.

**Definition 34.1.** A polynomial $f$ is **irreducible over a field** $F$ if there are no non-constant polynomials $g$ and $h$ (i.e., no polynomials of positive degrees) over the field $F$ so that $f = gh$. Otherwise $f$ is called **reducible over $F$**.

We emphasize that irreducibility or reducibility of a given polynomial depends strongly on what field $F$ it is over. For example, $f(x) = x^2 - 2$ is irreducible over the rational numbers $\mathbb{Q}$, but is reducible over the real numbers $\mathbb{R}$ since
$f(x) = (x - \sqrt{2})(x + \sqrt{2})$. Likewise, $g(x) = x^2 + 2$ is irreducible over the real numbers $\mathbb{R}$, but is reducible over the complex numbers $\mathbb{C}$ since
$g(x) = (x - i\sqrt{2})(x + i\sqrt{2})$.

A first question we might ask is if there are some simple criteria to determine irreducibility. Well, to start with, all linear (i.e., degree 1) polynomials over any field are automatically irreducible. Beyond that, the following is true.

**Theorem 34.1.** *Suppose $f$ is a polynomial of degree 2 or greater over a field $F$.*

(a) *If $f$ has a root in $F$, then $f$ is reducible.*

(b) *If the degree of $f$ is 2 or 3 and $f$ has no root in $F$, then $f$ is irreducible.*

**Proof.** Part (a) follows directly from Theorem 33.4. For Part (b), we assume that $f$ is of degree 2 or 3 and prove the contrapositive. Suppose $f$ is reducible; then by definition $f = gh$ where at least one of these factors, say $g = ax + b$, is of degree 1. But then $-\frac{b}{a}$ is a root of $f$ in $F$, which is a contradiction. ∎

**Example 34.1.** By simply plugging in some values, one can discover that $-3$ is a root of $f(x) = x^3 + x^2 - 2x + 12$. By Theorem 34.1 Part (a) then, $f$ must be reducible over the real numbers $\mathbb{R}$, in fact $f(x) = (x + 3)(x^2 - 2x + 4)$. It turns out, however, that in the case of polynomials over $\mathbb{R}$, *every* cubic (i.e., degree 3) polynomial is reducible since its graph must cross the $x$-axis at least once (why?), and that crossing point is a root. Concerning Part (b) of Theorem 34.1, the claim in that part fails for polynomials of degree 4 or greater since now its proper factors (if any) may all be irreducible of degree greater than 1. For example, over $\mathbb{R}$, $g(x) = x^4 + 3x^2 + 2 = (x^2 + 1)(x^2 + 2)$ is obviously reducible and just as obviously has no roots in $\mathbb{R}$.

Before being able to state the unique factorization theorem for polynomials, we need to settle an ambiguity that can arise with respect to the leading coefficients of a given polynomial and its irreducible factors. Looking back at the unique factorization theorem for integers (Theorem 14.4), you will see that it is stated entirely in terms of *positive* integers. This is to avoid the loss of uniqueness which would arise if, for example, we allowed $6 = (2)(3)$ but also $6 = (-2)(-3)$. We need to do this same kind of restriction in our polynomial setting, so we shall require that our unique irreducible factors be *monic*, i.e., have leading coefficients of 1. We will not require that the polynomial $f$ (which is to be factored) be monic, but we will place at the beginning of its unique representation its leading coefficient $\alpha$, a non-zero element of the field $F$. Given this set-up, we can now state our central theorem, in close analogy to Theorem 14.4.

**Theorem 34.2.** (**Unique Factorization of Polynomials**) *Suppose that $f$ is a non-constant polynomial over a field $F$ whose leading coefficient is $\alpha$.*

(i) (*Existence*) *The polynomial $f$ may be written as $\alpha$ times a product of (not necessarily distinct) monic irreducible polynomials $p_i$; i.e., $f$ may be written in the form*

$$f = \alpha p_1 \cdots p_r$$

*where each polynomial $p_i, i = 1, \ldots, r$, is monic and irreducible.*

(ii) (*Uniqueness*) *Moreover, this factorization is unique except for the order of the irreducibles; i.e., if we also have $f = \alpha q_1 \cdots q_s$ where each $q_i$ is a monic irreducible polynomial, then $r = s$ and (if necessary) upon re-ordering, $p_i = q_i, i = 1, \ldots, r$.*

We ask you to prove Part (i) of this theorem, using strong induction on polynomial degrees, in Exercise 34.8. The proof of Part (ii) is practically identical to that of Part (ii) of Theorem 14.4, being done by regular induction on the number of factors, so we urge you to simply go back over that proof carefully.

**Example 34.2.** Let $f(x) = 3x^4 - 12$. We illustrate that unique factorization into monic irreducibles depends heavily on what field the polynomial is viewed as being over.

(a) Over the rational numbers $\mathbb{Q}$, we get

$$f(x) = 3(x^2 - 2)(x^2 + 2).$$

(b) Over the real numbers $\mathbb{R}$, we get

$$f(x) = 3(x - \sqrt{2})(x + \sqrt{2})(x^2 + 2).$$

(c) Over the complex numbers $\mathbb{C}$, we get

$$f(x) = 3(x - \sqrt{2})(x + \sqrt{2})(x - i\sqrt{2})(x + i\sqrt{2}).$$

(d) Over the finite field $\mathbb{F}_7$, we get

$$f(x) = 3(x - 3)(x - 4)(x^2 + 2). \text{ (Check this!)}$$

Continuing our study of analogies between the integers and polynomial rings, we finish this chapter with a look at a possible answer to the following question: *Over what field or fields $F$ does the multiplicative structure of $F[x]$ most closely resemble that of the integers $\mathbb{Z}$?* Well, since the prime numbers and the irreducible polynomials are, by unique factorization, the "building blocks" of the multiplicative structures, we could ask our question in a different way: *Over what field or fields $F$ is the prevalence of irreducible polynomials in $F[x]$ most similar to the prevalence of primes in $\mathbb{Z}$?* For example, as you may well already suspect and as we shall state explicitly in the next chapter, the only irreducible polynomials over the complex numbers $\mathbb{C}$ are the linear polynomials! This is not likely to produce a multiplicative structure which is similar to the integers, where the prime numbers are scattered throughout $\mathbb{Z}^+$. It turns out that the clear answer to the question about prevalence of primes and irreducibles, as we shall now see, is that the *finite fields $\mathbb{F}_p$*
($p$ prime) provide the closest prevalence of irreducibles in $\mathbb{F}_p[x]$ to that of prime numbers in $\mathbb{Z}$. (Note: It could just as well be $\mathbb{F}_q$ where $q$ is a prime *power*, but for simplicity we'll stick with $\mathbb{F}_p$.)

Okay then, let's start with estimating the prevalence (or "density") of the prime numbers in $\mathbb{Z}^+$. We discussed this briefly at the very end of Chapter 14 when we stated the so-called Prime Number Theorem in two ways. We repeat here the second way since it's the one of interest to us now. This statement is imprecise, but is adequate for our purposes.

**Theorem 34.3. (The Prime Number Theorem - Density Version)** *Let N be a positive integer. In the vicinity of N (i.e., among all of the integers which are relatively near to N), about 1 out of every* $\ln(N)$ *integers is prime (where* $\ln$ *is the natural log function). Said another way, in the vicinity of N, the density of the primes is approximately* $1/\ln(N)$.

We must remark that this is an amazing fact. Somehow, the natural log function measures (approximately) the prevalence of primes in $\mathbb{Z}^{+}$!!

**Example 34.3.** Suppose $N = 10,000$. The Prime Number Theorem predicts that "nearby" $N$ approximately 1 out of every $\ln(10,000) \approx 9.21 \approx 9$ integers is prime. This would predict then, for example, that between 9000 and 11,000 there would be about $2,000/9.21 \approx 217$ primes. The *exact* number of primes in that range turns out to be 218. Pretty darn good estimate! Going higher, suppose $M = 1,000,000$; then we get a prediction of 1 out of about $\ln(1,000,000) \approx 13.83 \approx 14$ integers near $M$ being prime. Between 900,000 and 1,100,000 then, we are predicted to get about $200,000/13.83 \approx 14,472$ primes. The exact number is 14,440; once again a pretty amazingly good estimate.

So now we turn to the prevalence (or density) of monic irreducible polynomials in the finite field $\mathbb{F}_p$. (Recall that the best analogue of *positive* in integers is *monic* in polynomials.) We need a way of measuring the size (or "absolute value") of a polynomial over a finite field. Here is a natural way to do it.

**Definition 34.2.** If $f$ is a polynomial of degree $n$ over a finite field $\mathbb{F}_p$, then its *absolute value* $|f|$ is $p^n$.

For example then, over $\mathbb{F}_5$ the absolute value $|f|$ of $f(x) = x^3 + 3x^2 + 4$ is $5^3 = 125$.

We observe here that in fact *the absolute value* $p^n$ *of a polynomial of degree n over* $\mathbb{F}_p$ *is also exactly the number of monic polynomials of degree n over* $\mathbb{F}_p$. So, for example, there are 125 monic polynomials of degree 3 over $\mathbb{F}_5$ (why?), all having the same absolute value 125. (Note the difference here from integers, where only two integers have the same absolute value.)

We now come to our "density theorem" for monic polynomials over finite fields, in analogy to Theorem 34.3.

**Theorem 34.4. (Density Theorem for Polynomials over Finite Fields)** *Among the* $p^n$ *monic polynomials of degree n over the finite field* $\mathbb{F}_p$, *about 1 out of every n of them is irreducible.*

The proof of this theorem, though *much* easier than the proof of the Prime Number Theorem, is still beyond the scope of this text.

**Example 34.4.** Among the 125 monic polynomials of degree 3 over $\mathbb{F}_5$, the theorem predicts that about $125/3 \approx 41.67 \approx 42$ are irreducible. The actual number is 40. Moving to a higher degree, the theorem predicts that among the

$5^7 = 78,125$ monic polynomials of degree 7 over $\mathbb{F}_5$, about $78,125/7 \approx 11,161$ are irreducible. The actual number is 11,160.

We have not yet come to the "punch line" on how close these predicted densities in integers and polynomials are to each other. To repeat, the density of primes near a positive integer $N$ is about 1 out of $\ln(N)$, whereas the density of monic irreducibles "near" monic polynomials of degree $n$ over $\mathbb{F}_p$ is about 1 out of $n$. But now if $f$ any such polynomial, its absolute value $|f|$ is $p^n$, and $\log_p(p^n) = n$. Thus our Theorem 34.4 could have been stated as follows:

*If $f$ is a polynomial in the set $S$ of monic polynomials of degree $n$ over the finite field $\mathbb{F}_p$, then in $S$ approximately 1 out of every $\log_p(|f|)$ polynomials is irreducible.*

So, it turns out the *natural* log function measures the density of the primes in the integers, whereas the *base-p* log function measures the density of monic irreducible polynomials in $\mathbb{F}_p[x]$. We hope you can appreciate how alike the multiplicative (and hence probably algebraic) structures of the two rings $\mathbb{Z}$ and $\mathbb{F}_p[x]$ must be since the prevalence of their "prime" elements is so similar! To finish, let us give one example of this likeness.

In Chapter 14 we briefly discussed Goldbach's Conjecture, which proposes that every even number $n \geq 4$ can be written as a sum of two primes. Though that conjecture is as yet unproved, in 1934 the Russian mathematician Vinogradov proved that every "sufficiently large" *odd* number $n \geq 7$ can be written as a sum of *three* prime numbers ("sufficiently large" means greater than some fixed $N$). This is generally referred to as *The 3-Primes Theorem*. In analogy then, it was proved around 1990 that every monic polynomial of degree $k > 2$ over every finite field $\mathbb{F}_p$ ($p$ an odd prime) can be written as a sum of three monic irreducible polynomials, one of degree $k$ and the other two of lesser degree. That two results this similar could hold in the rings $\mathbb{Z}$ and $\mathbb{F}_p[x]$ is strong evidence of how alike their structures are.

**Example 34.5.** Illustrating The 3-Primes Theorem,

$$33 = 11+11+11 = 13+13+7 = 17+13+3 = 17+11+5 = 19+11+3 = \cdots.$$

In fact, there are 9 ways to write 33 as a sum of three primes. See Exercise 34.10.

Now illustrating our "3-Irreducibles Theorem," there are, let's just say, numerous ways to write the polynomial $f(x) = x^3$ over $\mathbb{F}_5$ as a sum of three monic irreducibles. Here are three of them, one using two quadratics, one using two linears, and one using one of each.

$$x^3 = (x^3 + 3x^2 + 4) + (x^2 + 3) + (x^2 + 3),$$

$$x^3 = (x^3 + 3x + 2) + (x + 2) + (x + 1), \text{ and}$$

$$x^3 = (x^3 + 4x^2 + x + 1) + (x^2 + 3x + 3) + (x + 1).$$

You need to take our word for it that all the polynomials on the right are indeed irreducible, but you should check that things add up properly over $\mathbb{F}_5$.

If you are interested in learning more about the similarities in the algebraic structures of the rings $\mathbb{Z}$ and $\mathbb{F}_p[x]$, see, for example, [4]. However, we shall now turn our attention solely to the more classic setting (and one which is certainly more familiar to you) of polynomials over the real numbers $\mathbb{R}$ and the complex numbers $\mathbb{C}$.

## Exercises

**34.1.** Factor the polynomial $x^3 + x^2 + x + 1$ into a product of monic irreducibles over
(a) the real numbers $\mathbb{R}$.
(b) the complex numbers $\mathbb{C}$.

**34.2.** Factor the polynomial $x^3 - 1$ into a product of monic irreducibles over
(a) the real numbers $\mathbb{R}$.
(b) the complex numbers $\mathbb{C}$.

**34.3.** Factor the polynomial $x^3 + x^2 + x + 1$ into a product of monic irreducibles over
(a) the finite field $\mathbb{F}_5$.
(b) the finite field $\mathbb{F}_7$.

**34.4.** Factor the polynomial $x^3 - 1$ into a product of monic irreducibles over
(a) the finite field $\mathbb{F}_5$.
(b) the finite field $\mathbb{F}_7$.

**34.5.** Write the polynomial $3x^2 - 4x - 2$ as the product of a real number and two monic irreducible polynomials over $\mathbb{R}$.

**34.6.** Show that over the finite field $\mathbb{F}_7$, the polynomial $x^4 + 3x + 2$ has no roots in $\mathbb{F}_7$ but is nonetheless reducible. (Note: This is an example of why the degree hypothesis of Theorem 34.1 Part (b) is necessary.)

**34.7.** (a) According to The Prime Number Theorem (Theorem 34.3), about how many primes are there between 90,000 and 110,000? (Note: The actual number is 1,740.)
(b) According to Theorem 34.4, among the set of all fourth degree monic polynomials over the finite field $\mathbb{F}_7$, about how many are irreducible? (Note: The actual number is 588.)

**34.8.** Using the proof of Theorem 14.4 as a guide, use strong induction to prove Part (i) of Theorem 34.2.

**34.9.** Students often think that unique factorization of polynomials is "automatic," but it is not automatic if the underlying ring is not a field. For example, show that $x^2 + 3x + 2$ can be factored in two different ways over the ring $\mathbb{Z}_6$. (Hint: See Exercise 33.4 Part (b).)

**34.10.** (a) In Example 34.5 we listed 5 of the 9 ways to write 33 as a sum of 3 primes. Find the other four.

(b) Concerning the 3-Irreducibles Theorem, given that $p(x) = x^3 + 4x + 2$ is irreducible over $\mathbb{F}_5$, write down a representation of $f(x) = x^3 + x$ as a sum of three monic irreducibles over $\mathbb{F}_5$.

# CHAPTER 35

# Polynomials over the Rational, Real and Complex Numbers

$$x^4 - 1 = (x - 1)(x + 1)(x - i)(x + i)$$

In the previous chapter we learned that over any field $F$, every polynomial in the ring $F[x]$ (i.e., every polynomial whose coefficients are in $F$) factors uniquely into a product of its leading coefficient and some collection of monic irreducible polynomials over $F$. We then asked the natural question: *Among the fields with which we are most familiar, what form do irreducibles over that field take?* We then finished that chapter with an examination of irreducibles over the finite field $\mathbb{F}_p$ ($p$ any prime), discovering that there are irreducibles of every positive degree $n$, and in fact we are easily able to get quite accurate estimates of how many such (monic) irreducibles there are of degree $n$ over $\mathbb{F}_p$ (specifically, there are about $p^n/n$).

In this chapter we shall explore this same question about irreducible polynomials over our "classical" fields: the rational numbers $\mathbb{Q}$, the real numbers $\mathbb{R}$, and the complex numbers $\mathbb{C}$. Because these fields are infinite, we will not be able to count irreducibles like we can in the finite field case, *but* we can certainly ask about what positive degrees $n$ allow irreducibles of degree $n$. The answer turns out to be quite different for our three fields of interest.

Before moving forward, we remind you that given a specific polynomial $f$, we need to make clear over what field we are viewing $f$. If the coefficients of $f$ are all integers and/or non-integral rational numbers, then $f$ can be viewed as being over $\mathbb{Q}$, $\mathbb{R}$ or $\mathbb{C}$. However, if $f$ has even one coefficient which is real but not rational, then it can only be viewed as being over $\mathbb{R}$ or $\mathbb{C}$. Finally, if $f$ contains a coefficient which is complex but not real, then $f$ can only be viewed as being over $\mathbb{C}$. The opening figures of this and the previous chapter illustrate this idea. In Chapter 34, the polynomial $x^4 - 1$ is being viewed as over either $\mathbb{Q}$ or $\mathbb{R}$ and is being factored into irreducibles as such, but in this chapter $x^4 - 1$ is clearly being viewed as over $\mathbb{C}$.

We also remind you here that over *any* field $F$ every linear polynomial (i.e., degree 1 polynomial) is automatically irreducible. In what follows we are curious about the existence of irreducibles of degree 2 or greater.

Okay, so let's start with the ring $\mathbb{Q}[x]$ of polynomials over the rational numbers $\mathbb{Q}$. Here it turns out that irreducibles of all degrees are quite plentiful. We can see this from the following theorem, whose proof we omit (feel free to seek a proof, for example via Google).

**Theorem 35.1. (Eisenstein's Irreducibility Criterion for $\mathbb{Q}[x]$)**
*Suppose $f(x) = a_n x^n + a_{n-1} x^{n-1} + \cdots + a_1 x + a_0$ where each $a_i$ in an integer. If there exists a prime number $p$ such that*
  *(i) $p$ divides $a_i$ for all $i \neq n$,*
  *(ii) $p$ does not divide $a_n$ (i.e., the leading coefficient), and*
  *(iii) $p^2$ does not divide $a_0$ (i.e, the constant term),*
*then $f(x)$ is irreducible over the rational numbers $\mathbb{Q}$.*

**Example 35.1.** Let $f(x) = 2x^4 - 3x^3 + \frac{9}{2} + \frac{3}{2}$. First, since two of the coefficients are not integers, we clear the fractions, getting

$$f(x) = \frac{1}{2}(4x^4 - 6x^3 + 9x + 3),$$

and we can apply Theorem 35.1 to the polynomial in parentheses. Using $p = 3$, we see that 3 divides $-6$, 0, 9 and 3, but does not divide 4. Moreover, $3^2 = 9$ does not divide 3, so we can conclude that $f(x)$ is irreducible over $\mathbb{Q}$ (since it is a constant times an irreducible).

Theorem 35.1 then provides us with many, many examples of irreducible polynomials of every possible degree over $\mathbb{Q}$. For example, we note that for every degree $n$ and every prime $p$, the polynomials $x^n + p$ and $x^n - p$ are irreducible, giving us an infinite supply of irreducibles of each degree $n$ (since there are infinitely many primes). And of course there are any number of other examples we could cook up using the theorem. Hence you can see that our statement above that irreducibles in $\mathbb{Q}[x]$ are "plentiful" is indeed true. Recall that in the previous chapter we showed that the "density" of irreducibles in the ring $\mathbb{F}_p[x]$ is comparable to the density of primes in the ring of integers $\mathbb{Z}$. Without being precise, we can say here that the density of irreducibles in the ring $\mathbb{Q}[x]$ exceeds each of those other two densities. However, we shall now see that this is very far from true for our two remaining rings, $\mathbb{R}[x]$ and $\mathbb{C}[x]$.

Let us now turn to the ring $\mathbb{R}[x]$ of polynomials over the field $\mathbb{R}$ of real numbers. This is of course the setting of polynomials in which you should be most comfortable, having worked with such polynomials for years when studying pre-calculus and calculus. In fact the very first topic we took up in this text, Example 1.1, was the famous Quadratic Formula, which we (implicitly) employed in the ring $\mathbb{R}[x]$. Concerning that formula, it involves two concepts, one of which we introduced just after Example 11.1, and one we introduce now. First, if $a$ and $b$ are real numbers, then the **complex conjugate** (or simply "conjugate") of the complex number $a + bi$ is the complex number $a - bi$. Second, given a polynomial $ax^2 + bx + c$ over $\mathbb{R}$ with $a \neq 0$, the expression $b^2 - 4ac$ is called the **discriminant** of $f$.

We remind you (though there is probably no need since *you* proved it in Chapter 1) that the Quadratic Formula applied on $\mathbb{R}[x]$ says that if $a \neq 0$, $b$ and $c$ are real numbers, then the roots of the polynomial $ax^2 + bx + c$ are

$$\frac{-b \pm \sqrt{b^2 - 4ac}}{2a}.$$

This then leads to our first theorem about the existence of irreducible polynomials in $\mathbb{R}[x]$.

**Theorem 35.2.** *The quadratic polynomial $f(x) = ax^2 + bx + c \in \mathbb{R}[x]$ is irreducible over $\mathbb{R}$ if and only if the discriminant $b^2 - 4ac$ is negative. More specifically,*
    (i) *if the discriminant $b^2 - 4ac$ is positive, $f$ has two distinct real roots and hence factors over $\mathbb{R}$ into the product of two distinct linear polynomials.*
    (ii) *if the discriminant is $0$, $f$ has one real root of multiplicity 2 and hence factors over $\mathbb{R}$ into the square of a single linear polynomial.*
    (iii) *if the discriminant is negative, $f$ has no real roots but two complex roots which are complex conjugates, and thus $f$ cannot be factored over $\mathbb{R}$.*

Speaking of pre-calculus and calculus, let us mention the geometric implications of the three cases above. In Case (i) the graph of $f$ in $\mathbb{R}^2$ crosses the $x$-axis at these two roots, in Case (ii) the graph of $f$ is *tangent* to the $x$-axis at this root, and in Case (iii) the graph of $f$ never touches the $x$-axis.

**Example 35.2.** Here are simple examples of each of the cases of Theorem 35.2.
    (i) If $f(x) = x^2 - x - 1$, the discriminant is 5 and the two real roots are $\frac{1+\sqrt{5}}{2}$ and $\frac{1-\sqrt{5}}{2}$.
    (ii) If $g(x) = x^2 - x + \frac{1}{4}$, the discriminant is 0 and the single real root (of multiplicity 2) is $\frac{1}{2}$.
    (iii) If $h(x) = x^2 - x + 1$, the discriminant is $-3$ and the two complex roots are $\frac{1+i\sqrt{3}}{2}$ and $\frac{1-i\sqrt{3}}{2}$.
    Hence, over $\mathbb{R}$, $f$ factors into two distinct linears, $g$ factors into the square of a single linear, but $h$ is irreducible.

Now we come to a very surprising fact which shows how different the multiplicative structure of $\mathbb{R}[x]$ is from those of $\mathbb{Q}[x]$ and $\mathbb{F}_p[x]$. That fact is that *we have already identified all the irreducible polynomials in* $\mathbb{R}[x]$. We have identified, of course, all the linear polynomials, and Theorem 35.2 identifies all quadratic polynomials whose discriminant is negative. That turns out to be it for irreducibles over $\mathbb{R}$! This is definitely important enough to state as a theorem.

**Theorem 35.3.** *There are no polynomials over the real numbers of degree greater than two which are irreducible.*

We shall be able to prove this result, but we must first develop some tools we will need for that proof. The key to all of this, however, turns out to be that if a non-real complex number $a + bi$ is a root of a polynomial $f$ of degree $n \geq 2$ with real coefficients, *then its complex conjugate* $a - bi$ *is also a root!* (This is what we need to prove.) But then, given this,

$$(x - (a + bi))(x - (a - bi)) = x^2 - 2ax + a^2 + b^2,$$

whose coefficients are real, so in $\mathbb{R}[x]$, we have

$$f(x) = (x^2 - 2ax + a^2 + b^2)(g(x))$$

for some polynomial $g$ of degree $n - 2$ (this follows from two applications of Theorem 33.4), and so $f$ is clearly not irreducible over $\mathbb{R}$. So let's work toward proving the above property (in italics) of polynomials in $\mathbb{R}[x]$. First, an example:

**Example 35.3.** As we did in Example 11.4, let's take another look at the polynomial $x^4 + 1$. At first glance it "sort of" looks irreducible over $\mathbb{R}$, but by what we have just said, it must not be. As we showed in that example, its four complex roots are

$$\frac{1}{\sqrt{2}}(1 + i), \frac{1}{\sqrt{2}}(1 - i), \frac{1}{\sqrt{2}}(-1 + i), \frac{1}{\sqrt{2}}(-1 - i).$$

Note that the first two are complex conjugates, as are the last two, so we get

$$\left(x - \frac{1}{\sqrt{2}}(1 + i)\right)\left(x - \frac{1}{\sqrt{2}}(1 - i)\right) = x^2 - \sqrt{2}x + 1, \text{ and}$$

$$\left(x - \frac{1}{\sqrt{2}}(-1 + i)\right)\left(x - \frac{1}{\sqrt{2}}(-1 - i)\right) = x^2 + \sqrt{2}x + 1, \text{ and so, over } \mathbb{R},$$

$$x^4 + 1 = (x^2 - \sqrt{2}x + 1)(x^2 + \sqrt{2}x + 1),$$

establishing that $x^4 + 1$ is indeed reducible over $\mathbb{R}$. (You definitely should check these three calculations. Remember, pencil and paper should always be handy when reading/doing math!)

The main tool we need to show that non-real complex roots of polynomials in $\mathbb{R}[x]$ always come in conjugate pairs is the following function:

**Definition 35.1.** The *complex conjugate function* $\sigma$ from $\mathbb{C}$ to $\mathbb{C}$ is given by $\sigma(a + bi) = a - bi$; i.e., $\sigma$ takes every complex number to its conjugate.

**Lemma 35.4.** *The function $\sigma$ has the following properties:*
(a) $\sigma$ *fixes elements of* $\mathbb{R}$; *i.e., if $a$ is real, then $\sigma(a) = a$.*
(b) $\sigma$ *"respects" complex addition, i.e.,*

$$\sigma((a + bi) + (c + di)) = \sigma(a + bi) + \sigma(c + di).$$

(c) $\sigma$ *"respects" complex multiplication, i.e.,*

$$\sigma((a + bi)(c + di)) = \sigma(a + bi)\sigma(c + di).$$

(d) *If $k$ is a positive integer, then $\sigma((a + bi)^k) = (\sigma(a + bi))^k$.*

Proof. The proofs of Parts (a) and (b) are very quick, and Part (d) follows from Part (c) since exponentiation is just repeated multiplication. However, Part (c) requires some work. We ask you to do all of the proof in Exercise 35.6. ∎

We can use all four parts of Lemma 35.4 to arrive at our desired result.

**Theorem 35.5.** *If $f(x) = \alpha_n x^n + \alpha_{n-1} x^{n-1} + \cdots + \alpha_1 x + \alpha_0$ is a polynomial in $\mathbb{R}[x]$ and if $a + bi$ is a root of $f$, then $a - bi$ is also a root of $f$.*

Proof. Let $\sigma$ be as defined in Definition 35.1. By Lemma 35.4, we have

$$0 = \sigma(0) = \sigma(f(a + bi))$$

$$= \sigma(\alpha_n(a + bi)^n + \alpha_{n-1}(a + bi)^{n-1} + \cdots + \alpha_1(a + bi) + \alpha_0)$$
$$= \alpha_n(\sigma(a + bi))^n + \alpha_{n-1}(\sigma(a + bi))^{n-1} + \cdots + \alpha_1\sigma(a + bi) + \alpha_0$$
$$= \alpha_n(a - bi)^n + \alpha_{n-1}(a - bi)^{n-1} + \cdots + \alpha_1(a - bi) + \alpha_0,$$

so $a - bi$ is a root of $f$. ∎

As we have pointed out before, when reading a proof like this one, it is an excellent idea to look at *every* equal sign and be sure you see what justifies it. For example, the very first one above ($0 = \sigma(0)$) is because 0 is real and $\sigma$ fixes all reals, the second is by definition of $a + bi$, and so on.

Between Theorem **35.5** and the argument given just following the statement of Theorem **35.3** (that the product $(x - (a + bi))(x - (a - bi))$ is quadratic with real coefficients), we have now proved Theorem **35.3**, and hence we are done with analyzing irreducible polynomials in $\mathbb{R}[x]$. The bottom line is that, relatively speaking, irreducibles in $\mathbb{R}[x]$ not very plentiful.

So now it is time to move to our final case, the ring $\mathbb{C}[x]$ of polynomials over the complex numbers. We suspect that you can guess what happens in this case, as we already have seen various examples of polynomials of degree $n$ which have $n$ roots in $\mathbb{C}$ (and never fewer, provided we adjust for multiplicities). Your likely guess is true, and is contained in a deep and very important theorem whose proof is well beyond the scope of this text. Here is that result.

**Theorem 35.6.** (**Fundamental Theorem of Algebra**) *Every polynomial $f$ in $\mathbb{C}[x]$ of degree $n > 0$ has exactly $n$ roots (counted by their multiplicities) in the field $\mathbb{C}$ of complex numbers.*

This of course gives us the information about irreducible polynomials in $\mathbb{C}[x]$ which we've been expecting.

**Corollary 35.7.** *The only irreducible polynomials over the complex numbers are the linear polynomials.*

C.F. Gauss (1777–1855), who was one of the most brilliant and prolific mathematicians of the early 19th century, was the first to prove Theorem 35.6, though several other proofs have since been devised. All of them depend in some way on the fact that the fields $\mathbb{R}$ and $\mathbb{C}$ are "complete," in the sense, loosely, that they contain "no holes." The proof by Gauss appeared in 1801 in his famous book *Disquisitiones Arithmeticae*. In [2] the authors provide an English version of this classic text.

Let's finish this chapter (and in fact this entire textbook) with an example showing how a relatively simple looking polynomial will factor into irreducibles in completely different ways over five different fields (three infinite and two finite). We shall ask you to verify some of these factorizations in the exercises below.

**Example 35.4.** Consider the polynomial $f(x) = x^4 + 2$. Here are its factorizations into irreducible polynomials over various fields:
(1) Over the rational numbers $\mathbb{Q}$, using Theorem 35.1 with $p = 2$: $x^4 + 2$.
(2) Over the real numbers $\mathbb{R}$:

$$\left(x^2 - \frac{2}{\sqrt[4]{2}}x + \frac{2}{\sqrt{2}}\right)\left(x^2 + \frac{2}{\sqrt[4]{2}}x + \frac{2}{\sqrt{2}}\right).$$

(3) Over the complex numbers $\mathbb{C}$:

$$\left(x - \frac{1}{\sqrt[4]{2}}(1+i)\right)\left(x - \frac{1}{\sqrt[4]{2}}(1-i)\right)\left(x - \frac{1}{\sqrt[4]{2}}(-1+i)\right)\left(x - \frac{1}{\sqrt[4]{2}}(-1-i)\right).$$

(4) Over the finite field $\mathbb{F}_3$:

$(x-1)(x-2)(x^2-2)$ (also can be written $(x+2)(x+1)(x^2+1)$).

(5) Over the finite field $\mathbb{F}_5$: $x^4 + 2$.

This brings us to the end of our three chapter investigation of some ideas about polynomials and the fields that compose their coefficients. More generally, it brings us to the end of our efforts to help you move from the largely computational mathematics of your past to the somewhat more theoretical basis of much of advanced mathematics. We hope you've enjoyed our journey together. As we said to you in the Preface, math is both important and fun, so enjoy immersing yourself in it!

**Exercises**

**35.1.** Factor the polynomial $x^2 + 5$ into a product of monic irreducible polynomials over
(a) the field $\mathbb{R}$ of real numbers.
(b) the field $\mathbb{C}$ of complex numbers.
(c) the finite field $\mathbb{F}_7$ (i.e., the integers modulo 7).

**35.2.** Factor the polynomial $x^2 + x + 3$ into a product of monic irreducible polynomials over
(a) the field $\mathbb{R}$ of real numbers.
(b) the field $\mathbb{C}$ of complex numbers.
(c) the finite field $\mathbb{F}_5$ (i.e., the integers modulo 5).

**35.3.** Factor the polynomial $2x^2 + 3x - 6$ into a product of a constant and monic irreducible polynomials over
(a) the field $\mathbb{Q}$ of rational numbers. (Justify your answer.)
(b) the field $\mathbb{R}$ of real numbers.

**35.4.** Factor the cubic polynomial $3x^3 - 4x^2 - 2x$ into a product of a constant and monic irreducible polynomials over
(a) the field $\mathbb{Q}$ of rational numbers. (Justify your answer.)
(b) the field $\mathbb{R}$ of real numbers.

**35.5.** Factor the cubic polynomial $x^3 + 5x - 6$ over
(a) the real numbers $\mathbb{R}$.
(b) the complex numbers $\mathbb{C}$.
(Hint: There is an obvious integer root. Use the Division Algorithm to find the remaining quadratic factor, then proceed.)

**35.6.** Prove Lemma 35.4.

**35.7.** Concerning Part (3) of Example 35.4, show that the factorization over $\mathbb{C}$ is correct by finding the four roots of $x^4 + 2$. This can be done by first using the Quadratic Formula on the polynomial $(x^2)^2 + 2$, then again using the formula on the two resulting quadratic polynomials in $x$.

**35.8.** (a) In Example 35.4, show that that the first two linear polynomials in Part (3) do multiply to give you the first quadratic polynomial in Part (2).
(b) Also in Example 35.4, show that the two quadratic polynomials in Part (2) multiply to give the original polynomial $x^4 + 2$.

**35.9.** Verify the claim of Part (4) by simply finding roots (if any) of $x^4 + 2$ over the finite field $\mathbb{F}_3$. (Note: Part (5) is harder to verify. We can check that our polynomial has no roots in $\mathbb{F}_5$, but as we have seen in various examples, this does not imply that our polynomial is irreducible. In this case we would need to identify the 10 (it turns out) monic quadratic polynomials in $\mathbb{F}_5$ and then use the Division Algorithm to verify that none of them divides $x^4 + 2$.)

**35.10.** Prove the following handy result about the possible existence of rational roots of a polynomial with integral coefficients:

If $\alpha$ and $\beta$ are non-zero integers and $\alpha/\beta$ is a rational root of the polynomial $f(x) = a_n x^n + a_{n-1} x^{n-1} + \cdots + a_1 x + a_0$, where all coefficients $a_i$ are integers (and $a_n \neq 0$), then $\alpha$ must divide $a_0$ and $\beta$ must divide $a_n$.

(Suggestion: Write out the equation $f(\alpha/\beta) = 0$, put the whole left side over the common denominator $\beta^n$, and clear the fractions, resulting in an equation which is all integers. Now notice what terms contain $\alpha$ as a factor, and the same with $\beta$.)

**35.11.** (a) According to the result in the previous exercise, there are only six possible rational roots of the polynomial $x^3 + x^2 + 4$. What are they?
(b) Is one of the answers to Part (a) in fact a root of $f$? If so, which one?
(c) List all the possible rational roots $g(x) = 2x^3 - 9$.

**35.12.** (a) Find all roots of the polynomial $x^4 - 1$ over the complex numbers $\mathbb{C}$.
(b) Sketch these four roots in the complex plane. What do you notice about their distance from the origin and how they are spaced?
(c) Find all roots of the polynomial $x^3 - 1$ over the complex numbers $\mathbb{C}$ and sketch them, making the same observations as in Part (b).
(d) Make a conjecture as to the placement of the roots of $x^n - 1$ in the complex plane for all $n \geq 2$. (Note: These complex numbers are called "$n$th roots of unity.")

# Suggested Solutions to Selected Examples and Exercises

## Chapter 1

**Example 1.3.**
$(a+b)^7 = a^7 + 7a^6b + 21a^5b^2 + 35a^4b^3 + 35a^3b^4 + 21a^2b^5 + 7ab^6 + b^7$
$(x-2)^6 = x^6 - 12x^5 + 60x^4 - 160x^3 + 240x^2 - 192x + 64$

**Example 1.4.** The second entry in the chart is, of course, $2/4$.

If $A$ is the measure of an interior angle of a regular $n$-gon, then the exterior angle is $180° - A$. These $n$ exterior angles add up to $360°$. Now solve for $A$.

### Exercises

**1.1.**
$$
\begin{aligned}
0 &= x^5 - 8x^2 + 2x^4 - 16x - 5x^3 + 40 \\
&= x^2(x^3 - 8) + 2x(x^3 - 8) - 5(x^3 - 8) \\
&= (x^3 - 8)(x^2 + 2x - 5) \\
&= x - 2)(x^2 + 2x + 4)(x^2 + 2x - 5),
\end{aligned}
$$
and you take it from there.

**1.3.** Your conjecture should be $\binom{n}{k} = \binom{n-1}{k-1} + \binom{n-1}{k}$. For the proof, write down what these latter two binomial coefficients mean in terms of factorials, put them over a common denominator, and add them.

## Chapter 2

**Example 2.1.** For example, if $f(x) = x^3$, then
$$
\begin{aligned}
f'(x) &= \lim_{h \to 0} \frac{(x+h)^3 - x^3}{h} \\
&= \lim_{h \to 0} \frac{x^3 + 3x^2h + 3xh^2 + h^3 - x^3}{h} \\
&= \lim_{h \to 0} \frac{h(3x^2 + 3xh + h^2)}{h} \\
&= \lim_{h \to 0} (3x^2 + 3xh + h^2) = 3x^2
\end{aligned}
$$

**Example 2.2.** To get you started:

$$(f(x)g(x))' = \lim_{h\to 0} \frac{f(x+h)g(x+h)-f(x)g(x)}{h}$$
$$= \lim_{h\to 0} \frac{(f(x+h)g(x+h)-f(x+h)g(x))+(f(x+h)g(x)-f(x)g(x))}{h}$$
$$= \cdots$$

**Example 2.3.** To get you started, if $F(x) = \int_a^x f(t)dt$, then

$$F'(x) = \lim_{h\to 0} \frac{\int_a^{x+h} f(t)dt - \int_a^x f(t)}{h}$$
$$= \lim_{h\to 0} \frac{\int_x^{x+h} f(t)dt}{h}$$

and now apply the Mean Value Theorem for Integrals.

**Example 2.4.** From the data on relative errors and their direction, the best conjecture would appear to be that Simpson's Rule gives (2 Mid + Trap)/3. Now work through the algebra.

**Exercises**

2.1(a). To start,

$$(1/g(x))' = \lim_{h\to 0} \frac{1/g(x+h)-1/g(x)}{h}$$
$$= \lim_{h\to 0} \frac{g(x)-g(x+h)}{hg(x)g(x+h)}$$
$$= \cdots$$

2.3 The Chain Rule says that when you *compose* functions (see Chapter 8 for a refresher if need be), their derivatives (i.e., their rates of change) *multiply*. In symbols, $((g \circ f)(x))' = g'(f(x))f'(x)$. For this exercise, we let $y = x^{n/m}$, so that $y^m = x^n$. Now use the Chain Rule on the left to take the derivative of both sides *with respect to* $x$. This gives $my^{m-1}y' = nx^{n-1}$. Now finish by (carefully) solving for $y'$ (replacing $y$ by $x^{n/m}$).

2.5. We solve the case $f(x) = x^2$ and leave you on your own for the case $f(x) = x^3$. By the Fundamental Theorem, $\int_a^b x^2 dx = (1/3)(b^3 - a^3)$. Now for Simpson's Rule, since $n = 1$, we have

$$\frac{(b-a)(a^2+4((a+b)/2)^2+b^2)}{6} = \frac{(b-a)(a^2+ab+b^2)}{3} = \frac{b^3-a^3}{3}.$$

# Chapter 3

**Example 3.1.** The function $f(x)$ crossing the $x$-axis means, of course, that $f(x) = 0$. Hence, by the Quadratic Formula, we have $x = \frac{-b\pm\sqrt{b^2-4ac}}{2a}$. Now use the hypothesis.

**Example 3.3.** We assume that $n = ad$ and $m = bd$. Then for any integers $x$ and $y$, we note that $xad + ybd$ is divisible by $d$ and hence cannot be 1.

**Example 3.6.** How about $x/2$?

Example 3.7.(a) $q = 6$, $r = 14$.

Example 3.8.(a) 21, (b) $(-2)(105)+(1)(231) = 21$

Example 3.9. For example,when $r = 3$, $n = 6q + 3$, which is divisible by 3, and $n + 1 = 6q + 4$, which is divisible by 2.

## Exercises

3.1. $x^2 - (r_1 + r_2)x + r_1r_2 = 0$, so $a = 1$, $b = -(r_1 + r_2)$ and $c = r_1r_2$. Hence $b^2 - 4ac = (-(r_1 + r_2))^2 - 4r_1r_2 = r_1{}^2 - 2r_1r_2 + r_2{}^2 = (r_1 - r_2)^2 > 0$.

3.3.(a) $1 + 3 = 4 = (9 - 1)/2$, $1 + 3 + 9 = 13 = (27 - 1)/2$,
$1 + 3 + 9 + 27 = 40 = (81 - 1)/2$, etc. Hence a good conjecture is that $1 + 3 + 9 + \cdots + 3^n = (3^{n+1} - 1)/2$.
(b) Briefly, using the inductive hypothesis, we have
$$(1 + 3 + 9 + \cdots + 3^k) + 3^{k+1} = (3^{k+1} - 1)/2 + 3^{k+1} = (3^{k+2} - 1)/2.$$

3.5.(a) Since the sequence of remainders is strictly decreasing and all are non-negative, the Well-Ordering Principle guarantees that we will eventually reach 0.
(b) The sequence of remainders is $\{15, 12, 3, 0\}$. The gcd of 102 and 87 is 3.

## Chapter 4

### Exercises

4.1. The first five numbers in this sequence are $\{1, 3, 7, 15, 31\}$. Note that indeed each is 1 less than a power of 2. Using induction on $n$, all of these supply base cases. For the inductive step, *assume* that for $n = k$ we have $a_k + 1 = 2^m$ for some positive integer $m$. Then
$a_{k+1} + 1 = (2a_k + 1) + 1 = 2(a_k + 1) = 2(2^m) = 2^{m+1}$, which is a power of 2.

4.3. The first four numbers are $\{0, 8, 80, 728\}$. All are clearly divisible by 8. Using induction, assume for $n = k$ that $3^{2k} - 1 = 8m$ for some positive integer $m$. Then $3^{2(k+1)} - 1 = 3^{2k}3^2 - 1 = 3^{2k}3^2 - 1 - 8 + 8$
$= (3^{2k} - 1)9 + 8 = (8m)9 + 8 = 8(9m + 1)$, and we are finished.

4.5. $2 + 4 + 6 + \cdots + 2n$
$= 2(1 + 2 + 3 + \cdots + n) = 2((n(n + 1)/2)$ (by Example 4.1) $= n(n + 1)$.

4.7. The first four sums are $\{2, 8, 20, 40\}$. They supply base cases. For the inductive step, assume now that for $n = k$

we have $1(2) + 2(3) + \cdots + k(k+1) = \frac{k(k+1)(k+2)}{3}$. Then for $n = k+1$ we have

$$1(2) + 2(3) + \cdots + k(k+1) + (k+1)(k+2) = \frac{k(k+1)(k+2)}{3} + (k+1)(k+2),$$

which can be simplified to obtain $\frac{(k+1)(k+2)(k+3)}{3}$.

4.9. The first 8 Fibonacci numbers are $\{1, 1, 2, 3, 5, 8, 13, 21\}$, and they supply bases cases for an inductive proof. Now assume that for $n = k$, $F_1 + F_2 + \cdots + F_k = F_{k+2} - 1$. Then for $n = k+1$ we have

$$F_1 + F_2 + \cdots + F_k + F_{k+1} = F_{k+2} - 1 + F_{k+1} = F_{k+3} - 1.$$

4.11. For $n = 1$ we have $1 = F_3 - 1 = 2 - 1 = 1$. Assume the result is true for $n = k$, so that $F_2 + F_4 + \cdots + F_{2k} = F_{2k+1} - 1$. Then for $n = k+1$ we have

$$F_2 + F_4 + \cdots + F_{2k} + F_{2k+2} = F_{2k+1} - 1 + F_{2k+2} = F_{2k+3} - 1.$$

4.13. For $n = 2$ both sides have the value $\frac{3}{4}$. Assume the result holds for $n = k$ so that

$$(1 - \frac{1}{2^2}) \cdots (1 - \frac{1}{k^2}) = \frac{k+1}{2k}.$$

Then for $n = k+1$ we have

$$(1 - \frac{1}{2^2}) \cdots (1 - \frac{1}{k^2})(1 - \frac{1}{(k+1)^2}) = \frac{k+1}{2k}(1 - \frac{1}{(k+1)^2}),$$

which after some simplification, becomes

$$\frac{k+2}{2(k+1)}.$$

4.15. For $n = 1$ both sides are 1. Assume that for $n = k$ we have

$$1 - 2^2 + 3^2 - \cdots + (-1)^{k-1}k^2 = \frac{(-1)^{k-1}k(k+1)}{2}.$$

Then for $n = k+1$, using the induction hypothesis we have

$$1 - 2^2 + 3^2 - \cdots + (-1)^{k-1}k^2 + (-1)^k(k+1)^2 = \frac{(-1)^{k-1}k(k+1)}{2} + (-1)^k(k+1)^2,$$

which after some algebraic simplification, becomes

$$\frac{(-1)^k(k+1)(k+2)}{2}.$$

**4.17.** For $n = 1$ both sides are 1. Assume that for $n = k$ we have

$$1 \cdot 1! + 2 \cdot 2! + \cdots + k \cdot k! = (k+1)! - 1.$$

Then for $n = k + 1$

$$1 \cdot 1! + 2 \cdot 2! + \cdots + k \cdot k! + (k+1)(k+1)!$$
$$= (k+1)! - 1 + (k+1)(k+1)!$$
$$= (k+1)!((1 + (k+1)) - 1 = (k+2)! - 1.$$

# Chapter 5

## Exercises

**5.1.** 360

**5.3.** An interval on the real line will be well-ordered if and only if it is closed on the left-hand edge.

**5.5.** (a) The set $\{a^2 + b^2\}$ of all squared norms in $S$ is a non-empty set of non-negative integers, so it is well-ordered. If $a_0^2 + b_0^2 \leq a^2 + b^2$ for all $a + bi \in S$, then $\sqrt{a_0^2 + b_0^2} \leq \sqrt{a^2 + b^2}$ for all $a + bi \in S$, i.e., $N(a_0 + b_0 i)$ is a least element in the set of norms of $S$.
(b) $1 = N(1 + 0i)$
(c) $\sqrt{26} = N(5 + i) = N(1 + 5i) = N(-1 + 5i)$. There are 8 such.

**5.7.** Let $S$ be the set of all positive integers $n$ which have the property that $1 + 2 + 3 + \cdots + n \neq \frac{n(n+1)}{2}$. By the Well-Ordering Principle, if $S$ is non-empty, then $S$ has a smallest element $n_0$. Thus we must have that $1 + 2 + 3 + \cdots + (n_0 - 1) = \frac{(n_0 - 1)(n_0)}{2}$. But then

$$1 + 2 + 3 + \cdots + n_0 = \frac{(n_0 - 1)(n_0)}{2} + n_0 = \frac{n_0(n_0 + 1)}{2},$$

which is a contradiction. Hence $S$ must be empty.

**5.9.** Suppose $\sqrt{2} = b/a$, a rational number in lowest terms. Square both sides and clear the fraction, creating an integer equation. Since one side is divisible by 2, it must be that $b^2$ is divisible by 2 and hence by 4. Now look back at the other side, and reach a contradiction.

**5.11.** The number $n_0$ cannot be 0, 1 or 2 since 8, 9 and 10 can be made by 3+5, 3+3+3 and 5+5. Hence $n_0 - 3$ is non-negative and cannot be in $S$, but then $(n_0 - 3) + 3 = n_0$ also cannot be in $S$, which is a contradiction. Thus $S$ is empty.

## Chapter 6

**Example 6.2.** The two sets are $\{\ldots, -12, -6, 0, 6, 12, 18, \ldots\}$ and $\{\ldots, -16, -8, 0, 8, 16, 24, \ldots\}$. $A \cap B = \{24n | n \in \mathbb{Z}\}$.

**Example 6.4.** An integer is odd if it is not divisible by 2, which will be true if and only if it can be written in the form $2i + 1$ for some integer $i$. If $2i + 1$ is in $C$, $2i + 1 = 2(i - 1) + 3$, which is in $D$ so $C \subseteq D$. On the other hand, $2k + 3 \in D$ is clearly odd, so $D \subseteq C$. Hence $C = D$.

**Example 6.5.** Clearly $P \cap B = \{5\}$ so they are not disjoint. However, $P \cap C = \emptyset$.

**Example 6.6.** For a given subset of $A$, there are two choices for each element of $A$: it is either in the subset or not in the subset. Hence there will be $2 \times 2 \times 2 = 2^3 = 8$ subsets of $A$ in all.

**Example 6.7.** $\mathcal{P}(B) = \{\emptyset, \{1\}, \{2\}, \{3\}, \{4\}, \{1, 2\}, \{1, 3\}, \{1, 4\}, \{2, 3\}, \{2, 4\}, \{3, 4\}, \{1, 2, 3\}, \{1, 2, 4\}, \{1, 3, 4\}, \{2, 3, 4\}, \{1, 2, 3, 4\}\}$. Conjecture: If $A$ has $n$ elements, then $\mathcal{P}(A)$ has $2^n$ subsets as its elements.

### Exercises

**6.1.** $A \cup B = \{a, b, 1, 2, c\}$, $A \cap B = \emptyset$ and $A \times B = \{(a, 1), (a, 2), (a, c), (b, 1), (b, 2), (b, c)\}$.

**6.3.** The set $A \times A = \{(2, 2), (2, 4), (2, 6), (4, 2), (4, 4), (4, 6), (6, 2), (6, 4), (6, 6)\}$. It has 9 ordered pairs as its elements. The set $\mathcal{P}(A \times A)$ has $2^9 = 512$ elements, which are all the subsets of $A \times A$.

**6.5.** To show that $B$ is a subset of $A$, we take an arbitrary element $b \in B$ and show that it's in $A$. But by hypothesis $b$ is in $A \cap B$, so by the definition of intersection, $b$ is also in $A$.

**6.7.** The sets $A \cap A$ and $A \cup A$ each have 10 elements, $A \times A$ has 100 elements, $\mathcal{P}(A)$ has $2^{10} = 1,024$ elements, and $\mathcal{P}(A \times A)$ has $2^{100}$ elements, which we choose not to expand out.

**6.9.** The set $A \cup B = \{1, 2, 3, 4\}$, so $(A \cup B) \times C = \{(1, a), (1, b), (2, a), (2, b), (3, a), (3, b), (4, a), (4, b)\}$. On the other hand, $A \times C = \{(1, a), (1, b), (2, a), (2, b), (3, a), (3, b)\}$ and $B \times C = \{(1, a), (1, b), (4, a), (4, b)\}$, and the union of these two sets is clearly equal to our first set above.

**6.11.** Using the sets from Exercise 6.9, $(A \times B) \cup C$ contains both ordered pairs (e.g., $(1, 1)$) and single elements (e.g., $a$), whereas $(A \cup C) \times (B \cup C)$ consists entirely of ordered pairs. Hence the two sets are not equal.

6.13. Suppose $x$ is in $(A \cup B)^C$, i.e., $x$ is not an element of $(A \cup B)$, so $x$ is not in $A$ and also $x$ is not in $B$, and so $x$ is in $A^C \cap B^C$. Hence $(A \cup B)^C \subseteq A^C \cap B^C$. On the other hand, if $y$ is in $A^C \cap B^C$, then $y$ is not in $A$ and also not in $B$, so it is not in $A \cup B$, i.e., it is in $(A \cup B)^C$. Thus $A^C \cap B^C \subseteq (A \cup B)^C$, and we are done.

## Chapter 7

**Example 7.1.** The relation $R_1$ satisfies all three, $R_2$ satisfies reflexivity and transitivity but not symmetry, $R_3$ satisfies only transitivity, and $R_4$ satisfies only symmetry.

**Example 7.2.** The numbers $-11, -6, -1, 4, 9, 14$ are all related to 4 in this relation. The relation is reflexive since for any integer $a$, $a - a$ is divisible by 5. It is symmetric since if $a - b$ is divisible by 5, then its negative $b - a$ is divisible by 5. Finally, it is transitive since if $a - b$ and $b - c$ are both divisible by 5, then so is $(a - b) + (b - c) = a - c$. Hence this is an equivalence relation.

**Example 7.3.** Just replace 5 by $n$ in the argument directly above.

**Example 7.4.** Since the condition on the ordered pairs involves equality, all three properties follow immediately. The equivalence class $[(0.6, 0.8)]$ is the unit circle centered at the origin. More generally, $[(a, b)]$ is the circle of radius $\sqrt{a^2 + b^2}$ centered at the origin.

**Example 7.5.** As in the argument just above, because the condition involves equality, the three properties follow. The equivalence class $[x^2]$ is $\{x^2 + c | c \in \mathbb{R}\}$. More generally, $[f(x)] = \{f(x) + c | c \in \mathbb{R}\}$.

### Exercises

7.1. Since "even" means divisible by 2, this is just a special case of the equivalence relation called congruence, introduced in Example 7.3. There are two equivalence classes, even numbers and odd numbers.

7.3. This is not an equivalence relation. It is symmetric since if $a + b$ is odd then so is $b + a$. However, it is not reflexive since, for example, $1 + 1$ is not odd. Also, it is not transitive since, for example, $1 + 2$ and $2 + 1$ are both odd, but $1 + 1$ is not.

7.5. The relation $R$ is reflexive since $(a, a), (b, b), (c, c)$ and $(d, d)$ are all present. It is not symmetric since $(a, c)$ is present but $(c, a)$ is not. It is also not transitive (this is the trickiest one to check) since $(a, b)$ and $(b, d)$ are present, but $(a, d)$ is not.

7.7. Though this relation is reflexive and transitive, it is clearly not symmetric (for example $\frac{1}{2} \leq \frac{3}{4}$, but not the reverse). Hence it is not an equivalence relation.

7.9. Since the condition on the complex numbers is equality of the "real" (i.e., the first) component, the three properties are easy to confirm, so this is an equivalence relation. The equivalence class $[3 + 4i]$ is the vertical line $\{3 + bi | b \in \mathbb{R}\}$.

7.11. (a) Let the set $X$ be $\{x, y\}$ and define a relation $R$ on $X$ by $xRx$. Then $R$ is symmetric and transitive, but it is not reflexive since $yRy$ is not present.
(b) The statement "If $xRy$ holds" assumes that some element $y$ besides $x$ is related to $x$. That may not be true, as we saw in Part (a), and $yRy$ may not hold.

7.13. Since the condition involves equality of the values of the integrals, the three properties follow easily, and so we do have an equivalence relation. The equivalence class $[x^2]$ is the set of all functions in the set $S$ for which the area under the curve between $x = 0$ and $x = 1$ is $1/3$. More generally, $[f(x)]$ consists of all functions in $S$ for which the area under the curve between $x = 0$ and $x = 1$ is $\int_0^1 f(x)dx$.

# Chapter 8

Example 8.2. The function $f : \mathbb{R} \to \mathbb{R}$ given by the rule $f(x) = x^3$ is both one-to-one and onto. Since the cube root of any real number exists and is unique, we see that $f$ is one-to-one since if $x_1^3 = x_2^3$, then $x_1 = x_2$. Likewise $f$ is onto since if $y$ is any real number, then $f(\sqrt[3]{y}) = y$. Concerning $g(x) = x^2$, however, square roots do not exist for negative real numbers, and square roots, when they do exist, are not unique. Hence $g$ is not one-to-one since, for example, $g(2) = g(-2) = 4$, and $g$ is not onto since, for example, $-4$ is not the image of any real number.

Example 8.3. Under this rule the image of 0 is not in the co-domain. If we change the co-domain to be all non-negative reals rather than all positive reals (i.e., make the co-domain $\mathbb{R} \cup \{0\}$), $f$ is now a function which is onto, but, as in Example 8.2, is still not one-to-one.

Example 8.4. The relation $R$ is not a function because
  (1) the element $c$ does not possess an image, and
  (2) the relation $R$ is "multi-valued" since both $(a, b)$ and $(a, c)$ are present.
We can turn $R$ into a function by replacing $(a, c)$ with $(c, a)$.

**Example 8.5.** The other seven are $\{(1,a),(2,a),(3,b)\}$,
$\{(1,a),(2,b),(3,a)\}$, $\{(1,a),(2,b),(3,b)\}$, $\{(1,b),(2,a),(3,a)\}$,
$\{(1,b),(2,a),(3,b)\}$, $\{(1,b),(2,b),(3,a)\}$ and $\{(1,b),(2,b),(3,b)\}$.

**Example 8.6.** Denote $h(x)$ by $y$. We have $y = 2x^3 - 4$, so $\frac{y-4}{2} = x^3$, and we arrive at $x = \sqrt[3]{\frac{y-4}{2}}$. Hence $h^{-1}(x) = \sqrt[3]{\frac{x-4}{2}}$.

**Example 8.7.**

$$(f \circ g)(x) = f(g(x)) = f(2x + 1) = (2x + 1)^2 = 4x^2 + 4x + 1.$$

**Example 8.8.** (a) No, because the co-domain of $g$, which is $C$, does not match the domain of $f$, which is $A$.
(b) Yes, because the co-domain of $f$, which is $B$, matches the domain of $g$, which is also $B$. The composite $g \circ f = \{(a, x), (b, z), (c, x)\}$.

**Exercises**

8.1. (a) An example of a one-to-one function from $A$ to $B$ would be $\{(a, 1), (b, 2)\}$. Since there are 3 choices for the images of $a$ but then only 2 choices for the images of $b$, there must be 6 one-to-one functions altogether.
(b) Because there are more elements in the co-domain than in the domain, there can be no onto functions.

8.3. (a) $\{(1,a),(2,a)\}, \{(1,a),(2,b)\}, \{(1,b),(2,a)\}, \{(1,b),(2,b)\}$ (4 in all).
(b) $\{(1,a),(2,a),(3,a)\}, \{(1,a),(2,a),(3,b\}, \{(1,a),(2,b),(3,a)\}$,
$\{(1,b),(2,a),(3,a)\}, \{(1,a),(2,b),(3,b)\}, \{(1,b),(2,a),(3,b)\}$,
$\{(1,b),(2,b),(3,a)\}, \{(1,b),(2,b),(3,b)\}$ (8 in all).
(c) $\{(a,1),(b,1)\}, \{(a,1),(b,2)\}, \{(a,1),(b,3)\}, \{(a,2),(b,1)\}$,
$\{(a,2),(b,2)\}, \{(a,2),(b,3)\}, \{(a,3),(b,1)\}, \{(a,3),(b,2)\}$,
$\{(a,3),(b,3)\}$ (9 in all).

8.5. (a) $4^7 = 16{,}384$.
(b) None.
(c) We help with the case 4-1-1-1 and leave the other two to you. Select a co-domain element to get the four images; there are 4 choices. Now select 4 of the 7 domain elements to send there; there are $\binom{7}{4}$ ways to do that (see Example 1.2). Finally there are 3! ways to arrange the remaining 3 domain elements into the remaining 3 co-domain elements. Hence we get $4 \cdot \binom{7}{4} \cdot 3! = 840$ onto functions of this type.

8.7. Since $k$ is odd, for any real number $x$, $(-x)^k = -(x^k)$. Hence if $x^k = y^k$, then $x$ and $y$ have the same sign. It follows that since

$$x^k - y^k = (x - y)(x^{k-1} + x^{k-2}y + \cdots + xy^{k-2} + y^{k-1}) = 0,$$

we must have $x = y$. Hence $g$ is one-to-one. Also, since each real number $y$ (positive or negative) has a $k$th root in $\mathbb{R}$, $g$ is also onto.

8.9. Since each domain element has a unique image, our only choice is that $f$ must be the identity function on $A$, i.e., for every $a \in A$, $f(a) = a$.

8.11. Suppose $f : A \to B$ and $g : B \to C$ are both one-to-one functions. Suppose that in $C$, $g(f(a_1)) = g(f(a_2))$, i.e., the images of $a_1$ and $a_2$ under the composite function $g \circ f$, are equal. Since $g$ is one-to-one, this implies $f(a_1) = f(a_2)$. But $f$ is also one-to-one, so we have $a_1 = a_2$, as desired.

8.13. There are so many! For example, $f(n) = n + 1$ and $g(n) = 2n$ both have this property.

8.15. Since $a$ divides $b$, there exist a $k \in \mathbb{Z}$ ($k \neq 0$) such that $b = ka$. Define $f : A \to B$ by $f(as) = kas$ for all $s \in \mathbb{Z}$. The function $f$ is onto since if $bt = kat$ is in $B$, then $f(at) = bt$. Moreover, if $kas_1 = kas_2$, then $as_1 = as_2$, so $f$ is also one-to-one.

8.17. Since $A$ and $B$ have 3 elements each, $A \times B$ has 9 elements. So the answer to Part (a) is $9^9 = 387,420,489$ and the answer to Part (b) is $3^9 = 19,683$.

# Chapter 9

Example 9.2. $\{0, 1 - 1, 2, -2, 3, -3, \cdots\}$.

Example 9.3. To go from a list of positive rationals to a list of all rationals, just mimic the idea in Example 9.2.

**Exercises**

9.1. The cardinality of both is 8.

9.3. Show that $f : E \to S$ given by $f(n) = 7n$ is a bijection.

9.5. If $A = \{a_1, a_2, \cdots\}$ and $B = \{b_1, b_2, \cdots\}$, then we can make the list $\{a_1, b_1, a_2, b_2, \cdots\}$.

9.7. Show that $f(x) = 1/x$ is a bijection between these two sets.

9.9. For example 231.56 is the "mixed number" $231\frac{56}{100}$, which is rational.

9.11. Suppose $x = 0.x_1x_2x_3 \cdots$ and $y = 0.y_1y_2y_3 \cdots$ are each arbitrary numbers in $(0, 1)$. Define $f : (0, 1) \times (0, 1) \to (0, 1)$ by $f((x, y)) = 0.x_1y_1x_2y_2x_3y_3 \cdots$. Show that $f$ is a bijection.

# Chapter 10

**Example 10.2.**
$\rho \circ \sigma(1) = \rho(\sigma(1)) = \rho(3) = 2,$
$\rho \circ \sigma(2) = \rho(\sigma(2)) = \rho(1) = 4,$
$\rho \circ \sigma(3) = \rho(\sigma(3)) = \rho(2) = 3,$
$\rho \circ \sigma(4) = \rho(\sigma(4)) = \rho(4) = 1.$

**Example 10.3.**

$$\begin{pmatrix} 1 & 2 & 3 & 4 \\ 4 & 3 & 2 & 1 \end{pmatrix} \begin{pmatrix} 1 & 2 & 3 & 4 \\ 3 & 1 & 2 & 4 \end{pmatrix} = \begin{pmatrix} 1 & 2 & 3 & 4 \\ 2 & 4 & 3 & 1 \end{pmatrix}.$$

**Example 10.5.** $(14)(23)$

**Example 10.6.** $(158)(27)(3)(46)$

**Example 10.8.** $(14)(23)(132)(4) = (124)(3)$

**Exercises**

10.1. (b) $\theta = (146)(23)(5)$, $\tau = (1643)(25)$

10.3. $\theta^{-1} = \begin{pmatrix} 1 & 2 & 3 & 4 & 5 & 6 \\ 6 & 3 & 2 & 1 & 5 & 4 \end{pmatrix} = (164)(23)(5)$

$\tau^{-1} = \begin{pmatrix} 1 & 2 & 3 & 4 & 5 & 6 \\ 3 & 5 & 4 & 6 & 2 & 1 \end{pmatrix} = (1346)(25)$

10.5. $\theta \circ \tau =$

$$\begin{pmatrix} 1 & 2 & 3 & 4 & 5 & 6 \\ 4 & 3 & 2 & 6 & 5 & 1 \end{pmatrix} \begin{pmatrix} 1 & 2 & 3 & 4 & 5 & 6 \\ 6 & 5 & 1 & 3 & 2 & 4 \end{pmatrix} = \begin{pmatrix} 1 & 2 & 3 & 4 & 5 & 6 \\ 1 & 5 & 4 & 2 & 3 & 6 \end{pmatrix}$$
$\theta \circ \tau = (146)(23)(5)(1643)(25) = (1)(2534)(6)$

10.7. (a) $(4)(2)(3) = (1)(2)(3)(4) =$ the identity permutation.
(b) the identity permutation.
(c) the identity permutation.

10.9. Suggestion: To get started, because $X$ is finite, it must be the case that for two non-negative integers $k > j$, $\sigma^k(x) = \sigma^j(x)$. This means that $\sigma(\sigma^{k-1}(x)) = \sigma(\sigma^{j-1}(x))$. But $\sigma$ is one-to-one, so $\cdots$. You complete the argument.

# Chapter 11

**Example 11.2.** $(3 + 4i) + (5 - 2i) = 8 + 2i$
$(3 + 4i) - (5 - 2i) = -2 + 6i$

$(3 + 4i)(5 - 2i) = 23 + 14i$
$(3 + 4i)/(5 - 2i) = (7/29) + (32/29)i$

Example 11.3. By the Binomial Theorem and because $i^3 = -i$,
$(1 + i\sqrt{3})^3 = 1^3 + 3(1^2)(i\sqrt{3}) + 3(1)(i\sqrt{3})^2 + (i\sqrt{3})^3$
$= 1 + 3i\sqrt{3} - 3(3) - 3i\sqrt{3} = 1 - 9 = -8$

**Exercises**

11.1. (a) $8 + i$   (b) $8 - i$   (c) $8 - i$

11.3. (a) $18 + 14i$   (b) $18 - 14i$   (c) $18 - 14i$

11.5. (a) $-46 + 9i$   (b) $-46 - 9i$   (c) $-46 - 9i$

11.7. (a) $(3/20) + i(1/20)$   (b) $(3/20) - i(1/20)$

11.9. (a) $(3/20) + i(11/20)$   (b) $(3/20) - i(11/20)$

11.11. $(-46/13) + i(178/13)$ (Had enough of that?)

11.13. $-2 \pm i$

11.15. $\pm 2$ and $\pm 2i$

11.17. (a) $(-i)^3 = (-i)(-i)^2 = (-i)(i)^2 = (-i)(-1) = i$.
(b) $(\cos(\pi/6) + i\sin(\pi/6))^3 = (\sqrt{3}/2 + i(1/2))^3$
    $= 3\sqrt{3}/8 + i(9/8) - 3\sqrt{3}/8 - i(1/8) = i(8/8) = i$.
(c) $(\cos(5\pi/6) + i\sin(5\pi/6))^3 = (-\sqrt{3}/2 + i(1/2))^3$
    $= -3\sqrt{3}/8 + i(9/8) + 3\sqrt{3}/8 - i(1/8) = i(8/8) = i$.
(Note: Concerning Part (a), we observe that
$\cos(9\pi/6) + i\sin(9\pi/6) = (\cos(3\pi/2) + i\sin(3\pi/2)) = 0 - i = -i$.
This is relevant to Exercise 11.18.)

## Chapter 12

Example 12.1. (a) Obviously $3 - 2 \neq 2 - 3$ and $3/2 \neq 2/3$. Also, $(4 - 3) - 2 = -1$
but $4 - (3 - 2) = 3$, and $(4/3)/2 = 2/3$ but $4/(3/2) = 8/3$.
(b) Composition of permutations *is* associative, but the proof is cumbersome.

Example 12.3.
(b) $BA = \begin{pmatrix} 1 & -9 \\ 9 & -16 \end{pmatrix}$
(d) $DC = \begin{pmatrix} 3 & 3 & 3 \\ 4 & -4 & 2 \\ 3 & -15 & -5 \end{pmatrix}$

**Example 12.5.** (a) $\mathbb{Z}^+$, $\mathbb{Q}^+$ and $\mathbb{R}^+$ do not possess an additive identity, but 0 is an additive identity for the other four sets.
(b) 1 is a multiplicative identity for all seven sets.
(c) The permutation $(1)(2)\cdots(n)$ in $S_n$ which fixes every element is an identity element.

**Example 12.6.** (a) None of the elements of $\mathbb{Z}^+$, $\mathbb{Q}^+$ and $\mathbb{R}^+$ possesses an additive inverse. For the other four sets, each element $a$ has an additive inverse of the form $-a$.
(b) Only one element of $\mathbb{Z}^+$ (namely 1) has a multiplicative inverse. Exactly two elements of $\mathbb{Z}$ (namely 1 and $-1$) have multiplicative inverses. Every *non-zero* element $a$ in each of the other five sets has a multiplicative inverse of the form $1/a$.

**Exercises**

12.1.
$$A + B = \begin{pmatrix} -2 & 6 \\ 3 & 1 \end{pmatrix}, A - B = \begin{pmatrix} 4 & 2 \\ 1 & -7 \end{pmatrix}$$

$$AB = \begin{pmatrix} 1 & 18 \\ -9 & -8 \end{pmatrix}, BA = \begin{pmatrix} 1 & -20 \\ 9 & -8 \end{pmatrix}$$

12.3. Both upper left-hand entries are $aej + afm + bgj + bhm$.

12.5. Both upper left-hand entries are $ae + aj + bg + bm$.

12.7. $AB = \begin{pmatrix} 0 & 11 & 3 \\ 3 & -1 & 0 \end{pmatrix}$.

12.9.
$$5A = \begin{pmatrix} 10 & 5 & -5 \\ -15 & 20 & 5 \\ 10 & -5 & 15 \end{pmatrix}, AB = \begin{pmatrix} 1 & -2 & 10 \\ 8 & -2 & -3 \\ 9 & -4 & -2 \end{pmatrix},$$

$$BA = \begin{pmatrix} 11 & -6 & 7 \\ 11 & -4 & 3 \\ 5 & 1 & -10 \end{pmatrix} \text{ and } A^2 = \begin{pmatrix} -1 & 7 & -4 \\ -16 & 12 & 10 \\ 13 & -5 & 6 \end{pmatrix}.$$

12.11. $\det(A) = 11$, $\det(B) = 0$, so only $A$ is invertible, and

$$A^{-1} = \begin{pmatrix} \frac{4}{11} & -\frac{1}{11} \\ \frac{3}{11} & \frac{2}{11} \end{pmatrix}.$$

12.13.
$$A^{-1} = \begin{pmatrix} \frac{1}{3} & -\frac{1}{2} & \frac{1}{6} \\ \frac{1}{3} & 0 & -\frac{1}{3} \\ 0 & \frac{1}{2} & \frac{1}{2} \end{pmatrix}.$$

## Chapter 13

**Example 13.4.** (b) $858 = 2 \cdot 3 \cdot 11 \cdot 13$ and $1092 = 2^2 \cdot 3 \cdot 7 \cdot 13$. Hence $\gcd(858, 1092) = 2 \cdot 3 \cdot 13 = 78$.

**Example 13.5.** (c)

$$
\begin{aligned}
420 &= 182(2) + 56 \\
182 &= 56(3) + 14 \\
56 &= 14(4) + 0,
\end{aligned}
$$

so the answer is 14.

**Example 13.6.** (b) Numbers to be replaced are in bold.

$$1 = 23 - \mathbf{22} = 23 - (45 - 23(1)) = \mathbf{23}(2) - 45 = (68 - 45(1))(2) - 45$$

$$= 68(2) - \mathbf{45}(3) = 68(2) - (249 - 68(3))(3) = 249(-3) + 68(11).$$

**Exercises**

**13.1.**
(a) $5|635$ is true since $635 = 5(127)$.
(b) $-5|635$ is true since $625 = -5(-127)$.
(c) $48|124$ is false since $48(2) = 96$ and $48(3) = 144$.
(d) $341|32871$ is false since the remainder is 125, not 0.
(e) $5|(15m - 10)$ is true since $15m - 10 = 5(3m - 2)$.
(f) $m|(-3m)$ is true since $-3m = m(-3)$.
(g) $(k + m)|(7k + 14m)$ is false since the remainder is $7m$, not 0.
(h) $k|(-6k^2 - k)$ is true since $-6k^2 - k = k(-6k - 1)$.

**13.3.** (a) $35 = 5 \cdot 7$, $180 = 2^2 \cdot 3^2 \cdot 5$, so $\gcd(35, 180) = 5$.
(b)

$$
\begin{aligned}
180 &= 35(5) + 5 \\
35 &= 5(7) + 0,
\end{aligned}
$$

so the answer is 5.

**13.5.** (a)

$$
\begin{aligned}
468 &= 224(2) + 20 \\
224 &= 20(11) + 4 \\
20 &= 4(5) + 0,
\end{aligned}
$$

so the answer is 4.

(b) $4 = 224 - 20(11) = 224 - (468 - 224(2))(11) = 468(-11) + 224(23)$.

**13.7.** In both (a) and (b), the hints should do the trick.

**13.9.** A proof by induction should work here. Note that
$4^{n+1} - 1 = (4^{n+1} - 4) + 3$.

**13.11.** An induction proof can work here, but it's a bit messy. Another approach is to observe that $n^3 - n = (n - 1)n(n + 1)$, i.e., it's the product of three consecutive integers.

**13.13.** (a) 3-5, 5-7, 11-13, 17-19, 29-31, 41-43, 59-61, 71-73.
(b) There are five such pairs. The smallest is 1019-1021.

## Chapter 14

**Example 14.1.** (b) $11 | 5357$.

**Example 14.3.** (b) $45,000 = 2^4 3^2 5^5$.

**Example 14.4.** $4307 = (59)(73)$.

**Example 14.5.** (b) In fact, $589 = (19)(31)$ and $899 = (29)(31)$, so their gcd is 31. Not easy to find "by hand" using this method.
(c)

$$
\begin{aligned}
899 &= (589)(1) + 310 \\
589 &= (310)(1) + 279 \\
310 &= (279)(1) + 31 \\
279 &= (31)(9) + 0,
\end{aligned}
$$

so $\gcd(589, 899) = 31$.

### Exercises

**14.1.** Since $p|n$, we have $n = pa$ for some $a \in \mathbb{Z}^+$. If $a \geq \sqrt{n}$, then $n = pa > (\sqrt{n})^2 = n$, which is a contradiction. Hence $a < \sqrt{n}$. Let $q$ be a prime which divides $a$, then $q < \sqrt{n}$, as desired.

**14.3.** (a) $384 = 2^7 3^1$
(b) $1,155 = 3 \cdot 5 \cdot 7 \cdot 11$

**14.5.** (a) Both; 3 only; 11 only; neither.

(b) We get you started on this part and leave Part (c) to you:

$$\begin{aligned} n &= 10^3 w + 10^2 x + 10y + z \\ &= (3^2+1)^3 w + (3^2+1)^2 x + (3^2+1)y + z \\ &= (3^6 + (3)(3^4) + (3)(3^2) + 1)w + \cdots \end{aligned}$$

Hence $n - (w + x + y + z)$ is divisible by 3. Now finish. (Note: This proof can easily, say by induction, be extended to integers with arbitrarily many decimal digits.)

14.7. $\gcd(2^5 3^8 5, 3^3 5^5 7^6) = \gcd(2^5 3^8 5^1 7^0, 2^0 3^3 5^5 7^6) = 2^0 3^3 5^1 7^0 = 3^3 5^1$.

14.9. Remember that an "if and only if" statement is two results, so it requires two proofs. Suppose first that $m$ is a square, so $m = n^2$ for some integer $n$. If the prime factorization of $n = p_1^{b_1} \cdots p_r^{b_r}$, then the factorization of $m = p_1^{2b_1} \cdots p_r^{2b_r}$, i.e., all the exponents are even. On the other hand, if all of $m$'s exponents are even, its factorization can be written in the form $p_1^{2b_1} \cdots p_r^{2b_r}$, so $m = n^2$ where $n = p_1^{b_1} \cdots p_r^{b_r}$.

14.11. Let $a$ and $b$ be positive integers greater than 1 and let $x$ be any variable. Then

$$x^{ab} - 1 = (x^a)^b - 1 = (x^a - 1)((x^a)^{b-1} + (x^a)^{b-2} + \cdots + x^a + 1).$$

Now replacing $x$ by 2, we see that if $n = ab$ is composite, then

$$2^n - 1 = (2^a)^b - 1 = (2^a - 1)((2^a)^{b-1} + (2^a)^{b-2} + \cdots + 2^a + 1),$$

and the right hand side is composite, being the product of two integers which are both greater than 1.

14.13. (a) For the integration by parts (i.e., $\int u\,dv = uv - \int v\,du$), set $u = 1/\ln(x) = (\ln(x))^{-1}$ and $dv = 1\,dx$. Then by the Power Rule and the Chain Rule, $du = (-(\ln(x))^{-2}/x)dx$, and $v = x$, so $uv = x/\ln(x)$ evaluated at $N$ and at 2. The result follows.
(b) This version of L'Hopital's Rule says that if two quantities are both going to infinity, then the limit of their ratio will be the same as the limit of the ratio of their derivatives. But the derivative of an integral is the original function, and so

$$\lim_{N \to \infty} \frac{\int_2^N \frac{1}{(\ln(x))^2}dx}{\int_2^N \frac{1}{\ln(x)}dx} = \lim_{N \to \infty} \frac{\frac{1}{(\ln(N))^2}}{\frac{1}{\ln(N)}} = \lim_{N \to \infty} \frac{1}{\ln(N)} = 0.$$

# Chapter 15

Example 15.1. (a) True since $43 - 17 = 2(13)$.
(b) False since $-16 - 5 = -21$ which is not divisible by 11.

Example 15.4. (c)

| + | 0 | 1 | 2 | 3 |
|---|---|---|---|---|
| 0 | 0 | 1 | 2 | 3 |
| 1 | 1 | 2 | 3 | 0 |
| 2 | 2 | 3 | 0 | 1 |
| 3 | 3 | 0 | 1 | 2 |

| · | 0 | 1 | 2 | 3 |
|---|---|---|---|---|
| 0 | 0 | 0 | 0 | 0 |
| 1 | 0 | 1 | 2 | 3 |
| 2 | 0 | 2 | 0 | 2 |
| 3 | 0 | 3 | 2 | 1 |

Because of the row for 2, $\mathbb{Z}_4$ looks more like $\mathbb{Z}_6$ than $\mathbb{Z}_5$.

Example 15.6. (b) In $\mathbb{Z}_8$, $1^{-1} = 1$, $3^{-1} = 3$, $5^{-1} = 5$ and $7^{-1} = 7$. (One might wonder how often every invertible element of $\mathbb{Z}_n$ is its own multiplicative inverse.)

**Exercises**

15.1. Parts (b), (c), (e) and (h) are true.

15.3. (ii) By assumption and by the definition of congruence, we have that $a - b = nk$ and $c - d = nj$ for some integers $k$ and $j$. Hence

$$(a - c) - (b - d) = (a - b) - (c - d) = nk - nj = n(k - j),$$

and the result follows.
(iv) Under these same assumptions, we have $ma - mb = m(a - b) = mnk$, and again the result follows.

15.5. (a) $\{\cdots, -19, -12, -5, 2, 9, 16, 23, 30, \cdots\}$.
(b) $\{x + kn | k \in \mathbb{Z}\}$.

15.7. In $\mathbb{Z}_{15}$, $1^{-1} = 1$, $2^{-1} = 8$, $4^{-1} = 4$, $7^{-1} = 13$, $8^{-1} = 2$, $11^{-1} = 11$, $13^{-1} = 7$, and $14^{-1} = 14$.

15.9. We choose to show this via factoring, but induction would also work well. We have

$$a^k - b^k = (a - b)(a^{k-1} + a^{k-2}b + a^{k-3}b^2 + \cdots + ab^{k-2} + b^{k-1}).$$

But $n$ divides $a - b$ by assumption, so $n$ divides $a^k - b^k$, as desired.

15.11. Reducing mod 5, we get $18^k - 13^k \equiv 3^k - 3^k = 0$ (mod 5), so 5 divides $18^k - 13^k$.

15.13. Running the Euclidean Algorithm forwards:

$$
\begin{aligned}
23 &= 16 + 7 \\
16 &= 2(7) + 2 \\
7 &= 3(2) + 1 \\
2 &= 2(1) + 0.
\end{aligned}
$$

Now running it backwards from the next to last line:

$$
\begin{aligned}
1 &= 7 - 3(2) \\
1 &= 7 - 3(16 - 2(7)) = 7(7) - 3(16) \\
1 &= 7(23 - 16) - 3(16) = 7(23) - 10(16).
\end{aligned}
$$

Thus in $\mathbb{Z}_{23}$, $16^{-1} = -10 \pmod{23} = 13$.

# Chapter 16

Example 16.1. (b) and (c) Modulo 8, we have $6(1) = 6$, $6(2) = 4$, $6(3) = 2$, $6(4) = 0$, $6(5) = 6$, $6(6) = 4$ and $6(7) = 2$. Hence (b) has no solutions and (c) has two solutions, 2 and 6.

Example 16.4. Starting with $16x \equiv 52 \pmod{20}$, we reduce modulo 20 to get $16x \equiv 12 \pmod{20}$. Since $\gcd(16, 20) = 4$ and since 4 divides 12, we can reduce to $4x \equiv 3 \pmod 5$. By inspection $x = 2$, so our four solutions in $\mathbb{Z}_{20}$ are 2, 7, 12 and 17.

## Exercises

16.1.
(a) The unique solution in $\mathbb{Z}_{11}$ is 10.
(b) No solutions.
(c) Four solutions in $\mathbb{Z}_{20}$ are 4, 9, 14 and 19.
(d) Reduce first to $14x \equiv 12 \pmod{18}$, then to $7x \equiv 6 \pmod 9$, giving $x = 6$, so the two solutions in $\mathbb{Z}_{18}$ are 6 and 15.

16.3. The two solutions are 3 and 8.

16.5. The solution is 60.

16.7. The solution is 32.

16.9. There are 23 things.

16.11. Below 96 the solutions to $x \equiv 2 \pmod{12}$ are $\{2, 14, 26, 38, 50, 62, 74, 86\}$. Reducing modulo 8, we get $\{2, 6, 2, 6, 2, 6, 2, 6\}$. Hence there are no solutions.

16.13. (a), (b) and (c) Every answer is 1.
(d) If $p$ is prime and if $a$ and $p$ are relatively prime, then $a^{p-1} \equiv 1 \pmod{p}$.

# Chapter 17

## Exercises

17.1. (a) 3, (b) 2, (c) 5, (d) 1

17.3. 20

17.5. Reduce to $3^3 \pmod{13} = 1$.

17.7. 4

17.9. (1) $18 = 16 + 2$
(2) $3^2 \pmod{50} = 9$
(3) $3^4 \pmod{50} = 31$
(4) $3^8 \pmod{50} = 11$
(5) $3^{16} \pmod{50} = 21$
(6) $3^{18} = 3^{16}3^2 = (21)(9) \pmod{50} = 39$.

17.11. $5^2 = 25$, $5^4 = 25^2 \pmod{100} = 25$, so $5^8 \pmod{100} = 25$.

17.13.
(a) Checking 2, 3, 5, 7, 11, 13, 17 and 19, none divide 499, so 499 is prime.
(b) $3^{503} \equiv 3^5 = 243 \pmod{499}$.

# Chapter 18

Example 18.2 (b)

| $a$ | 1 | 2 | 3 | 4 | 5 | 6 |
|-----|---|---|---|---|---|---|
| $k$ | 1 | 3 | 6 | 3 | 6 | 2 |

so 3 and 5 are primitive.

## Exercises

18.1. Modulo 13, $2^2 = 4$, $2^3 = 8$, $2^4 = 3$, and $2^6 = 12$, so 2 must be a primitive root (note that we only checked powers which are divisors of 12). Now Alice computes and sends $c = g^a = 2^3 = 8 \pmod{13}$, and then Bob computes $8^5 = 8 \pmod{13}$. Likewise Bob computes and sends $d = g^b = 2^5 = 6 \pmod{13}$, and then Alice computes $6^3 = 8 \pmod{13}$. So their secret shared key is 8.

18.3. Using *Mathematica*:

PrimeQ[577] = True

Primitive Root[577] = 5

Alice: PowerMod[5,52,577] = 428

Bob: PowerMod[428,34,5770 = 130

Bob: PowerMod[5,34,577] = 53

Alice: PowerMod[53,52,577] = 130

so the secret shared key is 130.

18.5. (a) 1 and 5 are relatively prime to 6, so the counts are both 2.

(b) The table for $p = 13$ (as in Example 18.2 Part (a) for $p = 11$) is:

| $a$ | 1 | 2 | 3 | 4 | 5 | 6 | 7 | 8 | 9 | 10 | 11 | 12 |
|---|---|---|---|---|---|---|---|---|---|---|---|---|
| $k$ | 1 | 12 | 3 | 6 | 4 | 12 | 12 | 4 | 3 | 6 | 12 | 2 |

so there are 4 primitive elements modulo 13. Now, 1, 5, 7 and 11 are relatively prime to 12, so again the counts match up.

(c) Since the elements of $\mathbb{Z}_{17}$ which are relatively prime to 16 are all the positive odd numbers below 16, we conjecture that there are 8 primitive roots modulo 17. In fact, the primitive roots modulo 17 are 3, 5, 6, 7, 10, 11, 12 and 14.

(d) You do it!

18.7. Modulo 13, we have $2^0 = 1$, $2^1 = 2$, $2^2 = 4$, $2^3 = 8$, $2^4 = 3$, $2^5 = 6$, $2^6 = 12$, $2^7 = 11$, $2^8 = 9$, $2^9 = 5$, $2^{10} = 10$ and $2^{11} = 7$.

Hence, $\log_2(1) = 0$, $\log_2(2) = 1$, $\log_2(3) = 4$, $\log_2(4) = 2$, $\log_2(5) = 9$, $\log_2(6) = 5$, $\log_2(7) = 11$, $\log_2(8) = 3$, $\log_2(9) = 8$, $\log_2(10) = 10$, $\log_2(11) = 7$ and $\log_2(12) = 6$.

Note that the function $\log_2$ is a bijection from $\mathbb{Z}_{13}^*$ onto the set $\{0, 1, 2, \ldots, 11\}$. See the following exercise.

18.9. (a) According to the chart in Example 18.2 Part (a), 9 is *not* a primitive root modulo 11.

(b) Modulo 13, $11^2 = 4$, $11^3 = 5$, $11^4 = 3$ and $11^6 = 12$, so 11 *is* a primitive root.

(c) Further hint: A difference between the arithmetic in $\mathbb{Z}_{11}$ and $\mathbb{Z}_{13}$ is that 10 is not divisible by 4, but 12 is.

# Chapter 19

Example 19.1. (b) $\phi(20) = 8$. $\phi(2)\phi(10) = (1)(4) = 4$; $\phi(4)\phi(5) = (2)(4) = 8$. $\gcd(2, 10) \neq 1$; $\gcd(4, 5) = 1$.

Example 19.3. (c) $\phi(363) = \phi(3)\phi(11^2) = (2)(11^2 - 11) = (2)(110) = 220$.

Example 19.5. (b) 5 and 75 are not relatively prime.
(c) $\phi(56) = \phi(7)\phi(8) = (6)(4) = 24$. By the Division Algorithm, $99 = (4)(24)+3$, so by Euler's Theorem

$$3^{99} = (3^{24})^4(3^3) \equiv (1^4)(3^3) = 27 \quad (\text{mod } 56).$$

## Exercises

19.1. (a) 18, (b) 12, (c) 28, (d) 8

19.3. (a) 6, (b) 1, 5, 7, 11, 13, 17

19.5. 240

19.7. 40,000,000

19.9. 51

19.11. (a) $\phi(n) = \phi(2^k) = 2^k - 2^{k-1} = 2^{k-1}(2 - 1) = n/2$.
(b) All the odd numbers in $\mathbb{Z}_n$.

19.13. 40

19.15. Hint: Just limiting ourselves to moduli $n$ between 2 and 10 inclusive and $a \in \mathbb{Z}_n$, there are five counter-examples available.

# Chapter 20

## Exercises

20.1. (a) $m = (17)(23) = 391$; $\phi(m) = (16)(22) = 352$
(b)
$$352 = (5)(70) + 2$$
$$5 = (2)(2) + 1.$$
Now working backwards:
$$1 = 5 - (2)(2)$$
$$1 = 5 - (2)(352 - (5)(70)) = (5)(141) - (2)(352),$$
so your private $d$, the multiplicative inverse of 5 in $\mathbb{Z}_{352}^*$, is 141.

20.3. 160

20.5 242

20.7 8

# Chapter 21

## Exercises

21.1. (a) No, because subtraction of integers is not associative. For example, $(8 - 5) - 3 = 0$, but $8 - (5 - 3) = 6$.
(b) No, because the identity element 0 is not present in $\mathbb{R}^+$.
(c) Yes. The product of two positive real numbers is a positive real number. Multiplication of real numbers is associative. Note that 1 is the identity element, and $1/a$ is the inverse of $a$.

21.3.

| + | 0 | 1 | 2 | 3 |
|---|---|---|---|---|
| 0 | 0 | 1 | 2 | 3 |
| 1 | 1 | 2 | 3 | 0 |
| 2 | 2 | 3 | 0 | 1 |
| 3 | 3 | 0 | 1 | 2 |

21.5. Because $e_1$ is an identity element, we know that $e_1 e_2 = e_2$. But likewise $e_1 e_2 = e_1$. Hence $e_1 = e_2$.

21.7. Using associativity, we have

$$(ab)(b^{-1}a^{-1}) = a(bb^{-1})a^{-1} = aea^{-1} = aa^{-1} = e,$$

so $b^{-1}a^{-1}$ is the "right-hand" inverse of $ab$. Similarly it is the "left-hand" inverse, and we are done.

21.9. Multiplying both sides of the first equation on the left by $a^{-1}$, we get $a^{-1}ax = a^{-1}b$, so $x = a^{-1}b$. Similarly, multiplying both sides of the second equation on the right by $a^{-1}$, we get $xaa^{-1} = ba^{-1}$, so $x = ba^{-1}$.

21.11.
Reflexivity: Choosing $g = e$, we get that $a$ is related to itself.
Symmetry: If $b = g^{-1}ag$, then multiplying on the left by $g$ and on the right by $g^{-1}$, we get $gbg^{-1} = a$. But $g = (g^{-1})^{-1}$, so $b$ is related to $a$ using the element $g^{-1}$.
Transitivity: If $b = g^{-1}ag$ and $c = h^{-1}bh$, then

$$c = h^{-1}bh = h^{-1}g^{-1}agh = (gh)^{-1}a(gh),$$

where we are using Exercise 21.7. Hence $c$ is related to $a$.

**21.13.** (a) Closure and associativity are exactly as in Example 21.13. The identity $e = -k$ and the inverse of $a$ is $-a - 2k$.
(b) $x = 43$.

**21.15.** (a) $(3.9, -31.2)$  (b) $35.1 - 11.24i$

**21.17.** First note that $f$ is a bijection of $G$ to itself since every element of $G$ has a unique inverse, so we need only check whether the "operation preserving" condition $f(a * b) = f(a) * f(b)$ holds if and only if $G$ is Abelian. Assuming $G$ is Abelian, use the fact that $b^{-1} * a^{-1} = a^{-1} * b^{-1}$ to establish the needed condition on $f$. On the other hand, assuming the condition holds, show then that $f(a * b) = f(b * a)$, so $a * b = b * a$ since $f$ is a bijection.

## Chapter 22

### Exercises

**22.1.** $x = b^{-2}a^{-1}bc^{-1}$

**22.3.** We wish to show that if $(ab)^{-1} = a^{-1}b^{-1}$, then $ab = ba$. On the one hand, $(ab)^{-1}(ab) = e$, but by assumption $(ab)^{-1}(ba) = a^{-1}b^{-1}ba = e$ also, so $(ab)^{-1}(ab) = (ab)^{-1}(ba)$, and canceling $(ab)^{-1}$ on both sides, we get $ab = ba$, as desired.

**22.5.** Every infinite cyclic group is countable, but according to Theorem 9.1, the three sets $\mathbb{R}$, $\mathbb{R}^+$ and $\mathbb{R}^*$ are all uncountable.

**22.7.** There are four generators, the smallest of which is 2 (the others are 6, 7 and 8).

**22.9.** (a) The set $U_{12} = \{1, 5, 7, 11\}$. Modulo 12, each of these squared is congruent to 1 modulo 12. So this is a Klein four-group.

(b) The set $U_{14} = \{1, 3, 5, 9, 11, 13\}$. Modulo 14,

$$3^1 = 3, 3^2 = 9, 3^3 \equiv 13, 3^4 \equiv 11, 3^5 \equiv 5, 3^6 \equiv 1.$$

Hence $U_{14}$ is cyclic.

## Chapter 23

### Exercises

**23.1.** (a) $2\mathbb{Z} \cap 3\mathbb{Z} = \{\ldots, -12, -6, 0, 6, 12, 18, \ldots\} = 6\mathbb{Z}$.

(b) $4\mathbb{Z} \cap 6\mathbb{Z} = \{\ldots, -24, -12, 0, 12, 24, 36, \ldots\} = 12\mathbb{Z}$.

(c) $n\mathbb{Z} \cap m\mathbb{Z} = (\text{lcm})\mathbb{Z}$ where lcm is the least common multiple of $n$ and $m$.

**23.3.** $2\mathbb{Z} \cup 3\mathbb{Z} = \{\ldots, 0, 2, 3, 4, 6, 8, \ldots\}$. This set is clearly not closed under addition.

**23.5.** (a) In $S_3$, (123) commutes with itself, with the identity (1), and with its inverse (132). However, it does not commute with (12) since $(12)(123) = (23)$ but $(123)(12) = (13)$, nor does it commute with (13) or (23). Hence $N_{(123)} = \{(1), (123), (132)\} = \langle(123)\rangle$, which of course is a subgroup of $S_3$.

(b) Closure: If $x, y \in N_a$, then $xya = xay = axy$, i.e., $xy \in N_a$.

Inverses: If $x \in N_a$, then $ax = xa$, Multiplying both sides of this equation on both the left and the right by $x^{-1}$, we get $x^{-1}axx^{-1} = x^{-1}xax^{-1}$, which simplifies to $x^{-1}a = ax^{-1}$. Hence $x^{-1} \in N_a$.

**23.7.** Subgroups of $S_3$:
$$\{(1)\}$$
$$\{(1), (12)\}$$
$$\{(1), (13)\}$$
$$\{(1), (23)\}$$
$$\{(1), (123), (132)\}$$
$$\{(1), (12), (13), (23), (123), (132)\} = S_3.$$

Yes, in all cases the size of the subgroup divides 6.

**23.9.** Let $G$ be an infinite cyclic group generated by $g$ and let $H$ be a non-trivial subgroup of $G$. By Theorem **23.2**, $H$ is cyclic, say generated by $h$. If $H$ were finite, then for some $n > 0$, $h^n = e$. But since $G$ is cyclic, $h = g^k$ for some $k \neq 0$, so we would have $g^{kn} = e$, implying that $kn = 0$, which is false. Hence $H$ must itself be infinite.

# Chapter 24

**Exercises**

**24.1.** $\langle 5 \rangle = \{5, 10, 0\}$
$1 + \langle 5 \rangle = \{6, 11, 1\}$
$2 + \langle 5 \rangle = \{7, 12, 2\}$
$3 + \langle 5 \rangle = \{8, 13, 3\}$
$4 + \langle 5 \rangle = \{9, 14, 4\}$.

**24.3.** To prove that $h * H \subseteq H$, let $h * h_1$ be a typical element of $h * H$. Since $H$ is closed under $*$, we get $h * h_1 \in H$, so $h * H \subseteq H$.

On the other hand, to show $H \subseteq h * H$, let $h_1$ be a typical element if $H$. Then $h_1 = e * h_1 = (h * h^{-1}) * h_1 = h * (h^{-1} * h_1) \in h * H$. Hence $H \subseteq h * H$, and we conclude that $H = h * H$. Note that the latter direction used existence of $e$, existence of inverses, closure, *and* associativity in $H$!

24.5. The sets $i + \mathbb{R}$, $2i + \mathbb{R}$ and $3i + \mathbb{R}$ are typical cosets. Each coset is a horizontal line in the plane.

24.7. Remember to compose cycles from right to left. Now $(12)\langle(12345)\rangle$
$$\begin{aligned} &= \{(12)(12345), (12)(13524), (12)(14253), (12)(15432), (12)(1)\} \\ &= \{(2345), (135)(24), (14)(253), (1543), (12)\} \end{aligned}$$

24.9. Every matrix in this coset has a determinant of $-5$.

# Chapter 25

## Exercises

25.1. $\{1, 2, 3, 5, 6, 9, 10, 15, 18, 30, 45.90\}$.

25.3. (a) $\{1, 2, 4, 7, 14, 28\}$.
(b) $\langle 1 \rangle = \mathbb{Z}_{28}$,
$\langle 2 \rangle = \{2, 4, 6, 8, 10, 12, 14, 16, 18, 20, 22, 24, 26, 0\}$,
$\langle 4 \rangle = \{4, 8, 12, 16, 20, 24, 0\}$,
$\langle 7 \rangle = \{7, 14, 21, 0\}$,
$\langle 14 \rangle = \{14, 0\}$,
$\langle 0 \rangle = \{0\}$.

25.5. Further hint: Since $h$ generates $H_1$, $h^k = e$, so $g^{rk} = e$ also. But $g$ generates $G$, so $\cdots$.

25.7. (a) 2 and 3.
(b) Of order 2: $\{(1), (12)\}, \{(1), (13)\}, \{(1), (23)\}$,
of order 3: $\{(1), (123), (132)\}$.

25.9. (a) $(12345) = (15)(14)(13)(12)$.
(b) $(12345)^2 = (12345)(12345) = (13524)$
$(12345)^3 = (12345)(13524) = (14253)$
$(12345)^4 = (12345)(14253) = (15432)$
$(12345)^5 = (12345)(15432) = (1)(2)(3)(4)(5) = (1)$.

# Chapter 26

## Exercises

26.1. (a) The set $2\mathbb{Z}$ is a ring, because
R1: The sum of two even numbers is even.
R2: Addition of integers is associative.

R3: The integer 0 is even.

R4: $-2n$ is the additive inverse of $2n$.

R5: Addition of integers is commutative.

R6: The product of two even integers is even.

R7: Multiplication of integers is associative.

R8: The distributive law hold in the integers.

(b) The set $2\mathbb{Z} + 1$ is not a ring, since it is not closed under addition (i.e., R1 fails), there is no additive identity (i.e., R3 fails), and hence there are no additive inverses (i.e., R4 fails).

26.3. The proof is very similar to that in Exercise 26.1 Part (a).

26.5. (a) The left hand side and right hand side are respectively

$$\begin{pmatrix} 1 & 1 \\ 0 & 0 \end{pmatrix} \qquad \begin{pmatrix} 1 & 2 \\ 0 & 0 \end{pmatrix}$$

so the formula fails.

(b) $(x + y)^2 = x^2 + yx + xy + y^2$

(c) Both sides are $\begin{pmatrix} 1 & 1 \\ 0 & 0 \end{pmatrix}$

26.7. By Theorem 26.2, the set of zero divisors in $\mathbb{Z}_{18}$ will be exactly the $17 - \phi(18) = 17 - 6 = 11$ non-zero elements which are not relatively prime to 18, i.e., the set $\{2, 3, 4, 6, 8, 9, 10, 12, 14, 15, 16\}$. Modulo 18, we have $(2)(9) = 0$, $(3)(6) = 0$, $(4)(9) = 0$, $(8)(9) = 0$, $(10)(9) = 0$, $(12)(6) = 0$, $(14)(9) = 0$, $(15)(6) = 0$, and $(16)(9) = 0$.

26.9. One example is $f(x) = x(x + 1)(x + 2) = x^3 + 2x$.

26.11. If $g$ generates $(R, +)$ and if $a \in R$, then for some integer $n$

$$a = \pm \underbrace{(g + g + \cdots + g)}_{|n| \text{ copies of } g},$$

depending on whether $n$ is non-negative or negative. If likewise $b = mg$ for some integer $m$, then by multiple applications of the distributive law and associativity of addition, we get

$$ab = ba = \pm \underbrace{(g^2 + g^2 + \cdots + g^2)}_{|nm| \text{ copies of } g^2}.$$

Hence $R$ is commutative.

26.13. (a) $2^4 = 16$ matrices over $\mathbb{Z}_2$.

(b) For example, $\begin{pmatrix} 0 & 1 \\ 0 & 0 \end{pmatrix} \begin{pmatrix} 1 & 0 \\ 0 & 0 \end{pmatrix} = \begin{pmatrix} 0 & 0 \\ 0 & 0 \end{pmatrix}$.

(c) For example, $\begin{pmatrix} 0 & 1 \\ 1 & 0 \end{pmatrix} \begin{pmatrix} 0 & 1 \\ 1 & 0 \end{pmatrix} = \begin{pmatrix} 1 & 0 \\ 0 & 1 \end{pmatrix}$, i.e, this matrix is its own inverse.

# Chapter 27

## Exercises

27.1. $\mathbb{Z}_{20} = \langle 1 \rangle = \{1, 2, 3, 4, 5, 6, 7, 8, 9, 10, 11, 12, 13, 14, 15, 16, 17, 18, 19, 0\}$,
$\langle 2 \rangle = \{2, 4, 6, 8, 10, 12, 14, 16, 18, 0\}$,
$\langle 4 \rangle = \{4, 8, 12, 16, 0\}$,
$\langle 5 \rangle = \{5, 10, 15, 0\}$,
$\langle 10 \rangle = \{10, 0\}$,
$\langle 0 \rangle = \{0\}$.

27.3. Using Theorem 27.1, because the sum and difference of any two even entries is even, and because the product of an even entry with any integer entry is even, this subset is an ideal.

27.5. (a) Using Theorem 27.1, if $a, b \in I \cap J$, then $a - b \in I$ and $a - b \in J$, so $a - b \in I \cap J$. Moreover, if $a \in I \cap J$ and $x \in R$, then $xa, ax \in I$ and $xa, ax \in J$, so $xa, ax \in I \cap J$, and we are done.
(b) Using $\mathbb{Z}_{20}$ (Exercise 27.1) as an example, $\langle 4 \rangle \cup \langle 5 \rangle = \{4, 5, 8, 10, 12, 15, 16, 0\}$, which is not even a subgroup of $\mathbb{Z}_{20}$, much less an ideal.

27.7. (a) Suggestion: Take your favorite irrational number and let it generate an additive cyclic subgroup of $\mathbb{R}$.
(b) Hint: Your irrational number has a multiplicative inverse in $\mathbb{R}$.
(c) Hint: Every non-zero real number has a multiplicative inverse.

27.9. (a) $\{0\}$, yes.
(b) $12\mathbb{Z}$, yes.
(c) $\mathbb{Z}_{20}$ and $\mathbb{Z}$ are both commutative. Look at the general case of closure under multiplication. (Note: The general case of closure under addition is okay, but not obvious. Take a look at it, recalling the "add and subtract trick" you used to prove the Product Rule in Chapter 2.)

27.11. (a) The other four elements of $\mathbb{Z}_{20}/I$ are $1 + I = \{6, 11, 16, 1\}$, $2 + I = \{7, 12, 17, 2\}$, $3 + I = \{8, 13, 18, 3\}$ and $4 + I = \{9, 14, 19, 4\}$.
(b) A typical element of $2 + I$ is $2 + 5a$ and of $3 + I$ is $3 + 5b$ for $a, b \in \{0, 1, 2, 3\}$, so we get that a typical element of $(2 + I)(3 + I)$ is $(2 + 5a)(3 + 5b) = 6 + 10b + 15a + 25ab \equiv 1 \pmod 5$, which is an element of $1 + I$, as desired.

# Chapter 28

## Exercises

28.1. (a) none, (b) 2/3, (c) 2, 6.

271

28.3. $2\mathbb{Z}$ inside $\mathbb{Z}$.

28.5. $\mathbb{Z}$ and $\mathbb{Z}_p$ for any prime $p$ have this property.

28.7. Among $2 \times 2$ matrices over $\mathbb{Z}_2$, we have

$$\begin{pmatrix} 0 & 1 \\ 0 & 0 \end{pmatrix} \begin{pmatrix} 1 & 0 \\ 0 & 0 \end{pmatrix} = \begin{pmatrix} 0 & 0 \\ 0 & 0 \end{pmatrix},$$

so we have zero divisors present. Beyond that,

$$\begin{pmatrix} 1 & 0 \\ 0 & 0 \end{pmatrix} \begin{pmatrix} 0 & 1 \\ 0 & 0 \end{pmatrix} = \begin{pmatrix} 0 & 1 \\ 0 & 0 \end{pmatrix},$$

which shows that the ring is not commutative.

28.9. The solutions are 9 and 10. Cancellation works because $\mathbb{Z}_{11}$ is an integral domain.

# Chapter 29

## Exercises

29.1. The element $a$ has a multiplicative inverse in $F$. Proceed.

29.3. To establish that $\mathbb{Z}[\sqrt{2}]$ is an integral domain, we must check closure of both operations, existence of additive inverses, existence of the unity element, and non-existence of zero-divisors. These are easy (but check them!) except for perhaps closure of multiplication and non-existence of zero-divisors. For the former, if $a, b, c, d \in \mathbb{Z}$, we have

$$(a + b\sqrt{2})(c + d\sqrt{2}) = (ac + 2bd) + (ad + bc)\sqrt{2},$$

which is in $\mathbb{Z}[\sqrt{2}]$. For the latter, because $\mathbb{Z}[\sqrt{2}]$ is a subset of $\mathbb{R}$, which contains no zero-divisors, $\mathbb{Z}[\sqrt{2}]$ can contain no zero-divisors. However, $\mathbb{Z}[\sqrt{2}]$ is not a field, since, for example, its element 2 has no multiplicative inverse.

29.5. All required properties to be an integral domain are easy to check (but again, do so) except non-existence of zero-divisors. But because $F$ contains no zero-divisors, the product of any two non-zero polynomials of degree 0 (i.e., constant polynomials) will be non-zero, and if one of the polynomials has positive degree, so will the product, and again will be non-zero. However, $F[x]$ is not a field since, for example, for any non-zero polynomial $f(x)$, $xf(x) \neq 1$, so $x$ has no multiplicative inverse.

29.7. Hint: When $p$ is odd, is it possible for a non-zero element of $\mathbb{Z}_p$ to equal its own additive inverse?

**29.9.** The polynomial $x^4 + 1 = (x+1)^4$. This is a special case of Exercise 29.11 Part (b).

**29.11.** (a) For $0 < k < p$, neither factor $k!$ nor $(p-k)!$ in the denominator of $\binom{p}{k}$ contains a factor of $p$. Hence $p$ divides $\binom{p}{k}$.
(b) Make an argument similar to (but more involved than) the one in Part (a).

**29.13.** (a) Does $\mathbb{F}_p$ have any non-trivial proper additive subgroups?
(b) $x^4 - 1 = (x^2 - 1)(x^2 + 1)$ and $x^4 - 1 = (x-1)(x+1)(x^2+1)$, so we have two factors, $x^2 - 1$ and $x - 1$ of the desired form. Setting $x = 2$, we get subfields which have $2^2 - 1 = 3$ and $2 - 1 = 1$ non-zero elements, i.e., subfields $\mathbb{F}_{2^2} = \mathbb{F}_4$ and $\mathbb{F}_{2^1} = \mathbb{F}_2$. Note that the allowable exponents are 2 and 1, the proper divisors of 4.
(c) Now, see if you can generalize (with justification) our solution to Part (b).

# Chapter 30

**Exercises**

**30.1.** (a) By the properties of 0 in $F$, property V4 of vector spaces, and the fact that $V$ is an additive group, we have $0\mathbf{v} = (0+0)\mathbf{v} = 0\mathbf{v} + 0\mathbf{v}$, so by cancellation in $V$, we get $0\mathbf{v} = \mathbf{0}$.
(b) The proof is similar but uses V5 instead.

**30.3.** (a) Yes, since this set is an additive group within $\mathbb{R}$.
(b) No, since this set is not closed under addition, nor contains a zero element.
(c) Yes, since all arithmetic in this set is done modulo 3, i.e., is done over $\mathbb{F}_3$. (Recall that additively $\mathbb{F}_{27} = \mathbb{F}_3 \times \mathbb{F}_3 \times \mathbb{F}_3$.)

**30.5.** This is not as simple as it might appear, e.g., we can't just "cancel" the $\mathbf{v}$. However, using the facts that $V$ is an additive group, that $F$ is a field, properties V2, V3 and V4, and Exercise 30.1(a), we can argue as follows: $\lambda\mathbf{v} = \mu\mathbf{v}$ implies that $(\lambda - \mu)\mathbf{v} = \mathbf{0}$. If $\lambda - \mu \neq 0$, then $(\lambda - \mu)^{-1}$ exists in the field $F$, and so

$$\mathbf{v} = 1\mathbf{v} = ((\lambda - \mu)^{-1}(\lambda - \mu))\mathbf{v} = (\lambda - \mu)^{-1}((\lambda - \mu)\mathbf{v}) = (\lambda - \mu)^{-1}\mathbf{0} = \mathbf{0},$$

which is a contradiction. Hence $\lambda - \mu = 0$, i.e., $\lambda = \mu$. Whew!

**30.7.** If $V = \mathbb{C}$ and $S = \mathbb{R}$, then $(i)(3)$ is not in $\mathbb{R}$, so scalar multiplication is not closed. Likewise with $\sqrt{2} \in V = \mathbb{R}$ and $5 \in S = \mathbb{Q}$, since $5\sqrt{2}$ is not in $\mathbb{Q}$.

**30.9.** There is more to check here than in most of our examples because we do not know from the start that our set of all linear mappings from $\mathbb{R}$ to $\mathbb{R}$ is an additive Abelian group. So establish that first (e.g., observe that the zero function is linear, etc.), then verify the five properties V1-V5 of scalar multiplication.

## Chapter 31

**Exercises**

**31.1.** Making use of Example 31.2, since $\det \begin{pmatrix} 2 & 3 \\ 4 & 9 \end{pmatrix} = 6$, we have that these vectors are independent. Alternately, since there are only two vectors, we could use Exercise 31.2.

**31.3.** Using the hint, there is clearly no scalar $\lambda$ such that $B = \lambda A$, so $A$ and $B$ are independent.

**31.5.** Using the determinant approach, we have

$$\det \begin{pmatrix} 2 & 3 & 4 \\ 3 & 4 & 5 \\ 4 & 5 & 6 \end{pmatrix} = 6\det \begin{pmatrix} 2 & 3 \\ 3 & 4 \end{pmatrix} - 5\det \begin{pmatrix} 2 & 4 \\ 3 & 5 \end{pmatrix} + 4\det \begin{pmatrix} 3 & 4 \\ 4 & 5 \end{pmatrix}$$

$$= 6(-1) - 5(-2) + 4(-1) = 0,$$

so these vectors are linearly dependent. If fact, we can see that

$$(2, 3, 4) + (4, 5, 6) - 2(3, 4, 5) = 0.$$

**31.7.** Call the given vectors $\{v_1, v_2, \ldots, v_{n+1}\}$. If the vectors in the subset $\{v_1, v_2, \ldots, v_n\}$ are linearly dependent, we have that the full set is likewise. If the vectors in the subset are linearly independent, they form a basis for our space. Hence $v_{n+1}$ can be written as a linear combination of the others, so again the full set in linearly dependent.

**31.9.** Since $\det \begin{pmatrix} 6 & 5 \\ 5 & 6 \end{pmatrix} = 11$, these vectors will be linearly independent over all $\mathbb{F}_p$ except over $\mathbb{F}_{11}$.

**31.11.** $V$ is an Abelian group since it is closed under matrix addition, the zero matrix is in $V$, and if $M \in V$, then $-M \in V$ also. Moreover, $V$ is closed under scalar multiplication, so $V$ is a vector space over $\mathbb{R}$. Its dimension over $\mathbb{R}$ is 3, and simple basis is $\begin{pmatrix} 1 & 0 \\ 0 & 0 \end{pmatrix}$, $\begin{pmatrix} 0 & 0 \\ 0 & 1 \end{pmatrix}$ and $\begin{pmatrix} 0 & 1 \\ 1 & 0 \end{pmatrix}$.

## Chapter 32

**Exercises**

**32.1.** (a) the $y$-axis
(b) and (c) you get the idea.

32.3. The vectors in $S$ are of the form $\mathbf{v} = (2y + 3z, y, z)$, so we can think of $y$ and $z$ as being independent and $x$ as being dependent on them. The sum of two vectors of this form are of this form, the zero vector is of this form, and $(-2x-3y, -y. -z)$ is the additive inverse of $\mathbf{v}$, so $S$ is an Abelian group. Also any scalar multiple of $\mathbf{v}$ is of this form, so $S$ is closed under scalar multiplication. Hence $S$ is a subspace of $\mathbb{R}^3$ of dimension 2. A simple basis is $(2,1,0)$ and $(3,0,1)$.

32.5. The set $S$ is not an Abelian group, nor is it closed under scalar multiplication. For example, $(1,1,\sqrt{2})$ is in $S$, but

$$(1,1,\sqrt{2}) + (1,1,\sqrt{2}) = 2(1,1,\sqrt{2}) = (2,2,2\sqrt{2})$$

is not in $S$ since it does not satisfy $x^2 + y^2 = z^2$. (Note: A picture of $S$ in $\mathbb{R}^3$ shows that it is, not surprisingly, curved. A picture of a subspace of $\mathbb{R}^3$, whether a line or a plane, will be "straight" or "flat", being a *linear* combination of vectors.)

32.7. (a) The zero subspace (i.e.,$\{\mathbf{0}\}$) and $V$ itself are distinct subspaces.
(b) The set $\mathbb{Z}_p \times \mathbb{Z}_p$, of dimension 2 over $\mathbb{Z}_p$, has $p$ subspaces of dimension 1, each having a basis $(1,\lambda)$ for some $\lambda \in \mathbb{Z}_p$. Together with the unique subspaces of dimension 0 and 2, as in Part (a), we get a total of $p + 2$ subspaces.

32.9. (a) You proved in Exercise 23.2 some time ago that the intersection $U \cap W$ of any two subgroups $U$ and $W$ of a group $V$ is itself a subgroup. Moreover, it is clear that if $\mathbf{v} \in U \cap W$ and $\lambda \in F$, then $\lambda\mathbf{v} \in U \cap W$ as well. Hence $U \cap W$ is a subspace of $V$.
(b) In $\mathbb{R}^3$ over $\mathbb{R}$, let $U = \{(a,0,0)|a \in \mathbb{R}\}$ and $W = \{(0,b,0)|b \in \mathbb{R}\}$ be two 1-dimensional subspaces. Then $U \cup W$ consists of all vectors of the form $(a,0,0)$ or $(0,b,0)$. However, $(1,0,0) + (0,1,0) = (1,1,0)$, which is not of the required form. Hence $U \cup W$ is not a group, so it is not a subspace of $V$. (Geometrically, $U \cup W$ is just two lines, but the subspace generated by those two lines is a plane.)

## Chapter 33

### Exercises

33.1. (a) $\{0\}$　(b) $\{0, \sqrt{3}, -\sqrt{3}\}$　(c) $\{0, \sqrt{3}, -\sqrt{3}, 2i, -2i\}$

33.3. (a) $\{2,4\}$　(b) none

33.5. $q(x) = 2x^2 - 5x + 15$, $r(x) = -27$

33.7. $x + \frac{1}{2}$

**33.9.**
Step 1: $x^3 + 3x^2 + 2x + 1 = (x^2 + x - 2)(x + 2) + 2x + 5$,
Step 2: $x^2 + x - 2 = (2x + 5)(\frac{1}{2}x - \frac{3}{4}) + \frac{7}{4}$, so gcd $= 1$.

# Chapter 34

## Exercises

**34.1.** (a) $(x + 1)(x^2 + 1)$   (b) $(x + 1)(x + i)(x - i)$

**34.3.** (a) Since 2, 3, and 4 are all roots of this polynomial over $\mathbb{F}_5$, it can be factored as $(x - 2)(x - 3)(x - 4)$. This same factorization can be put in the form $(x + 1)(x + 2)(x + 3)$ because, for example, in $\mathbb{Z}_5$, $1 = -4$.
(b) $(x + 1)(x^2 + 1)$

**34.5.** $3(x - \frac{2 + \sqrt{10}}{3})(x - \frac{2 - \sqrt{10}}{3})$

**34.7.** (a) $1,737$   (b) $600$

**34.9.** (a) $(x + 1)(x + 2)$ and $(x + 4)(x + 5)$

# Chapter 35

## Exercises

**35.1.** (a) $x^2 + 5$
(b) $(x - i\sqrt{5})(x + i\sqrt{5})$
(c) $(x - 3)(x - 4)$ (can also be written $(x + 4)(x + 3)$)

**35.3.** (a) $2x^2 + 3x - 6$ (by Theorem **35.1** with $p = 3$)
(b) $2(x - \frac{1}{4}(-3 + \sqrt{57}))(x - \frac{1}{4}(-3 - \sqrt{57}))$

**35.5.** (a) $(x - 1)(x^2 + x + 6)$
(b) $(x - 1)((x - \frac{1}{2}(-1 + i\sqrt{23}))(x - \frac{1}{2}(-1 - i\sqrt{23}))$

**35.7.** Applying the Quadratic Formula to $(x^2)^2 + 2$ gives us
$x^4 + 2 = (x^2 - i\sqrt{2})(x^2 + i\sqrt{2})$. Now apply the Quadratic Formula to each of these factors, arriving at $x = \pm\sqrt{i}\sqrt[4]{2}$. Since you now have $\sqrt{i}$ in your expressions, it will help to look at Exercise **33.10.**

**35.9.** If $f(x) = x^4 + 2$ over $\mathbb{F}_3$, then $f(1) = f(2) = 0$, so $(x - 1)(x - 2) = x^2 + 2$ must divide $x^4 + 2$. Do this to get the irreducible quadratic factor.

**35.11.** (a) $\{1, -1, 2, -2, 4, -4\}$
(b) $-2$
(c) $\{1, -1, 3, -3, 9, -9, 1/2, -1/2, 3/2, -3/2, 9/2, -9/2\}$

# Bibliography

[1] H. Anton, *Elementary Linear Algebra*, Eighth Ed., Wiley, New York, 2000.

[2] A.A. Clarke, W.C. Waterhouse, J. Brinkhuis, and C. Greiter, *Disquisitiones Arithmeticae*, reprinted English version, Springer, New York, 1986.

[3] W. Diffie and M.E. Hellman, New directions in cryptology, *IEEE Trans. Infor. Thy.* IT-22(1976), no. 6, 644–654.

[4] G. Effinger, K. Hicks and G.L. Mullen, Integers and Polynomials: Comparing the Close Cousins $\mathbb{Z}$ and $\mathbb{F}_q[x]$, *The Mathematical Intelligencer*, 27 (2005), vol. 2, 26–34.

[5] R. Hill, *A First Course in Coding Theory*, Oxford Appl. Math. and Comp. Sci. Ser., Clarendon Press, Oxford, 1986; reprinted 2009.

[6] G.L. Mullen and C. Mummert, *Finite Fields and Applications*, Student Math. Library, Vol. 41(2007), Amer. Math. Soc., Providence, RI.

[7] G.L. Mullen and D. White, A polynomial representation for logarithms in $GF(q)$, *Acta Arithmetica* 47(1986), 255–261.

[8] R.L. Rivest, A. Shamir, and L. Adelman, A method for obtaining digital signatures and public-key cryptosystems, *Communications of the ACM* 21(1978), no. 2, 120–126.

[9] J.H. Silverman, *A Friendly Introduction to Number Theory*, Fourth Edition, Pearson Prentice Hall, Upper Saddle River, NJ, 2013.

# Index

Abel, N., 144
Abelian group, 144
Adleman, L., 135
algebraic structure, 71
associative, 72
automorphism, 151

basis, 213
   standard, 214
bijection, 47
binary operation, 71
   associative, 72
   commutative, 72
   distributive, 72
binomial coefficients, 2
Binomial Theorem, 4
Box Principle, 18

canonical factorization, 91
Cantor diagonalization argument, 53
cardinality, 51
   countable, 52
   uncountable, 53
Cartesian product, 37
characteristic of a ring, 186
Chinese Remainder Theorem, 109
commutative, 72
commutative group, 144
commutative ring, 182
complex conjugate, 66, 239
complex numbers, 65
   conjugate, 66, 239
   imaginary part, 66
   real part, 66
composite number, 87

composition of functions, 48
congruent modulo $n$, 97
Continuum Hypothesis, 54
coset, 168
   left, 168
   right, 168
countable set, 52
cycle, 59
   $r$-cycle, 59
cycle notation, 59
cyclic group, 155

degree, 222
derivative, 8
determinant, 76
diagonal matrix, 218
Diffie, Whitfield, 122
Diffie-Hellman Key Exchange, 123
digital signature, 138
dihedral group, 147
dimension, 213
Discrete Logarithm Function, 125
Discrete Logarithm Problem, 124
discriminant, 239
disjoint cycles, 60
Disquisitiones Arithmeticae, 97, 242
distributive, 72
Division Algorithm, 17, 30, 81

Eisenstein's Criterion, 238
encryption, 121
Eratosthenes, 88
Euclid, 15, 17, 80, 81, 90
Euclid's Lemma, 90
Euclidean Algorithm, 17, 82